ABSOLUTE BEGINNER'S GUIDE

TO

Building Robots

Gareth Branwyn

800 E. 96th Street,
Indianapolis, Indiana 46240

Absolute Beginner's Guide to Building Robots

Copyright © 2004 by Que Publishing

International Standard Book Number: 0-7897-2971-7

Library of Congress Catalog Card Number: 20-03103655

Printed in the United States of America

First Printing: August 2003

06   05   04   03   4   3   2   1

Bulk Sales

Que Publishing offers excellent discounts on this book when ordered in quantity for bulk purchases or special sales. For more information, please contact

> **U.S. Corporate and Government Sales**
> **1-800-382-3419**
> corpsales@pearsontechgroup.com

For sales outside of the U.S., please contact

> **International Sales**
> **1-317-581-3793**
> international@pearsontechgroup.com

Trademarks

Warning and Disclaimer

Publisher
Paul Boger

Associate Publisher
Greg Wiegand

Executive Editor
Rick Kughen

Development Editor
Todd Brakke

Managing Editor
Charlotte Clapp

Project Editor
Andy Beaster

Copy Editor
Jessica McCarty

Indexer
Heather McNeill

Proofreader
Katie Robinson

Technical Editor
David Hrynkiw

Team Coordinator
Sharry Lee Gregory

Interior Designer
Anne Jones

Cover Designer
Dan Armstrong

Page Layout
Eric S. Miller

Graphics
Mark Frauenfelder
Blake Maloof

Photography
Jay Townsend/Primal Design

Reviewers
Carlo Bertocchini
Roger Gilbertson

Contents at a Glance

Introduction 1

Part I   **Robot.edu** **11**

1   What's in a Name? 13
2   Robot Evolution 33
3   Robot's Rules of Order 61
4   Robot Anatomy Class 75

Part II   **Gettin' Dirty with It** **111**

5   The Right Tools for the Job .113
6   Acquiring Mad Robot Skills .147
7   Project 1: Coat Hanger Walker 185
8   Project 2: Mousey the Junkbot 227
9   Project 3: Building a DiscRover 263

Part III   **Resources** **299**

10   Robot Hardware and Software 301
11   Robot Books, Magazines, and Videos 319
12   Robots on the Web 333

Glossary 345
Index 355

Table of Contents

Introduction 1

I Robot.edu 11

1 What's in a Name? 13

A Bot by Any Other Name... 13

Bots in Brief 19
Domo Arigato, Mr. Roboto 19
No More Rogue Robots! 21
The Real Rossum: Industrial Bots from Unimate to Co-Bots 22
One Small Dance for Robo-Critter Kind 23
Off to a Shakey Start 25

Great Moments in Robot History 28

The Absolute Minimum 30

2 Robot Evolution 33

Don't Call Me Chief! 33

Humanoid Robots 35

Robotic Presence 37

Biomimicry 40
Behavior-Based Robotics (BBR) 42
BEAM 46

Robot Sports 49
Combat Robotics 49
SUMO 51

Robotic Performance Art 52

Robot Warriors 54

Domestic Robots 55
Robo-Pets 55
Of Maids and Mowers 57
Rosie, Where Are You? 58

Robotic Evolution Through Open Networking 58

The Absolute Minimum 59

3 Robot's Rules of Order 61

Bots in Legal Briefs 61
Asmivo's Three (er...Four) Laws of Robotics 62
Tilden's Laws 62
Moore's Law 63
Ohm's Law 64
Moravec's Timeline 65
The Turing Test 66
Amdahl's First Law 67
Brooks's Research Heuristic 67
Braitenberg's Maxim 68
The Krogh Principle 68
The Sugarman Caution 68

The Rules for Roboticists 69

The Absolute Minimum 73

4 Robot Anatomy Class 75

Flesh and Steel 75

Robot Body Types 76
Industrial Manipulators 76
Utility Robots 77
Robo-Critters 78
Humanoids 79
Embedded Bots 79
The Development Platform 81

Robots: The Exploded View 82

The Frame 83
Wood 84
Steel 85
Aluminum Alloy 85
Plastic 86
Titanium 86
Carbon Composite 86
Printed Circuit Board (PCB) 87
LEGO Bricks 87

Actuators **88**
    Motors and Gears 88
    Hydraulic Cylinders 88
    Pneumatic Cylinders 89
    Muscle Wire 89

Drive Train **90**
    DC Motor 90
    Servo Motor 90
    Stepper Motor 91

Locomotion **91**
    Wheels 92
    Tank Tracks 93
    Legs 93

Power Systems **93**
    Batteries 94
    Pressure Systems 95
    Solar Cells 95
    Other Power Sources 96

Manipulators/End Effectors (Optional) **97**

Sensors **98**
    Pressure Sensors 98
    Light Sensors 98
    Sound Sensors 99
    Vision Systems 100
    Heat Sensors 101
    Gas Sensors 102
    Other Types of Sensors 102

Controllers **102**
    Microcontrollers 103
    Off-Board Control 103
    Nervous Nets 103
    Other Controllers 104

Communication (Optional) **104**
    Two-Way Wireless 104
    Speech 104
    Infrared 105

Outer Shell (Optional) **105**

Robot Soul (Software) **106**

A Word About Spaghetti **106**

Robot Pop Quiz **107**
    The Questions 108
    The Answers 108

The Absolute Minimum **109**

II Gettin' Dirty with It **111**

5 The Right Tools for the Job **113**

The Basic Bot Builder's Toolbox **114**
    Must-Have Tools 114
    Should-Have Tools 117
    Make-Life-Easier Tools 119
    Supplies 122

Thumbnail Guide to Digital Multimeters **126**
    Where's the Juice, Bruce? 129
    Resistance Is Futile 129
    You Light Up My Cadmium-Sulfide Sensors 130
    Resources 131

Your Robot Workshop **131**

Thumbnail Guide to Soldering **132**
    Desoldering 137
    Practice, Practice, Practice 137
    Resources 139

Building a Ventilation System **139**
    Parts List 139

Safety First **144**

The Absolute Minimum **145**

6 Acquiring Mad Robot Skills **147**

Killer Kits **148**
    Rockit Sound-Controlled Robot 148
    Spider 3 Walker 149
    WAO-G 149

Carpet Rover 2 Combo Kit 150

Hexapod Walkers 151

CyBugs 152

Build Your Own Robot 152

Sumo-Bot 153

BOE-Bot 154

SolarSpeeder 155

Photopopper 4.2b 156

ER1 Personal Robot System 157

Robot Building Systems **159**

LEGO MINDSTORMS Robotics Invention System 159

Fischertechnik 160

Robix Rascal Robot Construction Set 161

Robotix 162

Out-of-the-Box Bots **162**

Sony AIBO 163

Sony SDR-4X 164

B.I.O.-Bugs 165

Roomba 167

Robomower 168

AmigoBot ePresence 169

Retro-Robotics **170**

Androbots 170

Heathkit Heroes 171

RB5X 172

Other Cool "Antique" Robots **173**

Armatron/Super Armatron 173

Omnibot 174

Maxx Steele 174

Big Trak 174

Thumbnail Guide to Electronics **175**

The Greenie Theory 176

Making Greenies Dance 177

How to Become an Electronics Whiz in Four Easy Steps 181

The Absolute Minimum **182**

7 **Project 1: Coat Hanger Walker** **185**

Gathering the Parts **187**

The Parts List 187

The Tools and Supplies List 189

Freeforming the Bicore Control Circuit **189**
  Breadboarding the Bicore 190
  Hacking the Servo Motor 191
  Soldering Up the Circuit 194

Thumbnail Guide to Breadboards **200**

Building the Walker Body **203**
  Making the Gears 204
  Fabricating the Idler Shaft 205
  Making the Idler Gear 208
  Mounting the Drive Gear 211
  Creating the Leg Assemblies 212
  Testing the Fit 219

The Power Plant **219**
  Attaching the Battery Packs 219
  Adding a Power Switch 221

Final Assembly **222**

Further Experiments **225**

The Absolute Minimum **225**

8 Project 2: Mousey the Junkbot **227**

The Fine Art of Making "Frankenmice" **227**

Gathering the Parts **229**
  The Parts List 229
  The Tools and Supplies List 230

Building the Body **231**
  Alien Mouse Autopsy 232
  Motor and Switch Placement 234
  Installing the Control Switch 237

Understanding Mousey's Brain **238**
  A Brief Word About Mousey's Senses 239
  Breadboarding the Circuit 241
  Freeforming Mousey's Control Circuit 249
  Mousey Gets a Pair of Eyes and a Wag in Its Tail 255

It's a Slightly Anxious, Light-Seeking Robot! **258**
  Playing with Mousey 258
  Troubleshooting a Wayward Mousey 259

Further Experiments 260

The Absolute Minimum 261

9 Project 3: Building a DiscRover 263

AOL Just Got a Whole Lot Smarter 263

Gathering the Parts 265
    The Parts List 265
    The Tools and Supplies List 267

The Brains Behind the Bot 268
    OOPic: The "Hardware Object" 268
    Motorhead 271

Thumbnail Guide to Microcontrollers 273

Trash Recycling Through Bot Building 275
    DiscRover's Body 275

Programming Your DiscRover 285
    Our Motor Controller Code 285
    Bump Whisker Code 290

Load 'n Go 293

Troubleshooting the DiscRover 294
    The Mechanics 294
    The Code 295

Further Experimentation 296
    Add a Prototyping Area 296
    Put Another Layer on That Robo-Cake! 296
    Add Higher-Order Sensor Systems 296
    Make Two DiscRovers Talk to Each Other 297
    Build a CD-ROM Boxbot 297
    Take Your OOPic Elsewhere 298

The Absolute Minimum 298

III Resources 299

10 Robot Hardware and Software 301

Sourcing the Right Stuff 301

Robot Kits and Bot-Specific Parts 302

Building Materials and Supplies 305

Motors and Drive Train 307

Power Systems 309

Electronics 310

Remote Control (R/C) Equipment 312

Microcontrollers and Control Software 313

Tools 315

Hobby Stores 316

Surplus 317

11 Robot Books, Magazines, and Videos 319

Attention Bot-Obsessed Bookworms! 319

Books (General, Theoretical) 320

Books (Bot Building) 322

Books (Electronics) 326

Magazines 328

Videos 330

12 Robots on the Web 333

Portals 334

Robot-Specific Sites 336

Robot Communities 337

Robot Research Centers 339

Useful Bot-Building References 341

Glossary 345

Index 355

About the Author

Gareth Branwyn is a well-known technology journalist and self-proclaimed "reluctant deep geek." He is a contributing editor for *Wired* and writes for other national magazines such as *Esquire* and *I.D.* (*International Design*). He has written over a half-dozen books on technology, media, and culture, including *Mosaic Quick Tour* (the first World Wide Web book ever published), *Jargon Watch: A Pocket Dictionary for the Jitterati*, and *Jamming the Media: A Citizen's Guide*. Gareth is also owner and "cyborg-in-chief" of Street Tech (www.streettech.com), a popular watering hole for gadget enthusiasts.

Dedication

To my son Blake. You said it yourself: "We're fighters. That's what we do." Your boundless spirit, impressive talent, and emotional strength lend credence to the new agey notion of the "old soul."

Acknowledgments

This book happened during an extremely difficult time in my life. I wouldn't have made it through (my life or the book) if it wasn't for the help and support of so many people: Blake Maloof, Mark Frauenfelder, Dave Hrynkiw, Jérôme Demers, Sa'id Dahdah, Sean Carton, David Pescovitz, Xeni Jardin, the amazing Street Tech Irregulars (especially my masterful Web master Tim Tate), Nate Heasley, Roger Gilbertson, the Fulwilers, Mary Fenske, Christino, Linda, Tom, and Cathie, David and Lisa, Sande, and my parents. Especially loud shout outs go to Paula Bass (for critical life support), Steve Wiggins (ditto), Alberto Gaitán (my brother from another mother), Teresa Gaitán (the other mother), Jesse Hurdus (techno-roboto whiz kid), Nathan (for the greater good) Zimmerman, Peter Sugarman (my life editor), and Jay Townsend (for the tireless and brilliant photography work). And Pam Bricker ("'Cause in my mind…always").

I also have to acknowledge my amazing editors Rick Kughen and Todd Brakke (and everyone else I worked with at Que). Your patience during this project, and your faith in me, was impressive. Writers get lucky sometimes and get to work with the best. This was one of those times.

We Want to Hear from You!

As the reader of this book, *you* are our most important critic and commentator. We value your opinion and want to know what we're doing right, what we could do better, what areas you'd like to see us publish in, and any other words of wisdom you're willing to pass our way.

As an associate publisher for Que, I welcome your comments. You can email or write me directly to let me know what you did or didn't like about this book—as well as what we can do to make our books better.

Please note that I cannot help you with technical problems related to the *topic* of this book. We do have a User Services group, however, where I will forward specific technical questions related to the book.

When you write, please be sure to include this book's title and author as well as your name, email address, and phone number. I will carefully review your comments and share them with the author and editors who worked on the book.

Email: feedback@quepublishing.com

Mail: Greg Wiegand
Associate Publisher
Que
800 E. 96th Street
Indianapolis, IN 46240 USA

For more information about this book or another Que title, visit our Web site at www.quepublishing.com. Type the ISBN (excluding hyphens) or the title of a book in the Search field to find the page you're looking for.

How to Use This Book

A wise philosopher once said (I don't remember who it was, but trust me, he or she was really big-brained): "There are basically two types of people: Those who insist on shoving people into silly little categories, and those who don't." Wait, that's a different dichotomy than the one I was going for. (Accessing memory banks…) Here's the one: "There are basically two types of people: Those who think that today's digital technologies are actually intelligent (and are therefore conspiring to make us humans look dumb) and those who know that these technologies are quite stupid and are basically standing by, waiting for us to make them do something interesting."

If you're the former kind of person, fear not. This book is here to hold your hand as we lead you into the exciting realm of the robot kingdom. Trust us: It's really not as complicated as it might look. We've designed this book to be as user friendly, fun, and plain-spoken as possible. If we've done our job, you should come away from this journey with a conversational knowledge of all things robotic: where robots came from, where they're headed, and how to build them yourself. This knowledge might not do much for your cred at cocktail parties, but it might help you break the ice with that condescending IT guy at work. More important than that, this book should help boost your confidence in approaching seemingly intimidating high-tech subjects and give you the tools to understand one of the next big growth sectors of technology: robots! If you find the three projects in this book too intimidating, you can be perfectly happy building the robot kits and building sets detailed in Chapter 6, "Acquiring Mad Robot Skills." Many of these don't even require soldering.

If you're the latter type of person, this book should serve you well too. We've tried to load it with interesting facts, people, Web addresses, and other resources that will help you quickly get up to speed on your robot explorations. And you should find the projects fun and challenging (but hopefully not *too* challenging).

This book is divided into three parts:

- Part I, "Robot.edu," serves as a crash course in robot history, basic robot sciences, the various schools of thought driving robot evolution, and more. This is not only here to give you some background, but also to give you some conceptual tools for how to think about robots as you begin to build bots yourself.

- Part II, "Gettin' Dirty with It," is where you start putting greasy thumbprints all over this book. We run through all of the basic tools, supplies, materials, and techniques used in basic robot building, review kits and building sets, and offer a number of quick tutorials on need-to-know subjects. We call these *Thumbnail Guides*. We also present three robot-building projects that increase in level of difficulty, from a simple one-motor walker to a programmable robot platform you can use for many experiments beyond what's covered in this book.

■ Part III, "Resources," is not just a two-dimensional listing of books, Web sites, and so forth, but substantive reviews of the best material out there related to robots. There is a revolution in do-it-yourself robotics going on now, with thousands of amateurs collaborating and sharing their work over the Internet. We plug you directly into the heart of all this exciting activity. Don't mention it. It's what we live for.

I wrote this book because I love robots and wanted to share what I know with you. I hope you learn a lot from it and will end up sharing some of my passion for the subject. If you have any questions, comments, or suggestions about the book, please don't hesitate to email me at garethb2@streettech.com, or visit my Web site at www.streettech.com. On my site, I will include project bug reports (read these before starting the projects!), the CD templates and programs for the DiscRover (Chapter 9, "Project 3: Building a DiscRover"), larger versions of critical images, such as breadboard hookups, reader robot hacks, and more. So stop by and join in.

IN THIS CHAPTER

- How I got started in robotics (it's a cute story...no, really)

- Our changing perceptions of robots (from mechanical monster-men to fussy butlers to robotic tuna fish)

- What robots mean to us today and to our future

INTRODUCTION

I, Robot

I guess you could say that robots are in my bones. Wait, you *can* say that! In 2000, I had a total hip replacement. I am now part state-of-the-art, cobalt-chrome titanium with irradiated cross-linked polymers. Where HAL 9000, the robotic ship in *2001: A Space Odyssey*, was fond of telling people that he was built in the H.A.L. Plant in Urbana, Illinois, I can boast that part of me was built at DePuy Industries of Indiana.

I was obsessed with robots long before I started sporting my own robot-like body parts. In fact, my first spanking was over robotics (back in the early '60s when corporal punishment was all the rage). I was six years old. I'd been lusting after the latest Erector Set that was being heavily advertised for Christmas. As with all toy commercials, the one for this mechanical, motorized building set made it look far more sophisticated than it actually was. After sending my wish list off to Santa, I started fantasizing about the one thing I wanted to build most of all: a robot! My over-active imagination got the better of me, and by the time Christmas morning dawned, I was convinced that a full-blown humanoid companion would be mine within hours of ripping through the presents.

Opening up that Erector Set box was the first of many lifelong disappointments related to the fantastic perception of robots versus the far more sober reality. The Erector Set was cool all right, but it was little more than a few steel girders, some nuts and bolts, and a tiny wrench. It had one weak little motor (the infamous DC3) and barely enough parts to make a little crane, let alone a robot (or even a small model of one). I was so crestfallen that I couldn't hide my disappointment. As I started to sob, my parents got understandably angry at my apparent ungratefulness.

Then the *real* trouble started. My cousin lived a few towns over and we went to visit him and his family for New Year's Eve. Once he and I were alone, we started telling each other what we'd gotten for Christmas. In my enthusiasm (and probably my desire for one-upmanship), I suddenly found myself going with the fantasy of what I'd wanted rather than the reality of what I'd actually gotten. I told him I'd built this incredible robot that was now helping me out with chores around the house and was my new best friend. His excited response only egged me on and the tale became more and more fantastic. When I stopped to catch my breath, he dropped the bomb: "That's great! I'll see it next weekend when we come to visit your house!"

I knew right then that I was toast. I had no idea that his family had plans to visit, and even if I had, I'm not sure it would have stopped me from telling my Pinocchio-sized lies. We traveled home the next day, and in the car, I dedicated myself to figuring out a way of making my lie become true. I had a week to somehow magically transform that

note

If you were a child of the '50s and '60s (or the '10s, '20s, '30s, or '40s, for that matter), and were the proud owner of an Erector Set, you'll be happy to know that Erector is alive and well…at least in cyberspace. Erector World (www.erectorworld.com) has all sorts of material related to the mechanical building sets that launched the career aspirations of many an engineer. It has histories, galleries, and timelines. You can even buy refurbished antique building sets. The site also sells the "new" Erector Set, which is basically the French Meccano system being sold under the Erector name (it's *not* compatible with old Erector parts, by the way). If you want to get *real* Erector Sets, they're usually available (used) on eBay. And if you want to try your hand at incorporating Erector or other mechanical building components into your bots, check out the Mechanical Construction Set FAQ at Robotics.com (www.robotics.com/erector.txt).

lowly little Erector Set (and any other toys I could cannibalize) into something that could at least pass for a facsimile of a robot. Of course, it didn't take long, sitting there on our den floor, to realize that magic, the kind that transformed Pinocchio from sticks of wood and bits of string into a real boy, was the only thing that could come to my rescue. After an hour or so of mindlessly bolting pieces together, I gave up, put the set away, and began the countdown to my cousin's fateful arrival. He came. He saw (no robot). He had a hissy fit, and I got a spanking (for lying).

This event has always stuck with me, not only because it was so embarrassing, and one of the few times that I got spanked, but because it was the exact point at which my lifelong fascination with robots began. I think (if I may stretch out here on the couch for a moment) that my continued interest in robotics has been partially an attempt at making the magic happen that I couldn't make happen back then.

When my own son was six, we were watching a *Felix the Cat* cartoon that featured an evil scientist and his rampaging robot (one of the same episodes that I likely watched as a child). When the video was over, he said, "Hey dad, can we make a robot?" That was all the encouragement I needed. I figured that, in the long years between my first encounter with robots (at least of the daydream kind) and the early '90s, we certainly must be a lot closer to the robot I'd desired when *I* was six. Within months, we had surrounded ourselves with robot kits, Radio Shack robot toys (gotta love that Super Armatron), books, building sets, and microcontrollers.

As we explored all of the robots that were available at that time, I quickly realized that reality was still not even close to catching up to my boyhood fantasies. Even today, some 40 years from that fateful Christmas morning, there's still no robot companion kit to be found under the aluminum Christmas tree. Sure, today's kids have AIBO, LEGO MINDSTORMS, B.I.O. Bugs, and other nifty bots and building sets that are light-years beyond any Erector Set, but your kids shouldn't expect any of them to help make their beds or carry on a conversation about the freakish success of *SpongeBob SquarePants*. The robot I dreamt about, the one I was prepared to sell my soul to animate, is *still* generations away.

note

When LEGO introduced its MINDSTORMS Robotics Invention System in 1998, I got a huge kick out of the first TV commercial. A kid is asked to clean up the dinner dishes. In response, his MINDSTORMS robot is seen bulldozing all of the tableware onto the floor, followed by a second LEGO bot with a gripper arm planting a vase of flowers on the newly cleared table. I loved the way that the commercial played upon the perennial fantasy of building a robot helper, while realistically depicting what such bots could actually do at that time.

I know that all of the preceding is about robotic dreams unrealized, but happily, this is far from the whole story. I've spent 40-some years (never you mind the hard number) tinkering on and off with robot tech, and every step has been a fascinating one. Like the ancient alchemists, who searched for the illusive Philosopher's Stone—the element that would turn lead into gold—only to find that the transformative quest itself was the "gold" they sought (discovering modern chemistry in the process), my quest has also been extremely fulfilling. My son and I have built robots that draw, follow mazes, stir coffee, lift objects, play golf, send dirty dishes to the kitchen (though not very well), do light vacuuming (ditto), follow light, avoid light, navigate rooms, and much more. We're both diehard computer geeks, but it is robotics that's gotten us away from the sedentary computer desktop and allowed us to add some mechanical engineering, electronics, and rubber-meets-the-road problem solving to our skill sets. Robots are what happen when computers venture out into the real world.

Years ago, I had the good fortune to strike up an email conversation with pioneering musician, deep thinker, and multimedia artist Brian Eno. This was a time before the Web, when Apple's HyperCard program was the cutting-edge hypermedia tool. We were talking about the future of hypermedia, networking, and computers in general. He told me how much he hated computers. He loved what they could do, but hated using them (he also complained about computer geeks having foul body odor and disagreeable skin conditions, but that's another story). The thing he most objected to about computers was their nearly complete nonphysical nature. He was concerned about how increasing virtuality might affect our health, our social abilities, and our cultures. When I started to get more deeply into robotics with my son, I realized that it was at least a step in the direction of addressing Eno's concerns.

Look at a group of school children working at computers, and then look at a group working with robots. It's like night and day. Whereas computing is usually a solitary and physically passive activity, robot building in the classroom is very kinetic and usually collaborative. Kids are running back and forth, from the PC (where they're designing or programming their bots) to the floor (where they're

note

Okay, all you computer cowboys, don't write letters to me about how wrong this assertion is. I know that computer-based design and programming can be extremely creative, exciting, and mentally kinetic. I've been there myself. But until you can show me that you're burning real calories sitting at your desk and not forging a permanent butt-print on your task chair, my point stands.

building and testing them). There's great excitement and a sense of wonder that you don't see as much with solely computer-based projects.

My first fond memory of collaborating with my son on robot building is of him (at six), excitedly scooting around all over the floor in my home office watching, interacting with, and changing the pens on a drawing bot we'd just built out of LEGOs. The robot was tethered to our Mac by a serial computer cable, but I got this profound sense at that moment that I was witnessing the beginning of computers leaving the confines of the PC and leaking into everyday objects—a migration that's now become so commonplace, it's nearly invisible to most.

Robots: Fantasy Versus Reality

As this book and my reminiscences illustrate, there is often a great disparity between the dreams we have of robots—what they *should* be capable of—and their far more earthbound reality. Anyone who's both a technologist (whether a pro or an amateur) and a techno-romantic (one who dreams of a future made better by smarter machines and smarter people) must come to terms with this discrepancy and enjoy real robots for what they are (while continuing to be inspired by sci-fi visions of what they might become).

Science fiction has become the myth-making machinery of our age. Where cultures of old looked to their pasts to find heroes and golden ages to inspire them, we tend to look to our future. Ask a group of children what lessons are to be learned from Icarus or Midas, and they'll likely tell you little, except maybe that at Midas, you're not going to pay a lot for that muffler. Now ask them what they can tell you about Yoda and The Force, or what assimilation by the Borg means. You'll likely hear more than you bargained for. Right smack in the middle of these futuristic myths and morality tales stand robots: Lt. Commander Data, C-3PO, R2-D2, the Terminator, RoboCop, the B-9 from *Lost in Space*, Robby from *Fantastic Planet*. All of these robots have become part of our collective consciousness, our culture, and our modern mythology.

Humans have dreamt of creating humanoid companions for longer than you might think. The Greek poet Homer described mechanical maidens cast in gold by Hephaestus, Greek god of metalsmiths. The Hebrews had the Golem, a man-servant created from clay and animated by Kabalistic magic. Leonardo da Vinci even sketched out plans for a mechanical man in the fifteenth century. In many early tales of automata, as in numerous modern robot stories, things usually went horribly wrong, probably owing to human guilt and fear over playing God. This guilt reached a peak in the late nineteenth and early twentieth centuries when science and technology became the animating forces breathing life into bots (where magic had usually been before). It was perhaps the fact that technology and science made the creation of such creatures so possible, so within our grasp, that led to the nightmarish visions of Mary Shelley's *Frankenstein*, Karel Capek's play *R.U.R.* (*Rossum's Universal Robots*), and the 1927 Fritz Lang film *Metropolis*. These disturbed visions of robots were also influenced by the encroachment of science and industrialism in general, and the fear that its machinery threatened to overwhelm our humanity. As the twentieth century wore on, and technology became an increasingly essential and accepted part of our lives, many of the robots in sci-fi became more benign. Robots became steadfast companions (B-9), helpers of humanity (C-3PO, R2-D2), even our equals (Lt. Commander Data).

This growing comfort level with machines and machine intelligence might have done wonders for our improved relations with robots of the fictional kind, but it has had some negative effects on real-world robotics. In a word, it's made them dull by comparison (at least to the general public). Kids fed a steady diet of Saturday morning cartoons of building-size mech-warriors, sci-fi films and TV shows, video games, comic books, and other media, don't have much patience with the temperamental, decidedly dumb bots of today. I've seen this in my own home. When we get a new robot, my son is obsessed with it, but only for a few days, a week at the most.

When we got the Probotics Cye robot a few years ago (a two-wheeled bot with a cart and a vacuum attachment), he did nothing for about five days but play with it and teach it (using a map-based teaching tool wirelessly connected from a laptop to the robot). He had big plans for how it was going to serve us dinner and haul our dishes back to the kitchen every night. After creating a "CyeServe" routine (that would deliver food in Cye's cart from the kitchen to the family room), he tried for several nights to work the bugs out (until we just got up and retrieved our rapidly cooling dinners from wherever Cye had gotten himself stuck).

Frustrated and bored, he soon went back to drawing robots that do what they're told, watching them in the media, and constructing them on paper for role-playing games. I don't know that he would act any differently if he didn't have a fantasy world of robots to retreat to, but I'm sure it doesn't help. I can see in his eyes the disappointment in comparing real world robots with ones inside his head, and I recognize it as my own—that kid inside me that still struggles with the same disparity.

In the '80s and '90s, while exciting new robot stories were finding their way into all forms of media through the franchised fantasies of *Star Trek*, *Star Wars*, *The Terminator*, and other blockbusters, a quiet revolution was going on in the halls of academe and on research laboratory benches. Roboticists were rethinking the basic assumptions of what constitutes a robot, machine intelligence, means of mobility, and other issues with which such engineers must wrestle. In many areas of science, looking to biology for design inspiration was becoming fashionable, and robotics was no exception. Soon, the idea of gold-plated humanoids with English accents and persnickety dispositions gave way to far more bizarre visions of dung beetle–like insectoid bots that scuttle under furniture and wrangle your dust bunnies, squirmy snake-like robots that can wriggle into sewers and through rubble, even robotic fish (modeled on the tuna) that can swim the world's oceans collecting scientific data. A weird thing happened. While fictional robots continued to get more and more fantastic (culminating in 1999's *The Matrix*, in which robots take over the world and use humans as their batteries), real-world robots were becoming smaller, faster, cheaper, and better.

Robots: Today and Tomorrow

So far in the twenty-first century, we've seen a resurgence of interest in real-world robotics, thanks in part to the popularity of such TV shows as *Battlebots*, *Robotica*, and *Robot Wars*, and genuinely autonomous robo-pets such as the Sony AIBO. Another trend driving robotic development is the cheap and widespread availability of robotic "building blocks." Building sets (such as LEGO MINDSTORMS), sophisticated (and reasonably priced) microcontrollers, and other building components now allow anyone with the desire and a few bucks in his or her pocket to build a robot.

Contrary to Karel Capek's vision in his landmark 1921 play *R.U.R. (Rossum's Universal Robots)*, there still are no "universal robots." Today's bots, be they factory floor workers, bomb-retrieving police bots, robotic surgeons, or Martian rovers, are tailored to a particular task (or set of tasks). This will likely be the case until robots learn how to learn (and apply these smarts to changing situations). But in many quarters, this isn't even a priority. Biomimicry (taking design cues from nature) is still a chief inspiration to many of today's robot pioneers. In the coming decades, we're just as likely (some would say *more* likely) to see widespread adoption of robots based on swarms of bugs, lizards, flies, and crustaceans, as we are bots built upon human dimensions. And while some labor to create intelligent, adaptive robots with big brains like ours, others are keen on creating bottom-up bots with more modest (and robust) tools for directly interacting with their environments.

note

We'll be discussing robot combat sports and robo-pets in Chapter 2, "Robot Evolution," and reviewing popular robot building sets, kits, and parts in Chapter 6, "Acquiring Mad Robot Skills."

Few recent developments have done more to revitalize people's excitement about real-world robots than the Sony AIBO and the uncannily human-like ASIMO (which recently starred in a Honda car commercial). Robo-tunas, mechanical snakes, and insectoid walkers may hold great promise for, respectively, swimming the oceans, searching sewers, and climbing into bubbling volcanoes, but we humans are still most inspired by robots that look and act like us (or our best friends). What can we say? It's a weakness.

> **note**
>
> We'll look at the many branches of robot evolution and examine their strengths and weaknesses in Chapter 2.

It is likely that our robotic future will include both these branches of robo-evolution, and many more. We haven't even mentioned the R2-D2s of the world, bots who don't try to look like anything special, whose form follows their function. This is the realm of the robot vacuum, the robo-mower, the roving security camera and smoke detector, all coming soon to a home in your neighborhood (most likely yours, after reading this book).

Even today, while we sit in front of the TV, stunned by the futuristic vision of the Honda humanoid, wondering when we'll be able to park one in our garage, more modest devices that meet the criteria of "robot" (which we'll grapple with in the next chapter) are migrating to every corner of our world. Security systems that read faces, fingerprints, or other biometric input, and then open doors (or not) accordingly, are robots. Highway traffic systems that control signs and redirect traffic based on sensor data are robots. "Smart houses" that sense what's going on in and around them and take appropriate actions (like turning up the heat when you literally "phone home" from work) are robots. The cruise missiles used by our military are robots. Robots are already everywhere; you just have to know where (and how) to look.

This book is supposed to be about robots, but it's really about people. It's about our desire to make computers dance, to extend the reach of our technology and our intellects. At its heart, it's a how-to, get-dirty-and-gain-some-valuable-experience book, but it's also a book about how to *think* about robots. In my travels through the robot kingdom, I've learned that (to steal a shopworn street phrase) "it's all good." Some despair over the wide chasm between robots in fantasy and robots in reality, some turn their noses up at one arena of robotic development or another, but I've come to think of it all in exploratory and evolutionary terms. Humans explore. That's what we do. That's who we are. As an explorer, you can't really fail 'cause it's the quest that counts.

It's the same with evolution. An evolutionary dead end tells us something about the reasons for the road not taken (and therefore, something about the roads that were). It is with this spirit of openness and respect for all forms of robotic exploration that we'll set out on our journey into the world of robot building. We'll start off with some basics and the background (the "how-to-think" part), and then we'll delve into the honest-to-goodness how to do. If all goes well, you'll walk away from this book with a righteous knowledge of what robots are, how they work, and how to build 'em. You'll also be the proud owner of three cool little robots. They still won't make beds, do laundry, or help your kids with their homework, but at least one of them (Project 3) might be able to bulldoze your dishes off the dining table. Ah…progress.

PART i

ROBOT.EDU

What's in a Name? 13

Robot Evolution 33

Robot's Rules of Order 61

Robot Anatomy Class 75

İN THIS CHAPTER

- Defining the word *robot*
- Robot innovators adding their two cents to explain the term
- A brief history of bots and an even briefer timeline

1

WHAT'S IN A NAME?

A Bot by Any Other Name…

Before we delve into the wonders of robots, and you find yourself up to your lab coat lapels in gears, motors, and sensor arrays, it might be a good idea to try and gain some clarity (no small feat, as you'll see) about what a robot actually is and where robots come from.

So how *does* one define *robot*? Think you know? It's obvious, you say? Tell you what: You go off and do a quick poll among friends, family members, and online chatterbots, while I go fetch a stack of dictionaries. Because one of my other hats is tracking computer and online jargon and slang, I've got *lots* of dictionaries. Meet me back here in a few minutes…

Okay, so what did you find in your survey? I bet (since I've done it before) you probably got responses something like:

- A machine that can do the work of humans.
- A humanoid machine that can think and act on its own.
- A machine that's self-aware.
- A mechanical man that serves drinks at parties.
- I can't really define *robot*, I just know one when I see it.

These definitions aren't exactly right, are they? They speak to some aspects of robots, but they're either too exclusive ("a machine that can do the work of humans" completely ignores recreational robots such as Sony's AIBO) or too sci-fi (machines are not yet self-aware machines, so that definition would mean that no robots yet exist!). It doesn't take long for you to realize that the last statement ("I know a robot when I see one") is where most civilians stand on the subject.

Surely lexicographers, those word nerds who toil away in musty library stacks in an effort to preserve the rigor of our language, have come up with a more respectable definition? Let's have a look and see.

> **note**
>
> Being a "word worker" (editing *Wired*'s "Jargon Watch" column for nine years), I believe that understanding the names of things and where they come from often goes a long way toward figuring out what they are. These proper labels are also very helpful in doing effective research (using an index, looking up technologies online, and so forth). This might seem obvious, but you'd be surprised. This book contains a useful (I hope), mainly plain English glossary of technical terms and slang related to robots. It should help reinforce what you learn in the text.

The venerable (and heavy!) *Oxford English Dictionary* (okay, it's actually the *Oxford American Dictionary*, but we won't hold that against it) defines *robot* as

> *A machine capable of carrying out a complex series of actions automatically, especially one programmed by a computer.*

The more blue-collar *Merriam Webster Dictionary* offers this definition:

> *1. A machine that looks and acts like a human being. 2. An efficient but insensitive person. 3. A device that automatically performs esp. repetitive tasks. 4. Something guided by automatic controls.*

Finally, here's what the *American Heritage Dictionary* has to say on the subject:

> *An externally manlike mechanical device capable of performing human tasks or behaving in a human manner.*

I could go on (and on). Every dictionary has a slightly different definition, and every one of them is either slightly or way off the mark. Webster's definition is probably the closest to a decent one. It states that robots often mimic human form and function, are often designed to perform tasks that humans would rather pawn off on machines (such as assembly-line production), and are pre-programmed or guided by some type of control system. But as we'll learn later in this book, there are species of self-respecting robots that don't meet any of these criteria. They don't look or act like humans, they don't perform repetitive tasks (they refuse to do dishes!), and they have little in the way of brains (no pre-programmed "maps" of their world, and only the most primitive instructions on how to behave in it).

note

Apologies to amateur robotics guru Gordon McComb. The "I just know one when I see it," is actually a quote taken from his seminal 1987 book, *Robot Builder's Bonanza.* He dedicated it to his daughter Mercedes "who still believes that a robot is a mechanical man that serves drinks at parties."

As mentioned in the introduction, *biomimicry*—the study of nature's designs and processes in search of inspiration for creating machines and processes that solve human problems—is all the rage in robotics right now. The robots skittering and lumbering through today's university laboratories are more likely to look like ostriches or geckos than mechanical men. Embedded systems (also not covered in the preceding definitions) are another area of technological development that can involve robots. Here, computer brains and robot mechanics disappear into the very structure of homes, cars, office buildings, and so forth. Again, think of our pal HAL from *2001: A Space Odyssey.* HAL would be an example of an embedded system robotic spaceship. As you can see, both people on the street and dictionaries define the word *robot* in a way that's too *anthropocentric* (human-centered), and therefore too limiting for our purposes.

So what does the robotics industry itself have to say on the subject? Here's the definition as presented by the Robot Institute of America (circa 1979):

> *A reprogrammable, multifunctional manipulator designed to move material, parts, tools, or specialized devices through various programmed motions for the performance of a variety of tasks.*

Even though this definition is Jurassic on modern technology's timetable (and obviously focused on industrial robots), it is still widely cited and has unfortunately influenced many dictionary definitions.

The Japanese Industrial Robot Association (JIRA) is also chiefly concerned with industrial robotics, but it still went so far as to create a whole robot classification system:

- **Manually operated manipulators**—Machines slaved to a human operator (think Ripley strapped into the robotic exoskeleton loader in the film *Aliens*).

- **Sequential manipulators**—Devices that perform a series of tasks in the same sequence every time they're activated (a phone switching system).

- **Programmable manipulators**—An assembly-line robotic arm.

- **Numerically controlled robots (also known as a *playback robot*)**— Robots that are instructed to perform tasks through the receipt of information on sequences and positions in the form of numerical data. Such robots are typically used for making precision machinery.

- **Sensate robots**—Robots that incorporate sensor feedback into their circuitry—touch sensors, proximity sensors, vision systems, and so forth. (The robots we will construct in Projects 2 and 3 fit into this category.)

- **Adaptive robots**—Robots that can change the way they function in response to their environment. Today's most sophisticated robots fit into this category.

- **Smart robots**—Robots that are considered to possess *artificial intelligence (AI)*. Whether or not any robots like this yet exist depends on how you choose to define AI (and you think *robot* is hard to define!).

- **Intelligent mechatronic systems**—*Mechatronics* ("mechanics" plus "electronics") is a fancy word coined by the Japanese in the 1960s to refer to the intersection of mechanical and electrical engineering and computer control systems. Here it refers to smart devices and embedded systems, such as the highway traffic control "robot" mentioned in the introduction.

The Japanese are certainly on the right track, and it's nice to have such a robot taxonomy to refer to, but if somebody asks me at a cocktail party how to define *robot* and I respond with the preceding list, he's going to think that I'm a bigger nerd than he does already! And committees undoubtedly generated both of the preceding association definitions, and we all know what happens when committees draft things.

Since it's likely dawning on you by now that a concise and widely agreed upon definition of *robot* might not be in your immediate future, let's try a different tack. Let's ask some of the guys and gals in the trenches, those who work with robots every day, how they define the term.

Carlo Bertocchini (Battlebots champion, owner of RobotBooks.com):

> *Deciding if a machine is or is not a robot is like trying to decide if a certain shade of greenish blue is truly blue or not blue. Some people will call it blue while others will vote not blue. It is very difficult to call one person's judgment correct and another's incorrect. The same is true with robots today, and I suspect it will remain true for a long time to come.*

Roger Gilbertson (muscle wire guru and owner of The Robot Store):

> *I define a robot as any autonomous sensor-processor-actuator system that functions in a specific world.*

Fred G. Martin (Assistant professor of computer science at UMass, creator of the infamous 6.270 Autonomous Robot Design Competition at MIT):

> *The term robot, while accurate, is too mentally confining. By my definition and thinking, systems such as musical light shows (which incorporate environmental sensing or sound processing), highway monitoring systems, hydroponic farms (with chemical sensors and environmental controllers), and even exercise machines which incorporate body sensors and performance monitoring features, can all be considered robots.*

Hans Moravec (robotics researcher and author of *Mind Children*):

> *How about: a machine with sensory and behavioral abilities once found only in animals and humans.*

Marc Thorpe (*Robot Wars* mastermind):

> *I would define a robot as any mechanical creation that has an identity apart from its functionality. Control and artificial intelligence are factors that define a robot to some extent. And, though I think control and AI are the determining factors in the development of robotics, I feel that neither is the defining factor. For me, paradoxically, a robot is a mechanical lifeform. In this sense, an automobile is not a robot, but a wind-up tin "robot toy" is. To me, a machine is a tool;*

note

Most of the roboticists whose definitions are included here are profiled in our *Heroes of the Robolution* trading cards in Chapter 2, "Robot Evolution." If some of the ideas and terms here are foreign to you, don't worry—they won't be by the time you've finished this book.

a robot is a mechanical being—usually with a unique identity and often a name.

Rodney Brooks (Director of the Artificial Intelligence Lab at MIT and author of *Flesh and Machines: How Robots Will Change Us*):

> *A robot is a machine which senses the world, computes, and then decides on some action in the world which has a physical reach beyond itself. The reason for the last clause is to make it so that a washing machine doesn't fit the definition. At the same time, I don't want to say that a robot has to be mobile, as there are lots of robots with fixed locations—as long as they can reach out somewhere beyond themselves. As with all definitions, it is possible to push on this one until it breaks.*

As you can see, there's enough ambiguity about what constitutes a robot to fry our circuits and make smoke shoot out of our ears! We could try to deconstruct all of these definitions, look for common threads and logical inconsistencies, and then extend to you our attempt at a super-duper ironclad definition. But I won't bother. I will, however, tell you how *I* define the term.

While I was working on this chapter, I got a copy of a book called *Mad Professor*, written and illustrated by Mark Frauenfelder—the same Mark Frauenfelder who illustrates this book. *Mad Professor* is a book of weird and wonderful kitchen chemistry experiments for little kids. In a chapter on robotics, Mark defines a robot as "a machine designed to do things that humans and animals can do." To the ears of babes—that's it! It makes it clear that a robot is a machine, a real-world mechanism, not a software program, and that it is designed by humans to mimic the behaviors of humans and animals (not just humans, as it is so often defined). This definition doesn't specify the need for computer control, artificial intelligence, or sensors. It also doesn't limit the definition only to robots that work, nor does it exclude things such as remote-controlled bots, BEAM robots (which we'll discuss throughout this book), and robots created solely for entertainment purposes, such as AIBO.

Our childish definition might be a bit too simplistic for some. We could fortify it a bit more for adult tastes. How about this: "A machine designed to mimic the behaviors of humans and animals, often in an effort to do jobs that are undesirable, dangerous, or inaccessible to humans, and often accomplished through the use of sensors, actuators, and computer control programs." Anyway, that'll serve as our tied-with-a-slipknot definition of robot. What's yours?

Bots in Brief

Part of the confusion over what constitutes a robot has to do with the many realms in which the term has gained currency. We won't spend too much time discussing robots in fiction, but because much of the concept comes from the fevered imaginations of science fiction writers, there are several milestones in that genre that should be mentioned (and one widespread misconception to clear up). After that, we'll take a look at three significant moments in real-world robot history, and then run down a few more milestones on the road to the robots of today.

Domo Arigato, Mr. Roboto

The term *robot* comes to us from the Czech word *robota*, which means forced labor or servitude. In Czech, a *robotnik* is a peasant or serf. The term was first introduced in Karel Capek's play *R.U.R. (Rossum's Universal Robots)*. Capek wrote *R.U.R.* in 1920, and it premiered in Prague in 1921. The play was introduced to the West when it was performed in New York in 1922, and was subsequently published in English in 1923. *R.U.R.* was controversial, widely debated in intellectual circles, and the term *robot* quickly replaced the earlier term *automaton*.

In the play, a mad scientist-type, "old Rossum," has discovered the secrets of creating artificial life (see Figure 1.1). His goal is to usurp the powers of God by proving that man has now acquired the intelligence and technology to create life. Rossum's far more pragmatic nephew, an industrialist, sees in artificial humans the perfect worker, a tireless laborer who doesn't complain, doesn't require health insurance, and doesn't demand a paycheck.

FIGURE 1.1

In later productions of *R.U.R.* (this image is from a U.K. production in the 1930s), Rossum's robots became menacing metallic men.

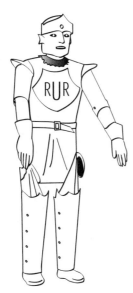

Capek's play is an industrial age morality tale that introduces many of the key questions concerning artificially intelligent robots that we still chew on today. Would a robot with true AI be considered a "person" or a possession? Would signs of an emerging free will be grounds for liberating the robots or sending them back to the factory to have their software re-installed? Is it just a matter of time before technology supercedes biology as the vehicle of evolution? And does that *have* to be a bad thing?

Ironically, Mary Shelly's *Frankenstein* is, in some ways, more of a robot story than *R.U.R.* It describes a man being constructed of assembled components (albeit fleshy dead human ones) and animated with electricity. Rossum's robots are more like clones or some other product of genetic engineering, grown in vats with a chemical process. It was only in later productions of the play in the 1930s that the robots began to be depicted as gleaming metal constructions.

It's also worth noting that the man who introduced the word *robot* to the world was always befuddled by the attention he got for it (he considered *R.U.R.* a minor work). In fact, he found the whole idea of (real-world) robots to be ridiculous. In an essay in 1935, he wrote:

> It is with horror, frankly, that [I] reject all responsibility for the idea that metal contraptions could ever replace human beings, and that by means of wires they could awaken something like life, love, or rebellion. [I] would deem this dark prospect to be either an overestimation of machines, or a grave offence against life.

There are many misconceptions about Capek's play, but the most common one is the belief (found in many dictionaries, text books, and books on robotics) that Karel himself coined the term *robot*. In fact, his brother Josef, also an accomplished

note

Some other interesting tidbits about *R.U.R.*:

- The name "Mr. Rossum" roughly translates to "Mr. Brain" or "Mr. Intellectual."

- A friend of Capek's claimed that Karel got the idea for *R.U.R.* while riding in a car (still a rather new experience). Looking out the window, the people he passed looked artificial, strangely nonhuman. This "vision" became the inspiration for the play.

- The 1973 Paul Selver translation of *R.U.R.* for Simon and Schuster is heavily edited and has a watered-down ending. Look for a version prior to this, or the Capek collection *Toward the Radical Center* (Catbird Press, 1990).

- If brother Josef hadn't suggested the word *robot*, this book might have been called *The Absolute Beginner's Guide to Building Labori*. That's the name Karel first considered for his creatures.

writer, dreamed it up. Josef had actually coined the word for a short story in 1912. When his brother was racking his brain for a name for his artificial workers, Josef suggested the term. The English spelling of it appealed to Karel, and he decided to use English for the title of his play to call attention to the globalizing effects of science and technology. Pretty advanced stuff for 1920, don't you think?

No More Rogue Robots!

By the 1930s, the widespread popularity of *R.U.R.* and the growing sci-fi pulp fiction movement led to a profusion of short stories about robots. Nearly all of them copied the basic template of Capek's tale (and *Frankenstein* before it). This formula came to be nicknamed the "rogue robot plot" (man makes bot, bot turns on man, either bot survives and man doesn't, or man survives and bot doesn't). A young science fiction writer named Isaac Asimov got tired of rogue robots and wanted to explore robots from a different, more scientifically thoughtful perspective. In 1940, the first of what came to be known as his *Positronic Robot* series was published. The story, "Strange Playfellow," appeared in Frederik Pohl's *Super Science Stories*. It had actually been written and widely submitted in 1939, under the title "Robbie," but no magazines were interested.

The highly intelligent and restlessly curious Asimov was probably the first writer to do any deep thinking about how robots might actually work (Capek and others were more concerned with the moral and religious implications). Annoyed at the unscientific nature of previous efforts, Asimov thought that surely a society advanced enough to create an intelligent machine would be smart enough to program it so that it would be incapable of harming its makers. In "Strange Playfellow," a robot (Robbie) is a nursemaid for a little girl. Her mother exhibits many of the same fears found in rogue robot stories, but her more rational father points out that Robbie "can't help being faithful and loving and kind. He's a machine—made so." This idea of being programmed not to do harm was the beginning of what Asimov would later develop into his "Three Laws of Robotics" (which we'll explore in Chapter 3, "Robot's Rules of Order").

> **note**
>
> In a nod of appreciation to Asimov, Gene Roddenberry, creator of *Star Trek*, gave his android character Data a "positronic brain." Although both men were keen on a high degree of scientific believability in their work, this invention of Asimov's was pure fiction. The positron (a positively charged subatomic particle) had recently been discovered by Carl Anderson (in 1932) and Asimov needed a sexy-sounding name for his robots' noggins.

Throughout the 1940s, Asimov wrote more robot stories for *Super Science*, *Astounding*, and other pulp sci-fi magazines. In 1950, the best of these stories was published in the collection *I, Robot*. It was this book that introduced a wider audience to Asimov's ideas about robots and his "laws" for robot behavior. Asimov continued to explore his thinking about robot behavior, design, and the robot's impact on society for the rest of his career.

It is fitting that, because Asimov's robot stories had such a huge impact on the engineers destined to bring robots into the real world, he should be responsible for coining the term *robotics* (for the science of studying robots). He did so in his 1942 short story "Runaround." Amazingly, the coinage was accidental. Asimov had no idea the word didn't already exist and was somewhat surprised to discover that he was its author.

The Real Rossum: Industrial Bots from Unimate to Co-Bots

The fantasy: a legion of vat-grown androids that can do all manner of menial labor without complaint. The reality: A bulky metal arm with hydraulic actuators that looks like an artillery piece. This real-world analog to Rossum's robots came about because of a chance meeting, and because of Isaac Asimov.

Legend has it that, in 1956, Joseph Engleberger, now considered the father of industrial robotics, met George Devol over cocktails. Devol was an inventor and entrepreneur; Engleberger, an aerospace engineer. The two discovered they had a mutual love of sci-fi, especially the works of Isaac Asimov, particularly his robot stories. Devol told Engleberger of a patent he'd been issued on a process called "programmable manipulation." Engelberger immediately saw the potential for this technology, the two of them sought funding, and soon they founded Unimation (Universal Automation), the world's first robot maker.

> **note**
>
> The well-known computer networking hardware manufacturer, U.S. Robotics, took its name from US Robotics and Mechanical Men, Inc., a company that appeared in many of Asimov's robot stories.

General Motors was the first company to install a Unimate in 1962 (see Figure 1.2). The arm was used for extracting hot parts from a die-casting machine, a dangerous job for a human worker. Just as fictional accounts of robots had captured the public's imagination before, massive publicity over the Unimate fueled a new wave of robot stories, a lot of green lighting of robot projects and purchases in boardrooms, and ushered in the so-called "Age of Automation." Once again, there was a backlash when reality didn't meet people's expectations, and growth of the industrial robotics industry had faltered in the U.S. by the 1980s.

FIGURE 1.2

The arrival of Joe Engleberger and George Devol's Unimate robotic arm on factory floors in the 1970s ushered in the "Age of Automation."

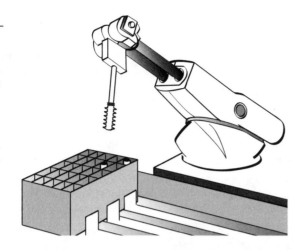

Although industrial robotics floundered in the United States, it took off in Japan, thanks to its '80s economic boom. Where we have had a love/hate relationship with robots in the West, it's all about the love in Japan. This has been attributed to many things, from their animistic religious background (no Judeo-Christian guilt), to their insularity (no immigrant labor force), to their overall passion for techno-toys. Whatever the reason, today robots saturate Japanese industry and many other sectors of life. In 1998, there were some 280 industrial robots for every 100,000 humans working in industry there.

In the United States, industrial automation has regained its lost ground in the last decade (although Japan still dominates in the actual making of robots). A new popular feature on American factory floors is a device called a *co-bot*, a "cooperative robot" that is manipulated or guided by a human (the bot does the heavy lifting; the human does the navigation and decision-making).

One Small Dance for Robo-Critter Kind

Before Unimate made inroads into humanoid robotics (okay, so it was just an arm and a gripper-hand, but hey, that was a start!), a renowned American neurophysiologist named W. Grey Walter was quietly exploring robots of an entirely different nature. In 1948, Grey conducted a series of experiments in mobile, autonomous robotics. Keep in mind, this is the late '40s—we're talking vacuum tube technology! Walter was interested in using his mobile robots to model simple brain functions. He wanted to explore his theory that complex behaviors arise from the simple firing of neurons.

Walter built a series of small mobile robots he called tortoises (we loutish Americans usually refer to his bots as turtles). The first two robots Walter built were named Elmer and Elsie (see Figure 1.3). These mousey-looking machines might seem crude by today's standards (with their plastic shells removed, they look like little steam engines), but for their time, they were a marvel. The most revolutionary thing about them was that they didn't have any brains or pre-programming. They used basic analog electronics of the time, two vacuum tubes (two "neurons" per bot), a touch sensor, and a light sensor to create robots that could avoid large objects, seek and avoid light, and even recharge their own batteries. The ideas underlying these robots (no program, rudimentary circuits, and built on animal rather than human inspirations) were very unconventional, and very ahead of their time.

FIGURE 1.3

This version of one of Grey Walter's ground-breaking robot "tortoises" is a restoration done in 1995. It is housed at the Smithsonian Institution. Photo courtesy of The National Museum of American History, Smithsonian Institution.

Walter was convinced the experiments would prove his point, and Elmer and Elsie did not disappoint. Walter found the interactions between his two robots and their environment, and between each other, to be extremely suggestive of living beings. They never followed the same path twice. They seemed to dance around each other, like two animals might—attracted at one point, repulsed the next.

In his observations of Elmer and Elsie's scuttlings, Walter described what today would be called *emergent behavior* (apparently spontaneous and complex behavior that arises from simple entities—whether real or artificial—interacting with each other or their environment). About these seemingly lifelike patterns, Walter wrote: "[I] noticed uncertain, randomic, free-will or independent characteristics [which constituted] aspects of animal behavioral and human psychology...despite being crude [the tortoises] conveyed the impression of having goals, independence and spontaneity."

Many budding robot engineers, and others in emerging postwar sciences such as cybernetics, were fascinated by Walter's experiments (first published in *Scientific American* in 1950) and they continue to have a significant impact today. Robot pioneers Hans Moravec, Rodney Brooks, and Mark Tilden (all of whom we'll learn more about in future chapters) were inspired by, and have built upon, Grey Walter's work. The robot we will build in Project 2, Mousey the Junkbot, is also a "two-neuron" mobile robot. It has the same basic components as Elmer and Elsie and will exhibit similar behaviors to its famous ancestors.

Off to a Shakey Start

Big, dumb robotic arms and little scurrying robot-critters are great, but as we've already established, we humans have a weakness for trying to cast robots in our own image, and that means brains. We humans have big, fancy brains (or, as the author Tom Robbins once pointed out, that's what the brain tells us). In the late '60s, researchers at Stanford Research Institute (now called SRI International) began work on what was to become the first "intelligent" robot.

tip

If you want to build your own version of a Walter Tortoise, you can! An enterprising LEGO roboticist named Michael Gasperi has re-created one of Walter's robo-critters out of LEGO blocks (and a homemade sensor). Details can be found at www.plazaearth.com/ usr/gasperi/walter.htm.

Their robot, Shakey, was designed to be a mobile robot that could reason on its own, and it did that...sorta. The robot was built on a four-wheeled base (two drive wheels and two caster wheels). This type of simple two-drive-wheel robot development platform has become *de rigueur* in robot research (we'll build a mini version of our own in Project 3). Needing to control only two drive wheels eliminates the need for a sophisticated steering mechanism. Shakey also sported wire fender-type bump sensors, a navigation system that used laser range finders to triangulate its position, and a bulky (by today's standards) TV camera (see Figure 1.4). The robot got its name because, laden with all of its large '60s/'70s-era technology, and designed with a tall camera and radio mast, it swayed and shuddered as it moved. All of the data to and from Shakey went over a radio link to DEC PDP-10 and PDP-15 minicomputers. (Ah...the DEC PDP. Memories, memories.)

For Shakey's noodle, a program was developed at SRI called *STRIPS (STanford Research Institute Problem Solver)*. Shakey's programming worked in three distinct layers. The first layer dealt with basic moving, turning, and navigating. The second layer took that information and strung it together to build routines that it could execute. The third layer was designed to receive instructions, plan the best course of action, and then execute the plan.

note

If some of these terms such as two-wheeled drive mechanism, range finder, and bump sensor befuddle you, don't sweat it. We'll look under the hood of these nifty technologies in upcoming chapters.

FIGURE 1.4

Shakey, SRI's intelligent robot, contemplates his next move. Note the solid-shaped, colored objects that constituted Shakey's "world." Used by permission of SRI International.

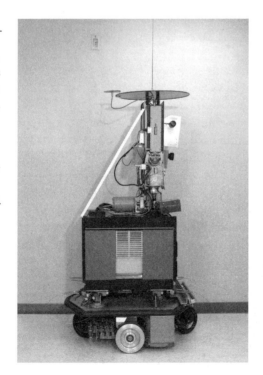

Armed with its planning software, Shakey was supposed to be capable of taking its camera, bump sensor, and range data, and making decisions to achieve a goal given to it by researchers. Unfortunately, Shakey was a temperamental sort, and spent a lot of time on the lab floor contemplating his metallic navel (decisions and actions took hours), or on the repair bench. Work first began on Shakey in 1966 and various versions and upgrades appeared until 1972. On a good day, Shakey offered a thrilling glimpse into the future of artificially intelligent machines. Isaac Asimov, in his 1985 book, *Robots: Machines in Man's Image*, recounts a fascinating Shakey tale. The robot was "asked" to push a box off a platform, but it couldn't get to the box. Shakey eventually found a ramp, pushed it against the platform, rolled it up, and shoved the box off onto the floor.

note

Shakey had an intriguing way of looking at the world. Probably owing to the limited computer processing power of the time, Shakey's program didn't analyze objects—it looked only at the place where the lines of object planes intersected, their vertices. So, a box was seen as nothing more than the points where its angles met. It was the same with triangles and other shapes.

In 1984, Shakey's descendant, Flakey, was "born" at SRI. With what-a-difference-a-decade-makes flamboyance, Flakey carried its own workstation-grade computer, had 12 sonar range finders on board, and could navigate many environments, not just an idealized laboratory one. Flakey was also the first bot to use Saphira, a software development tool for robots that's still in use today in commercial robot maker ActivMedia's (www.activmedia.com) robots (see Figure 1.5).

FIGURE 1.5

Shakey's great, great, great grandchild, the commercially available AmigoBot from ActivMedia. Used by permission, ActivMedia Robotics, www. MobileRobots.com.

Great Moments in Robot History

The events in the evolution of robo-kind that we've discussed so far only scratch the surface. There are more, so many more. We'll detail some of these other advances in later chapters. But in the meantime, let's fast forward at 4x speed through some other significant milestones along the road to our robotic future.

- ■ **Third Century B.C.**—Aristotle gets the robo-ball rolling by penning these words: "If every instrument could accomplish its own work, obeying or anticipating the will of others…if the shuttle could weave, and the pick touch the lyre, without a hand to guide them, chief workmen would not need servants."

- ■ **Eighteenth Century**—Automatons, clockwork "robots" built by inventors and watchmakers in France, Germany, and elsewhere, become a courtly rage. These mechanically animated dolls can play musical instruments, draw, and quack like a duck (well, the duck one could, anyway). Some of these automatons even had different "programs" (interchangeable cam sets) that would alter their actions.

- ■ **1801**—Joseph Jacquard invents a "programmable loom" that operates via punch cards.

- ■ **1898**—Nicola Tesla, inventor of the induction motor, alternating current (AC) transmission, and the *actual* inventor of radio (and a bunch of other stuff he's not given credit for) patents the *teleautomaton*. Children around the world will come to know it as the remote-controlled toy. The stage is set for a future of *Battlebots*.

- ■ **1940**—Westinghouse's Electro and Sparko spokesbots entertain crowds at the World's Fair in New York. These animatronic wonders marked the first time that electric motors were used to power and actuate robots.

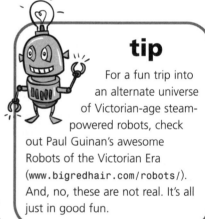

tip

For a fun trip into an alternate universe of Victorian-age steam-powered robots, check out Paul Guinan's awesome Robots of the Victorian Era (www.bigredhair.com/robots/). And, no, these are not real. It's all just in good fun.

- ■ **1956**—MIT's John McCarthy coins the term "Artificial Intelligence" during a Dartmouth computer workshop.

- **1960**—The Johns Hopkins "Beast" begins prowling the university's hallways. Similar in functionality to Grey Walter's Elmer and Elsie, it sports more complex sonar navigation and can seek out wall sockets when its Beastly batteries get that empty feeling.

- **1973**—Tokyo's Waseda University, under the direction of Ichiro Kato, develops WABOT-1, the first humanoid walking robot. Honda is impressed and secretly begins work on its own humanoid (which will become the infamous P3).

- **1976**—Robot arms in space! Viking 1 and 2 Martian landers are outfitted with robotic arms first developed at Stanford University.

- **1979**—The Stanford Cart is rebuilt by a young Hans Moravec (see *Heroes of the Robolution* trading cards, Chapter 2). Work had actually begun on the cart in 1965. It is billed as the first computer-controlled autonomous vehicle after it successfully crosses a chair-filled room without human intervention.

- **1982**—The Heath Company begins sales of their HERO 1 personal robot kit.

- **1983**—Androbot, Inc., a company founded by Nolan Bushnell of Atari Computers fame, releases TOPO I, a personal home robot, followed by BOB (Brains on Board), a surprisingly sophisticated robot for its time. Androbots' futuristic good looks, shared by many imitators that followed, spur a small home robot boom. But ultimately, given computer tech of the time, the bots do not live up to expectations and the boom goes bust.

- **1984**—Red Whittaker's team from Carnegie-Mellon University (CMU) sends its RRV robot into the irradiated landscape of the Three Mile Island nuclear plant site and the "field robotics" industry (mobile robots doing actual work) is born.

- **1994**—CMU's six-legged extreme environments robot, aptly named Dante II, successfully climbs into the Mt. Spurr volcano in Alaska.

tip

If you're interested in finding out more about any of the robots listed in this timeline, pop their names into Google (`www.google.com`) or your favorite search engine, and you're likely to get a motherload of info.

- **1997**—IBM's Deep Blue computer beats world chess champion Garry Kasparov in a match.

- **1997**—Shockingly, the Mars Pathfinder actually lands. Its Sojourner Rover delights and amazes armchair astronauts around the world, who watch near real-time coverage on TV and the Internet. Suddenly, kids think robots are cool again.

- **1997**—Honda stuns the world with its P3 humanoid robot. Unnerved skeptics swear it must be a child in a plastic space suit.
  It isn't.

- **1999**—Sony begins selling its AIBO robotic pet. The product is almost affordable and extremely sophisticated. Anti-cutesy Web site critics the world over shudder at the thought of personal AIBO pet pages (there are now *hundreds*).

- **2000**—*Battlebots* premieres on *Comedy Central*. Gearheads all over America head to the garage to cannibalize the family mower.

These are only a few of the significant stories to be found in the rapidly growing annals of robot history. If you see glaring omissions, it might be because we'll be detailing these bots in other sections of the book (for instance, the release of LEGO MINDSTORMS and the creation of Rodney Brook's COG).

THE ABSOLUTE MINIMUM

This chapter got our feet wet in grappling with some of the issues surrounding the very word *robot*. Many groups of people, from scientists and industrialists to artists and toymakers, call it their own. We also touched on some of the watershed moments in robot history, bringing us up to speed with the robots covered in the rest of this book. Here's what you would have learned if you'd been paying attention:

- Defining *robot* is hard. We gave up. You'll know one when you see one.

- Don't believe all of those *other* robot books and dictionaries! The Czech playwright Karel Capek might have brought the word *robot* into the world, but it was his brother Josef who coined it. Brothers never get any credit.

- Capek's dystopian play led to many bots-gone-wild sci-fi stories in the 1930s and '40s. These came to be known as "rogue robot stories."

■ Pulp sci-fi author Isaac Asimov got sick of rogue robots and began publishing his highly influential "positronic robot" series, which came to define much of how we think about robots. If you haven't read *I, Robot* yet, do it now! Go ahead, we'll wait.

■ We identified three significant moments in real-world robot history: the introduction of the Unimate robotic arm (1962), the earlier, lesser-known, but equally significant autonomous mobile robot experiments of Dr. Grey Walter (1948), and the...ah...shakey emergence of Shakey, the first robot with any brains. So, we had brawn, we had mobile autonomy, and we had basic brainpower—robots were off to a good start. And so are you...

In This Chapter

- A little sci-fi "micro-fiction"
- The many branching paths of robot evolution
- Humanoids R Us
- Is Mother Nature the ultimate robot builder?
- Robotic evolution through human competition
- *Heroes of the Robolution* trading cards—collect the whole set

2

ROBOT EVOLUTiON

Don't Call Me Chief!

The year is 2008. You're in your home office working feverishly on your latest book and you're in dire need of a major snack break. "Dexter, can you come here?" you shout. From the living room, you hear the sound of the vacuum cleaner powering down and the increasingly loud whine of servo motors as your Sony Roboman 3000 lumbers in. "Can you get me a diet soda, the guacamole from the fridge, and that bag of corn chips from the kitchen counter?" you ask. "Hey, not a prob, Chief," your bot responds, in disturbingly casual English. You sigh. "I thought I told you not to call me 'Chief'," you scold, as Dexter heads for the kitchen. You make a mental note to complain to the service company about the slacker attitude of today's "brain." Dexter brings back your snacks, along with the day's hard mail. "Well, that's the end of my shift, Chief. Call me history," he announces, after he sees that you're happily munching away. Without waiting for a reply, he walks over to the recharging station and snaps himself in. The activity lights on his chest and head slowly wink out as your R3000 goes into "stand-by" mode.

Across the country, in a nondescript California office building that once housed an endless parade of dot-com start-ups, your "brain" peels off his virtual reality (VR) headset and steps out of his motion capture suit. "What a load that guy is," he snorts, fluffing up his multicolored dreadlocks. "A writer? Ha! And I thought my gig offered poor job security." He hops onto his gyro-stabilized scooter and buzzes down a hallway lined with cubicles of other people in VR suits hired to inhabit the world's robots, to serve as the higher intelligence that, so far, still eludes the world's machines.

Is this the vision of our near-term robotic future? Rodney Brooks, of MIT's Artificial Intelligence (AI) Lab and cofounder of iRobot Corporation, thinks it may be (or at least something like it). Dubbed *robotic presence* (the operating of a robot body by a human), it represents a clever "workaround" to a daunting problem facing robot engineers: How can we give robots big brains sooner rather than later? As recent innovations such as the Honda P-series, the Honda ASIMO, and the Sony SDR-X have shown, we've come a long way in terms of the mechanics of robotics, even working out such significant problems as bipedal balance and walking gaits. But as futuristic as these prototype robots look, they're still relatively stupid. The Honda robots are only smart from the waist down. An "off camera" operator is actually controlling the top half and making decisions for the robot. And the Sony bot, although extremely sophisticated by previous standards, is basically an upgrade of the same technology developed for the AIBO robo-pets loaded into a small humanoid body. None of these robots, even when they hit the market, are going to be serving anybody chips and guacamole anytime soon.

note

Because this is a book about *building* robots, keep in mind as you read how these various approaches might come into play when actually building robots of your own. We'll point out when what's being discussed here relates to the robot kits and building sets discussed in Chapter 6, "Acquiring Mad Robot Skills," and the robot projects we'll be building later in the book (Chapters 7, 8, and 9).

In the last few decades, robot research and development has peeled off in a number of divergent directions. The goal is basically the same: to create autonomous artificial life forms that can perform useful work, but the paths to getting there are as diverse as the eccentric cast of characters forging them (see the *Heroes of the Robolution* "trading cards" throughout this chapter).

In this chapter, we'll explore some of these evolutionary paths to a roboticized future. Few people argue that robots won't be an important part of our future, but there are major schoolyard scraps over which developmental approaches are going to get us there. As in all areas of academia and scientific investigation, careers and egos are at stake, so there's a lot of high-domed, one-upmanship and the slinging of fancy multisyllabic mud, in books, published papers, and lectures. But we as amateur robot builders and armchair investigators don't have to soil ourselves with any of this. It's fun, fascinating, and very enlightening to stand back and consider all of the various schools of robotic thought and development. We can speculate where these varies tracks might lead, and even how it is likely that the future will be a hybridized version of these varied approaches. Let's take a brief look at some of today's most promising areas of robotic development.

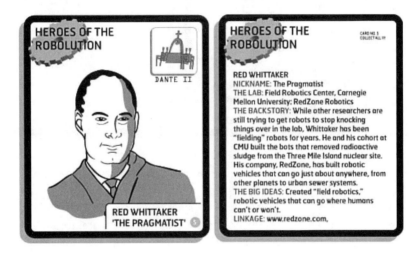

Humanoid Robots

As we've already discussed, we humans love a robot fashioned in our own image. Science fiction has pushed this idea of what a robot should be, almost to the exclusion of all others. So it's no wonder that anytime a robot builder—amateur or pro—creates a robot with two legs, two arms, and a head, the world beats a path to the laboratory door. The world has been so taken by Honda's flashy humanoid robots that few people (including otherwise skeptical journalists) have bothered to ask what sort of brains control these bots. The answer is: human. Honda spent over a hundred year's worth of development hours (and untold yen) getting the walking technology perfected, and figuring out how to cram all of the electronics, servos, and batteries into a humanoid shell. The P-series and the ASIMO are really proof-of-concept models for robotic mobility. The progenitor of the line (the E1) was actually *only* a set of legs, with a weight above the hips to approximate a full body weight. It was only after Honda engineers got the walking gait and balancing technologies working that they built the rest of the robot.

There's nothing wrong with this approach; in fact, as we'll see in "Behavior-Based Robotics (BBR)" later in this chapter, a bottom-up approach (in this case, literally) makes a lot of sense. But Honda appears to want people to think that its robots are more advanced and a lot smarter than they actually are. In other words, it isn't going out of its way to point out that a human operator is calling the shots. Honda, and other Japanese robot makers (like Sony), and the world's humanoid robot research labs, are thrilled that these bipedal bots have so thoroughly captured the public's imagination. Now, while we're all busy watching ASIMO deftly walk out of the family garage to fetch the morning paper on Honda's TV ads (note that we never actually see little 'MO bend over and *pick up* the paper), the company's hard at work trying to fill this fantasy container they've built with some honest smarts.

HEROES OF THE
ROBOLUTION

SAIKA 3

HIROCHIKA INOUE
'OLD MAN OF JAPANESE ROBOTICS'

HEROES OF THE
ROBOLUTION

CARD NO. 6
COLLECT ALL !!!

HIROCHIKA INOUE
NICKNAME: Old Man of Japanese Robotics
LAB: Inoue-Inaba laboratory, University of
Tokyo
BACKSTORY: One of the directors of Japan's
humanoid robot initiative, a massive effort to
create an intelligent humanoid robot in a
decade. Specializes in robotic vision and
learning by seeing.
BIG IDEAS: Humanoid robotic form makes
sense because environments are already
designed for human manipulation (access
heights, machine controls, transportation,
etc.). Inoue calls this "softness," the ability for
bots to fit in without big changes to the
environment.
LINKAGE: www.jsk.t.u-tokyo.ac.jp

The Honda P-series, ASIMO, the Sony SDR-4X, and many other not-as-photogenic robots are all part of a countrywide humanoid robot initiative in Japan. Robot researchers there are hard at work trying to create autonomous bipedal robots that can cope with changing environments and perform complex tasks. Although the Japanese plan on doing this within a decade (the project clock started ticking in 1998, by the way), you realize just how far we still have to go when you hear Sony making a big deal over the fact that the SDR-4X can walk from a wooden floor onto semi-thick carpet without falling over. The idea behind the Japanese humanoid project is also to create robots that are not necessarily universal, but can at least handle multiple tasks. One application commonly talked about is care for the elderly. One can only imagine how many different and changing circumstances *that* might entail.

tip

Building a humanoid in your spare time is not a project undertaken lightly, but that doesn't stop some hearty amateurs from trying. The site Android World (www.androidworld.com) chronicles their efforts, details available hardware, and links to the latest R&D projects.

Robotic Presence

To robot pragmatists like Rodney Brooks, we don't need to wait decades for robots to develop sophisticated enough noggins to be able to separate laundry and iron dad's dress shirts. For years now, Brooks's company iRobot has been experimenting with, and even marketing, machines that a human can "robot in" to. Its first prototype, which it began showing off in August 2000, was called the iRobot-LE. It had eight wheels—six drive wheels and two additional wheels raised up in the front that enabled the robot to travel up stairs (a significant innovation). But the main purpose of the LE was to demonstrate the concept of robotic presence. Although the LE had no manipulators or anything else on it that would enable it to do useful work (as in the sci-fi story that started this chapter), it did enable the user to be in two places at once.

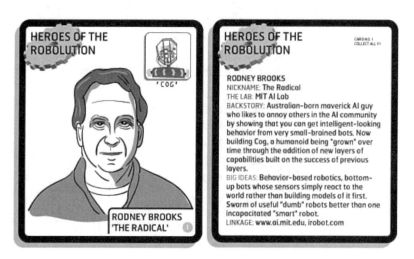

HEROES OF THE ROBOLUTION

'COG'

RODNEY BROOKS
'THE RADICAL'

HEROES OF THE ROBOLUTION

CARD NO. 1
COLLECT ALL 11!

RODNEY BROOKS
NICKNAME: **The Radical**
THE LAB: **MIT AI Lab**
BACKSTORY: Australian-born maverick AI guy who likes to annoy others in the AI community by showing that you can get intelligent-looking behavior from very small-brained bots. Now building Cog, a humanoid being "grown" over time through the addition of new layers of capabilities built on the success of previous layers.
BIG IDEAS: Behavior-based robotics, bottom-up bots whose sensors simply react to the world rather than building models of it first. Swarm of useful "dumb" robots better than one incapacitated "smart" robot.
LINKAGE: www.ai.mit.edu, irobot.com

Imagine planning a trip to your parents' house for Thanksgiving. At the last minute, your trip has to be canceled when the airport gets snowed in. "That's okay," says mom over the phone. "You can spend the day with the family anyway by *robotting in* via our new remote presence rover." Your dreams of a rare Thanksgiving away from your loudmouth brother-in-law and all-hands Aunt Edna fade away as Mom gives you the appropriate Web address. You log in and soon are seeing what the robot sees (through a high-quality color camera in its head). Microphones on the robot let you hear the endless prattle about who's in the hospital, who just got divorced, and who just bought a vacation home at Myrtle Beach. You confound everybody into silence when, desperate to participate, you ask (through the speaker on the bot's neck) if anybody's read a good book lately. After more chitchat and hors d'oeuvres (them: Brie and cranberry tarts, you: Ritz Crackers and Cheese Whiz),

Mom tells everyone to move into the dining room. You're stunned when you realize that, for a second, you forget you aren't really there. You find yourself "walking" down the hallway, asking your niece and nephew what they've been up to. As you "sit" at the Thanksgiving table—your robotic eyestalk head raised up to human head height—looking over your extended family (and the extended feast), a feeling of well-being and…well…presence comes over you. You're glad you came, after all. Your family's not so bad, and you're thankful that advanced robot and networking technology has made such a virtual visit possible. Then you come to your senses and realize that you've probably just inhaled too many fumes from the aerosol cheese.

note

Although the iRobot-LE never made it into the home market, iRobot does sell a robot designed for remote presence. Called the Co-Worker, it's basically a Web conferencing system on wheels that enables a remote business associate to robot in on meetings, building walk-throughs, to do "on-site" troubleshooting, and so forth. Information can be found at `www.irobot.com`.

A robot like the one described previously (which is basically how the iRobot-LE worked) is a baby step toward full-blown working humanoids with remote-login brains. In Brooks's 2002 book *Flesh and Machines* (ISBN: 0-3757-2527-X), he offers an argument for why such a near-term humanoid future makes so much sense. Besides the domestic "killer apps" of vacuuming and lawn mowing, what other dreadful chores would you like to pawn off on mechanical domestics? Dishwashing? Laundry? Ironing? Cooking? Now imagine what a huge developmental leap will be required to automate these jobs, which not only require a great degree of manual dexterity, but also require significant decision-making smarts. A robot would have to discern colors, textures, shapes, weights, measures, pressures, and make complicated, even potentially dangerous, decisions (by AI standards, anyway). But what if you hired a human to robot in to work, to inhabit some future generation of Honda robots that had more sophisticated vision systems and more dexterous limbs, but weren't really that far from today's P3 model? The hired brain could make all of the choices required to do these domestic tasks. And you'd have a nearly universal robot. You wouldn't need a separate bot for the yard work, the dusting and vacuuming, the laundry, and the kitchen. One human/robot "symbiot" could do it all. Brooks imagines people from third world countries, with little local work opportunities, being hired to robot in from home (or from a well-wired workplace) to anyplace else in the world where robot brains are needed. This might have unsettling social and political implications to some, but as Brooks points out, this type of remote labor (away from heat, smoke, fumes, and physical exertion) is probably a lot more desirable than doing the same work in person.

HEROES OF THE ROBOLUTION

BUSH ROBOT

HANS MORAVEC
'THE TRANSHUMANIST' ②

HEROES OF THE ROBOLUTION

CARD NO 2
COLLECT ALL 11!

HANS MORAVEC
NICKNAME: The Transhumanist
THE LAB: Robotics Institute, Carnegie Mellon University
BACKSTORY: Moravec believes that computers and robots are our progeny, our Mind Children (title of his 1988 book). In Moravec's view, technology will one day replace biology as the vehicle of evolution.
BIG IDEAS: The robotic bush, a fractal branching robot "tree" with molecule-sized end effectors that can take any shaped required to perform tasks. The Moravec timeline (see Chapter 3), a way of predicting when machine intelligence will become possible.
LINKAGE: www.frc.ri.cmu.edu/~hpm/

It is likely that, given the immense complexities of true AI, Japan won't be meeting its 2008 deadline for truly autonomous and intelligent humanoid robots that can be tasked with taking care of Grandma by themselves. While we wait for the break-throughs that will enable such big-brained bots to take their place beside us on the commuter train, robotic presence might be just the ticket.

Biomimicry

The idea behind *biomimicry* (from *bios*, meaning life, and *mimesis* meaning to imitate) seems painfully obvious. Nature has had millions of years to evolve organisms and ecosystems that work, and work under many different circumstances. So, why not try to *reverse engineer* that natural design intelligence in human-made devices and systems? Of course, biomimicry is nothing new. Humans have been imitating nature from the very beginning. Ancient peoples observed and copied how other predators hunted, the airplane was obviously inspired by nature's flight technologies (though we were sidetracked for awhile by that whole flapping thing), and the telephone was inspired by the workings of the human ear. But the atomic and petrochemical ages made us humans a little cocky, and we began to think that we could "easily" outdo nature's designs. With the emergence of greater environmental awareness in the 1960s and '70s, and then the advent of computers and computer modeling in the '80s and '90s, suddenly, natural systems started to be more closely examined and effectively deconstructed, reconstructed, and "hacked" in high-tech laboratories.

A newfound appreciation arose for biological systems and their potential applications in human technologies and even social systems. Whereas physics had always been the foundational discipline in the realm of the hard sciences, the influence of biological sciences began to be felt in many quarters. In the early '50s, with the birth of AI as a branch of computer science, researchers were confident that, as soon as the computational power was there, they could build smart machines. As a testament to this hubris, we see HAL 9000, the intelligent robotic ship in the film *2001: A Space Odyssey*. Marvin Minsky, one of the founders of the science of AI, and the AI Lab at MIT, helped director Stanley Kubrick create this character (based on research projections of the time).

tip

If you'd like to know more about how the study of biological systems is affecting human systems, and how human systems (read: technology) are affecting the study of biology, check out Kevin Kelly's book, *Out of Control: The Rise of Neo-Biological Civilization* (ISBN: 0-2014-8340-8). Written in 1994, at the dawn of the Internet revolution, it chronicles how the born (the biological) is converging with the made (the technological). Its hubris is a bit dated, but it does an astounding job of detailing the degree to which biological models and processes were affecting technology (and the other way around) by the end of the twentieth century. The chapter, "Machines with Attitude," about behavior-based robotics and robot performance art, is especially recommended.

As computing power increased, advances in AI didn't keep pace. Frustrated, some scientists began looking for other approaches to building thinking machines. They started to look at how brains and bodies might have evolved in the first place. Rather than building a complete brain and then switching it on, how about letting the brain evolve over time?

HEROES OF THE
ROBOLUTION

SOCIETY
OF MIND

MARVIN MINSKY
'GRANDPAPPY OF AI'

HEROES OF THE
ROBOLUTION

CARD NO. 3
COLLECT ALL !!!

MARVIN MINSKY
NICKNAME: Grandpappy of AI
THE LAB: MIT Media Lab, MIT AI Lab
BACKSTORY: Co-founded MIT Artificial
Intelligence Lab in 1959. Helped Stanley
Kubrick create the HAL 9000 AI for the film
2001: A Space Odyssey.
BIG IDEAS: Intelligence emerges from a
"society of mind," a complex inner ecology of
smaller, interrelating (and competing) thought
agents. These agents focus on different goals
(reproduction, securing food, eating, security),
and from their complex negotiations, emerges
what looks like unified intelligence.
LINKAGE: web.media.mit.edu/~minsky

Behavior-Based Robotics (BBR)

Rodney Brooks, now director of MIT's AI Lab, stunned the AI community in the mid-1980s when he started talking about ditching the traditional *cognition box*—the complex central control and planning systems—the "big brain" that most AI researchers were struggling to build into their bots. Rather than having to build maps of the external world and then plan what to do in that world (while constantly updating the map), Brooks asked: Why not let the world represent itself? In other words, why not create a much more direct link between a robot sensing its world and reacting to it?

Up to that point, control systems for robots were all *sense-plan-act*. The robot, using video vision, touch sensors, sonar, or other sensors, would send collected data to its "brain" (at that time, usually an off-board computer). Software in the brain would build a map based on the input data, and then send the map to another part of the program that would try to make sense of it, and then, figure out what to do next. When it arrived at a decision, it would send signals to the robot's *actuators* (motors that move the robot itself, or an arm, or whatever). As we discussed with Shakey in Chapter 1, "What's

note

Bob Full and his students at U.C. Berkeley's Poly-PEDAL Lab don't like the term *biomimetics* because it implies a slavish imitation of nature's designs. They use the term *biological inspiration* because it doesn't constrain engineers solely to how nature has chosen to address a given design problem.

in a Name?" more often than not, this led to more contemplation than actuation. Brooks proposed a fundamentally more direct approach that amounted to a *sense-act* control system. He used insects as an inspiration. They have puny brains, not nearly big enough to handle the computational demands of the world-modeling robots of the day, and yet, they can deftly navigate their world and overcome many obstacles to their survival.

At first, it probably appeared to the rest of the robot/AI community that Professor Brooks wasn't properly ventilating the lead-based fumes from his soldering iron. But when he started showing up at conferences with working sense-act robots, doing with few brains what other bots struggled (and failed) to do with (then) state-of-the-art computing muscle, his colleagues became royally ticked off. Surely this wasn't *serious* robot research. Where was the higher-order complexity? Where were the thousands of lines of computer code? Where was all the fancy computer hardware? As he says in his 2002 book, *Flesh and Machines*, "It was this talk [Second International Symposium on Robotics Research in 1985] that split me off from the mainstream of robotics research. The arguments that started that day continue to today."

The control approach that Brooks created was initially called *subsumption architecture*. Today, the concepts behind it are more generally referred to as *behavior-based robotics (BBR)*. To understand BBR in general, let's use subsumption architecture as an example.

Subsumption Architecture

Subsumption architecture is built upon virtual "devices," programmed in software, called *augmented finite-state machines (AFSM)*. Basically, these amount to simple, independently operating software "agents" that focus exclusively on a control function of the robot (say, lifting a leg).

These AFSMs talk to other AFSMs as needed, taking input from related "machines," and sending output from others. They also frequently have "state timers" built into them that execute an action at a given clock time. These AFSMs are networked into layers, with "bottom" layers handling basic functions, such as walking, while "higher" layers handle such things as sensing heat, light, and sound. It's called *subsumption* architecture because higher-level behaviors subsume lower-level functions when they wish to take control. So, if we were to build a robot using a subsumption model, it might look something like this:

Layer 1: Avoid Obstacles

This layer would include all of the input sensors (such as "bump" contact sensors) and output actuators (such as motors) required so that the robot could move around a room and not get trapped by obstacles. The "Avoid Obstacles" routine (coded into a microcontroller) would tell the robot's motors to move forward until the bump sensors were triggered. If triggered, the motors would be told to back up, with one motor turning a bit faster so that the robot would turn away from the obstacle. Then it would motor forward again until the bump sensors registered another hit, or a higher-order behavior subsumed layer 1.

Layer 2: Wander Aimlessly

On top of the layer 1 network of AFSMs would be a routine that kicked in periodically (based on a given time interval coded into the software) to *subsume* it. This behavior would cause our robot to dart off in a random direction to further explore its world. If the robot's bump sensors detected an obstacle, control would be given back to layer 1. Wander behaviors are often useful for getting a robot out of a jam.

Layer 3: Seek Ye the Light!

After another set interval, a third layer of behavior could kick in: light seeking. A light sensor on our robot would send signals to our motors to direct the robot toward the brightest light source in the room. We could also add a light-avoiding sub-behavior that, again at a time interval we specify in the control program, would make the robot suddenly become a recluse and head for the darkest corner of the room. After another set time period, it would

note

In computer science, a *finite-state machine (FSM)* is a software "machine," mainly used to study computational dynamics and logic functions. It's called "finite-state" because it's designed with a set number of states (such as "enough" and "not enough"). The machine also has inputs and outputs. So, if the input triggers the "complete" state, an output occurs. (To use a real-world analogy: The vending machine drops your potato chips to their doom at the bottom of the retrieval box because you input the correct change. If in the "not enough" state—you're one quarter shy of a bag of Doritos—the machine waits for more input.)

Brooks dubbed his little parallel programs "augmented" FSMs because they include internal timers that augment input/output. (If AFSMs could speak, they'd say something like: "This idiot's taking too long getting that last quarter out of his skin-tight Levis—let's just trigger the output and toss him the chips anyway.")

pass control back to a lower behavior, such as "wander aimlessly," or a higher layer could subsume it.

Layer 4: Hot Stuff, Baby

After we have all of the "machines" of our three-layer bot working smoothly, we might add another behavioral layer, such as pyroelectric (heat) sensing, which would make our robot motor toward a heat source in the room, like us! Our robot could then follow us around the room like a mechanical robo-puppy.

The BBR Recipe

One of the hallmarks of subsumption architecture (and all BBR) is not only simplicity of control (asking yourself what's the least amount of control I need to get a useful behavior?), but also building in layers, and building upon previously successful layers. At MIT, Brooks and his students came up with this "recipe" for BBR:

1. Do simple things first.
2. Learn to do them flawlessly.
3. Add new layers of activity over the results of simple tasks.
4. Don't change the simple things.
5. Make new layers work as flawlessly as the previous ones.
6. Repeat ad infinitum.

tip

Obviously, mutation is a critical ingredient in the rich gumbo of biological evolution (insert your own X-Men analogies here). Robot engineers have realized that programming in rarely occurring behaviors, or the potential for random events into robotic control, can be both useful (as in the wandering behavior outlined here) and in creating the illusion that the robot is "alive." When possible, program such random events into your robot control programs. You'll be surprised by the results.

Over the years, since Brooks built his first robots that demonstrated subsumption architecture (such as the now-famous Genghis pictured in Chapter 1), he and his students at MIT have continued to expand upon the BBR approach. Their recent robots, such as Kismet and Cog, are of humanoid design, but they still function using the same layered architecture approach and the idea of building upon previous behavioral successes. Their approach has been dubbed *evolutionary AI*, the idea that robot intelligence might be best grown, mimicking the growth stages of biologi-cal evolution, rather than completely engineering a robot brain in one ambitious super-project.

HEROES OF THE ROBOLUTION

KISMET

CYNTHIA BREAZEAL
'THE SOCIAL ROBOTICIST'

HEROES OF THE ROBOLUTION

CARD NO. 4
COLLECT ALL !!!

CYNTHIA BREAZEAL
NICKNAME: The Social Roboticist
THE LAB: MIT AI Lab
BACKSTORY: Builds robots at MIT's AI Lab that interact with people in very human-like ways. The robots, such as the famous Kismet, are not smart, but people interact with them as if they are because they mimic human movement and emotional gestures.
BIG IDEAS: The more a robot looks and acts like a human, the more humans will interact with it as if it was one. Robots can be programmed to learn from human interactions.
LINKAGE:
www.ai.mit.edu/people/cynthia/cynthia.html

In Information Technology (IT) circles, there's an expression: "Don't try to boil the ocean," mean-ing, don't try to do everything at once. You can boil a pot of water, but not an entire ocean. Evolutionary AI and BBR are all about building the future of robotics one boiled pot of water at a time.

> **tip**
>
> See "The Rules for Roboticists" in Chapter 3, "Robot's Rules of Order," for more sound principles for robot construction.

BEAM

In 1991, Rodney Brooks was traveling around the world showing off his behavior-based robots at uni-versities and robotics conferences. At the University of Waterloo, an engineer named Mark Tilden was thrilled by the radical idea of these bottom-up, bio-inspired bots. He was impressed by the minimal use of computer control, but he couldn't help wondering if this was as minimal as robot control could get. Could robots be built with little more than analog electronics (no computer chips required)? Could one *hardwire* (in conventional electronic circuits) similar sense-act behavior networks, rather than using software programming? Tilden began to experiment and soon answered with a resounding *yes*. He came up with an approach to robotics he dubbed *BEAM (Biological Electronic Aesthetics Mechanics)*.

HEROES OF THE
ROBOLUTION

SOLAR
WALKER

MARK TILDEN
'BIG GOD OF BEAM'

HEROES OF THE
ROBOLUTION

CARD NO. 7
COLLECT ALL 19

MARK TILDEN
NICKNAME: Big God of BEAM
THE LAB: Wow Wee Toys/Hasbro
BACKSTORY: If Brooks is the robot radical, Tilden
is the unrepentant heretic. What Brooks did with
minimal microcontrollers, Tilden does with
analog electronics. Simple just got simpler.
Tilden also insists that robots be autonomous, so
most BEAM bots use solar power.
BIG IDEAS: A human is a way for a robot to build
a better robot. Robots can evolve through
human competition. Life-like behaviors can arise
through the use of analog circuits designed to
create sense-act nervous nets (as opposed to
the sense-plan-act neural nets of AI).
LINKAGE: www.solarbotics.com

Tilden likes to say that what BEAM stands for depends on what sort of mood he's in, but on most good days, it stands for *Biology, Electronics, Aesthetics, and Mechanics*. The *B*, the *E*, and the *M* are obvious, but the *A* is worth exploring. There are two ideas behind Tilden's insistence that BEAMbots be aesthetically pleasing. One is the belief that beauty arises from good design. If your robot is well thought out, well designed, and well built, chances are, it'll be aesthetically interesting too. The aesthetic element also, Tilden contends, helps BEAM to evolve through biological rules of attraction. If your robot looks really cool, people are going to want to check it out; they're going to be inspired by it; they're going to talk about it, and want to make their own. Like a beautiful, sweet-smelling flower in a meadow, your handsome robot is going to help spread its seed, to pollinate other humans. Here are a few other key tenets of the BEAM philosophy:

- The *M* in BEAM could also stand for *minimalism*. In Tilden's early experiments, he proved that you could dispense with microcontrollers altogether. He was able to use basic analog electronics (transistors, resistors, capacitors, diodes, and integrated circuit chips) to orchestrate similar behaviors to what Brooks was getting from his bots (walking, light following, light avoiding, heat following, and so forth). Tilden dubbed his control architecture *nervous nets* (a play on the brainy *neural networks* being developed in AI circles).

- A human is a way in which robots build better robots. Tilden created BEAM games (robot competitions) as a way of accelerating BEAM's evolution. Human builders would create robot competitors and enter them in various

Olympic-like events. The winning robots would be celebrated and admired, and in true hardware-hacker style, winning innovations would be shamelessly "borrowed" by other builders for next year's robot designs. This combination of healthy competition and open source cooperation would allow BEAMbots to evolve while everyone was having a jolly good time and learning a lot. As with Brooks's recipe for BBR (see "The BBR Recipe" previously), BEAMbots would start as simple as possible and build upon previous design triumphs.

tip

BEAM robot competitions are held every year all over the world. Check out www.solarbotics.com and www.solarbotics.net for news of upcoming event announcements.

- Tilden also advanced the idea that robots should be environmentally hearty and autonomous (see "Tilden's Laws" in Chapter 3). In practical terms, this has led to most BEAMbots using solar energy as their power source.

- Another tenet of BEAM concerned the use of techno-junk, using recycled parts from cast-off consumer electronics devices such as pagers, portable stereos, answering machines, and the like. One of Tilden's most famous robots is a stick bug–looking critter, called the VBug 1.5, which was made with parts from four cannibalized Sony Walkmans. The use of techno-junk and inexpensive, readily available electronics, makes BEAM a very accessible entry into amateur robot building.

Tilden's work has inspired an entire movement of robot builders dedicated to his BEAM approach. If Brooks is dismissed in *hard AI* circles (which he usually is), Tilden is barely a blip on anybody's radar. But like Brooks, his influences (or at least the core concepts that inform both men's work) are widely evident. In labs like JPL, Los Alamos (where Tilden worked for a time), other defense department labs, and elsewhere, there is much talk of robo-critters, swarms of autonomous, resilient, skittering robots that can clear mind fields, clean toxic waste tanks, explore other planets, and more. Small (and cheap and disposable and bug-brained) has definitely become beautiful in the emerging robot kingdom.

Robot Sports

Tilden definitely hit on something with the idea of using human competition as a way of accelerating robot evolution. Other robot researchers, such as Fred Martin and Randy Savage at MIT, were also using games as a way of inspiring students to "go for the gold" in robot design. MIT's infamous 6.270 electrical and computer engineering design course turned into a robot design competition in 1992. Its *6.270 Robot Builder's Guide* helped inspire a generation of robot builders (including your humble author). For the MIT competition, LEGO blocks were used as building components (years before LEGO released its MINDSTORMS building sets), ushering in the age of quick, cheap, and easy prototyping in the building of small robots. Robot sports and academic competitions have grown over the years, to the point where robotic sports have been embraced by a mainstream audience. Here are a few other arenas of robotic competition:

Combat Robotics

You can say what you want about Battlebots, Robot Wars, Robotica, and combat robotics in general, but no one can deny that the attention this emerging sport has gotten has raised the general public's interest in all things robotic. Even though some argue that these machines aren't robots at all, but rather, remote-controlled (R/C) vehicles with seriously bad tempers, there have been real innovations in the field. As we'll delve into in Chapter 4, "Robot Anatomy Class," robots are constructed of numerous subsystems (structural frames, actuators, power systems, controllers, and so forth). All of these systems can benefit from tireless devotion to designing, redesigning, building, and rebuilding—learning from previous successes and failures. Although most combat robots have

note

In William Gibson's groundbreaking cyberpunk stories of the 1980s (*Neuromancer*, *Count Zero*, and *Mona Lisa Overdrive*), he coined the phrase "the street finds its own uses for things." This idea was at the heart of the "cyberpunk" ethos, the idea that technology could be appropriated in ways that its makers never could have imagined (think of the scratching turntables of hip-hop). BEAM is a prime example of "street tech." The integrated circuits used in these bots (often rescued from techno-junk) were designed as lowly signal amplifiers, or to help direct data traffic on a computer circuit board, or as simple clock timers. BEAM builders hack these buck-a-piece chips to turn them into robot brains, controlling motors, sensors, and orchestrating lifelike behaviors.

no sensors, and the main controller is usually a geek with an R/C radio, these robots continue to get more impressive in their mechanical and electronic sophistication. As with BEAMbots, in which survival and a robust attitude are key, combat robots have to be powerful, sturdy, and resilient. Builders spend inordinate amounts of time studying and testing different construction materials, structural forms, building techniques, and trying to maximize power while minimizing weight (which, as we'll see in Chapter 4, is a constant battle for all robot builders). An impressively large number of combat robot builders work in the movie special effects industry (the inventor of the sport, Marc Thorpe, worked on *Star Wars* at Lucasfilm). Several builders, such as one of the sport's pioneers, Mark Setrakian, claim that they've applied technologies developed for their robots to their work in movie animatronics.

HEROES OF THE ROBOLUTION

ROBOT WARS

MARK THORPE
'FATHER OF ROBOTIC COMBAT'

HEROES OF THE ROBOLUTION

CARD NO. 11
COLLECT ALL 19

MARK THORPE
NICKNAME: Father of Robotic Combat
THE LAB: The Stupid Fun Club
BACKSTORY: Was working at Lucasfilms when he dreamt up idea for Robot Wars (Star Wars. Robot Wars. Get it?). Got mired in combat of the legal kind. Had to sit out emergence of the sport he created.
BIG IDEAS: Combat robotics as a legitimate sport. The idea of using sponsorships of robotic competitors, a la NASCAR, to support builders and to help build bigger and better bots.
LINKAGE: www.marcthorpe.com

From the very beginning of the idea for Robot Wars, creator Mark Thorpe saw its great entertainment potential. He likes to describe combat robotics as the first sport of the twenty-first century. Like Tilden, he saw the potential for competition and builder cooperation via robot games, but he added a new vehicle of robot evolution: commercialism. He knew that if robot sports became popular, there would inevitably be spin-off products (toys, games, kits, T-shirts, how-to books, and the like) and he saw this as a very good thing. He would cut the builders in on the licensing royalties of products based on their bots, and they could use this money to become robot sports professionals. Also, money from the inevitable corporate sponsorships (a la NASCAR) could go toward building better bots. All of this has happened in the case of Battlebots: Robot evolution through human consumerism!

As of this writing, the future of Battlebots is in limbo. The *Battlebots* TV show on Comedy Central was canceled in 2002 and it is unknown whether it will ever air anyplace else. *Robot Wars*, a BBC production, is cablecast on TNN in the U.S., and *Robot Wars Extreme* (its U.S. counterpart) can now be seen on the TechTV network. Although the future of these shows is shaky, the future of robotic combat appears rock-solid. Those who've gotten a taste for it (as both builders and viewers) want more. To find out about these and other robot combat sports, check out the Web sites listed in Chapter 12, "Robots on the Web."

note

Although robot combat shows struggle to impress advertisers and network executives, a new, related genre of engineering-challenge shows seems to be booming on cable. As of this writing, there are shows airing on cable and satellite networks such as *Junkyard Wars*, *Operation Junkyard* (JYW for kids), *Full Metal Challenge*, *Escape from Experiment Island*, *Monster Garage*, and *Panic Mechanics*. All of these shows are a tinker-geek's dreamcast, with challenges to build crazy contraptions, pitted against rival teams or the clock. As in the digital revolution of the '90s, where all of a sudden, geek became chic, is it now sexy to be a gadget-lovin' gearhead? Nah. But we can dream.

SUMO

There is probably nothing more inherently Japanese than sumo, a form of wrestling in which two gargantuanly fat men in disturbingly small loin clothes try to steamroll each other out of a small ring. The stars of the sport are heroes in Japan, sex symbols, even. In the late 1980s, the chairman of Fuji Software, a sumo fan himself, dreamt up the idea of replacing the human wrestlers with robotic ones. The first official tournament, held in Japan in 1990, drew nearly 150 competitors. In 2001, some 4,000 bots participated in countrywide tournaments. Not wanting Japan to have all the bot-shoving fun, the sport was introduced in the U.S. soon after it began and has spread steadily, here and throughout the world.

The goals of the game in robot sumo are the same as in its human predecessor. Two robots (either R/C-controlled or autonomous) face off in a raised ring (60 5/8 inches or 154cm). There are three matches. The first robot to get two points (one point per match) wins. To score a point, a robot has to push its opponent out of the ring (or a point is scored against a bot that drives itself out of the ring). That's about it. There are two weight classes. A regular robot sumo can weigh up to 3 kilograms (6.61 lbs). A mini-sumo competitor can weigh up to 500 grams (16.64 ounces).

Obviously, a lot of the fun of robot sumo is building the bot. It is something of a Zen art to create a robot that has serious traction and pushing power but that stays within the weight restrictions. It might sound like *all* of the fun is in the building and that watching the sport is boring, but you'd be surprised. Given this chapter's framing concept of robot evolution, it's fascinating to see how there can be so many solutions to such a simple design problem (power pushing).

Robotic Performance Art

No one can deny that technological and scientific innovation accelerates during wartime. Human conflict and the desire to kill people and take their stuff is, most unfortunately, a prime accelerator of machine evolution.

The apparent inextricable link between human conflict and advanced killing machines has always fascinated San Francisco artist Mark Pauline. In 1979, he created an art collective called Survival Research Labs (www.srl.org) and began staging bizarre underground events in which scavenged military-industrial junk was reanimated as actors in a symphony of wanton bot-on-bot destruction. Pauline once described his mechanical performances as a cross between a demolition derby, a modern battlefield, and a three-ring circus.

note

One of the robot events popular among high school students is the FIRST Robotic Competitions. FIRST (For Inspiration and Recognition of Science and Technology) was started in 1992 by Dean Kamen (inventor of the Segway scooter). Teams from Canada, Brazil, the U.K., and almost every state in the U.S. are challenged to build robots (in six weeks) that solve a given design problem (for example: robots that can stack storage containers). Winners of regional events move on to a final international showdown. If you'd like to get involved as a sponsor, mentor, or participant, surf on over to www.usfirst.org.

tip

There are many more robotic performance artists than are covered here. To find others, do a Web search on "robot performance art," "robotic performance art," and "robot art."

HEROES OF THE ROBOLUTION

FIRE BELCHER

MARK PAULINE
'BAD BOY O' BOTS'

HEROES OF THE ROBOLUTION

CARD NO. 9
COLLECT ALL !!!

MARK PAULINE
NICKNAME: The Bad Boy of Bots
THE LAB: Survival Research Labs
BACKSTORY: Godfather of all robot combat
sports and mechanized performance arts.
Does shows, with gigantic (and dangerous)
teleoperated machines, that are part
demolition derby, part post-apocalyptic
battlefield, part three-ring circus.
BIG IDEAS: Using machines as performers.
Transferring the destructive impulse in humans
from human-on-human violence to machine-
on-machine violence.
LINKAGE: www.srl.org

Although the deeper artistic messages behind SRL shows might be lost on many who attend them, the brilliantly designed machines, and the idea of machines killing each other (rather than being used to kill human beings) has inspired legions of geeks, cyberpunks, and gearheads. SRL's work has spawned an entire branch of robo-evolution: robot (or mechanical) performance art. SRL and its killer robots were also one of the inspirations for the sport of robotic combat (see "Combat Robotics," previously).

Today, there are dozens of artists and art collectives exploring robot performance art. Like Pauline, Chico MacMurtie (www.amorphicrobotworks.org) creates robot performers, but they usually don't kill each other. His radio-controlled techno-puppets dance, play instruments, and do acrobatics. Christian Ristow (www.christianristow.com), a former SRL member, continues in the grand tradition of artistic bot-bashing with ominous mechanical critters that sport names such as "The Subjugator," a giant 5,000 lb. three-bladed claw arm that also breaths fire. Ristow also helped build robots for the films *AI* and *Bicentennial Man*. Kal Spelletich's

note

Even DARPA (Defense Advanced Research Projects Agency) is getting in on the robot evolution-through-competition game. It's now holding the DARPA Grand Challenge, a contest with a $1 million prize to be awarded to anyone who can build an autonomous robot vehicle that can travel from L.A. to Las Vegas without human intervention. Check out the details at www.darpa.mil/grandchallenge.

(www.seemen.org) machine performances focus on audience participation. One goal of his shows is to have the audience overcome their initial fear of these dangerous mechanical beasts by letting them literally take control. Kal is currently working on a robot that a human can control via brainwaves.

As with robotic sports, some roboticists turn their bespectacled noses up at robotic performance art, not recognizing it as having anything to do with the science and technology of real robots. Maybe. Maybe not. Who cares? Both of these forms of robot-inspired activity keep the notion of robots in the public's consciousness, get people interested in engineering, computer sciences, math, industrial design, and "real" robots. They speak to the degree to which robots now "saturate" all corners of our social fabric. What's not to like?

Robot Warriors

While art-bot builders such as Mark Pauline and Christian Ristow like their robots to make artistic statements about human conflict, our fear of machines, the animal impulse to kill, and so forth, other builders are creating bots to *serve* human conflict, to *generate* fear of machines, and to assist soldiers on the battlefield. This is the realm of *real* combat robotics. Recent conflicts in Afghanistan and Iraq have been proving grounds for such robotic warriors as the Predator *UAV (Unmanned Arial Vehicle)*. First fielded as a spy drone, Predators are now being used as remotely controlled warplanes, carrying laser-guided Hellfire missiles. A "pilot" can sit in an army command center in the U.S. and attack a target halfway around the world. We won't get into the chilling implications of this type of warfare here, but feel free to ruminate on your own.

So far, most of the military robot development has been focused on extending the eyes and ears of soldiers through "man portable" machines that can scout ahead of a unit. One such robot type, called a *throwbot*, looks like a toilet-tank float. The soldier throws it toward the area he or she wants to explore, and the robot deploys a spiky set of wheels and rolls toward the target area. A laptop or handheld computer shows what the throwbot sees through its onboard camera. IRobot is one of the leading developers of battlefield robotics. Its PackBot is another man-portable system. It is designed to function as a platform to accommodate a number of different applications, from doing reconnaissance such as a throwbot, to sensing biochemical weapons, to carrying ordinance onto the battlefield, and even as a robot ambulance for carrying injured combatants to safety.

The Defense Department's latest project reads like a page right out of the Rodney Brooks handbook for bottom-up bots. Called the DARPA Distributed Robotics Program (www.darpa.mil/mto/drobotics/), it seeks to develop the next generation of miniature and micro robots. It's exploring robots whose intelligence is built upon behavioral layers, that use biological inspirations, that are cheap enough to be deployed in large numbers, and that can act together like an insect colony. Sound familiar? Some proposed uses of such bots include minefield detection, reconnaissance (through city sewerage systems, under doors, and even key-holes), pathfinding, and even acting as decoys ("Hey, look at that swarm of mechanical insects! What the...?").

caution

Robot performances are dangerous! These are machines that are designed to chew, stamp, bash, blast, and incinerate anything in their path. The builders want the *illusion* of danger and destruction without hurting anybody of the meatbot persuasion, but they also like pushing the envelope. If you go to a show, be on your guard at all times, have an exit plan, and for heaven's sake, bring *good* earplugs! These events are usually deafeningly loud.

Domestic Robots

Just as nearly every robot engineer and pundit has a slightly different theory about which robot evolutionary path will be taken in the end, they also have different takes on where and how the first permanent settlement of robo-kind will be established. On the battlefield? In the corporate office? In every nook and cranny where humans fear to tread (toxic waste dumps, inside volcanoes)? In the gladiator arena? How about your living room?

Robo-Pets

Sometimes it comes out of nowhere. Who would have thought that Furby, a big-eyed mechanized fuzzball, would become history's first household robot. In the winter of 1998, otherwise reasonable adult humans went feral over Furby, rushing toy store shelves to scoop up limited quantities of the interactive doll. Furby was ingeniously designed, with simple sensors under its fur, and a motor that animated its eyes and mouth. It didn't learn, but it gave the *impression* of learning by using a clock timer in its microcontroller brain to create seemingly random behaviors, unique personality quirks (every Furby was different), and speaking skills that appeared to improve over time. Legions of Furby owners, both children and adults, were awestruck by the apparent sophistication of the critter, ascribing to it all sorts of lifelike behaviors it did not possess. And, just when the proverbial fur was settling over Furby, in scampered AIBO.

Sony released AIBO, its high-priced mechanical pet (at U.S. $2,500), in June 1999, not having any idea what the response would be. It had only made 5,000 units. The gadget- and pet-loving world was stunned by AIBO's level of sophistication. It really did appear to be alive. Like the Furby, the AIBO used random events, the appearance of growth stages, and personality quirks to create the illusion of life. But unlike the Furby, this was no crude toy. The AIBO displayed extremely lifelike and sophisticated movements. It could play with a ball, beg, right itself when it fell over, and other amazing feats. The first 3,000 units sold in Japan in 20 minutes. The remaining 2,000 sold out in the U.S. within four days. The phenomenal success of the first version of AIBO has led to several new generations, each with more interactive features, more lifelike behaviors, improvements on its mechanical and digital technologies, and much more. Sony also now sells an LM series, a cheaper AIBO (for $599) with fewer capabilities.

The phenomenal success of the Furby (one of the fastest-selling toys in history) and the AIBO has created a new breed of pet owner. What? You don't consider these mechanical creatures pets? Try telling that to a rabid AIBO owner and see what happens. In true robot-evolution-through-commercialism fashion, Sony has taken what it has learned from the AIBO, improved upon it, and is planning on unleashing its SDR-4X humanoid...er...pet sometime in the near future.

note

Although most art-bot builders want their creations to move and act like self-respecting robots, some are content to treat them as pure art objects. Since 1997, sculptor and ceramicist Clayton Bailey has been building life-size robot sculptures out of junk and found objects. His bots are not completely static, though. Electronics built into them makes their eyes and displays blink and flash, their needle dials flutter, and other animations happen that make Bailey's creations look alive. See his robot gallery at `www.claytonbailey.com/robogroup.htm`.

note

For more information on the Sony AIBO and the SDR-4X, see the section "Out-of-the-Box Bots" in Chapter 6.

Of Maids and Mowers

Although the Sony AIBO appears to be the closest thing the robot world has gotten to a "killer app" (an application that can "bootstrap" an entire industry), everyone knows that, one day, the real revolution in home robotics will be one in which robots wield mops and feather dusters. For years now, we've seen an endless parade of vacuum manufacturers showing off prototype robo-vacs. They always make great "cutting-edge tech" segments on the evening news, but nothing ever seems to make it as far as real-world dirty carpets. That changed in 2002, with the introduction of iRobot's affordable and reasonably efficient Roomba (see "Out-of-the-Box Bots," Chapter 6). The Roomba uses many of Brooks's ideas about robot control and engineering. It's a "set it and forget it" machine that wanders around a room on its own, using soft bump sensors for obstacle avoidance. Although it's not flawless (the company claims it's for "upkeep vacuuming" and that you might need to drag out the ol' canister vac now and again), it's still a significant achievement. At only $199, it falls into that "why *shouldn't* I get it?" category for any gadget geek or housework hater. The coolest thing about the Roomba is how it fits into iRobot's vision of the future in which fast, cheap, and out-of-control tiny bots go about their work nearly unseen, cleaning floors, washing windows, mowing lawns, and gobbling up the food on your dinner dishes while they stay on the table (okay, that last one's kind of creepy).

Next to Hoovering the house, the next domestic drag in need of robotic rescue is mowing the lawn. Scrubbing toilets and washing windows probably trumps vacuuming and mowing, but robots that can do these more…uh…sophisticated chores are a long ways off. The gnarly problems of manipulators and end effectors (as we'll discuss in Chapter 4) come into play with these tasks. Vacuums and mowers are relatively straightforward by comparison. Like the robo-vac, we've been catching glimpses of robotic mowers for years now. Friendly Robotics released its Robomower in 2000, and it seems to have the most staying power in the market (like the Roomba, it's reasonably functional, and at $750, it's *almost* affordably priced). The Robomower uses a similar navigation scheme to the Roomba, with bump sensors for obstacle avoidance and a combination of set movement patterns and random wandering. Also as with the robo-vac, you might need to do some touch-ups. To make this manual mowing more couch potato–friendly, the Robomower comes with a remote control, so you can trim those trouble spots from poolside.

Rosie, Where Are You?

It is perhaps a sad testament to early twenty-first century technology that this is where domestic robots stand to date. We probably spent too much of the early computer revolution mesmerized by a machine vision of ourselves, C-3PO–like humanoids that are still likely decades away. And even if such smart machines are *ever* available to help out around the house, the price for such multitasking (or "universal") robots will likely bankrupt anyone whose name is not Bill Gates. Hopefully, with the more modest goals of bots like Roomba, Robomower, and half a dozen copycat robo-maids and mowers that should be on the market by the time you read this, we'll see a flowering of domestic robotic development in the coming decade.

> **note**
>
> For more information on the Roomba and the Robomower, see the section "Out-of-the-Box Bots" in Chapter 6.

Robotic Evolution Through Open Networking

The most revolutionary development in robotics in the last decade is not a robot at all, but a gigantic human-machine hybrid brain called the Internet. The Internet was originally developed as a means for scientists, academics, and researchers to share information and ideas. The World Wide Web (a hyper-linkable interface to that Internet) was also created with this academic information and idea sharing in mind. In amongst today's online superstores, mailbox spam, red light districts, and dancing hamster fan pages, that sharing and collaborative spirit lives on. The Internet has become a place where robot builders can create or subscribe to e-lists, build Web sites, participate in Yahoo! discussion groups, read (or publish) technical papers, and use other powerful Net tools, many of them free of charge. In this global forum, professionals, academics, and amateur roboticists discuss building bots; exchange design ideas, schematics, and robot control programs; and show off pictures and movies of their creations like proud new parents. Not only are these groups communicating all of this amazing stuff among themselves, but also across their respective groups. Amateurs now have unprecedented access to the work of academics and professionals, and these groups are gaining newfound respect for work being done by nonprofessionals. There is nothing more inspiring than being involved in a heady (and heated) online discussion about robot minutia, only to discover months into it that the group you've been learning so much from includes a high school student, a rocket scientist, an artist, several academics, a lawyer, and assorted computer geeks. In cyberspace, nobody knows (or cares) if you're an amateur as long as you have something meaningful to contribute.

> **tip**
>
> To tangle yourself up in the World Wide Web of robotic goodness, skitter on over to Chapter 12.

THE ABSOLUTE MINIMUM

Our branching tree of robot evolution split off in so many directions in this chapter, it's starting to look like a Moravecian robot bush. If you have absolutely no clue what that last sentence means, you weren't paying attention, were you? Not a problem, but we are going to have to ask you to put down the GameBoy and flip back through the chapter until you find the reference we just dropped (hint: look at the pretty pictures). After you're done, come back here and check out these other chapter tidbits, predigested for your gastrointestinal pleasure:

- Bipedal humanoid robots are darn hard to build! All of Japan's uber-nerds are trying to complete such a mechanical marvel by 2008. Making robots walk? Hard (but doable). Making robots think and reason and understand language? Well, let's just say someone's going to blow a really high-profile deadline.

- In the last few decades, biology and looking to nature for design inspiration (called *biomimicry*) has had a huge impact on robotics R&D. This has given rise to *behavior-based robotics (BBR)*, a scheme for creating bots that react directly to their environment (sense-act) rather than building maps of it first (sense-plan-act).

- Robot competitions—from wrestling robots, to combat robots, to robot "Olympics"—are becoming a major way in which robots "evolve." Or, as builder Mark Tilden says: "A human is a way in which a robot builds a better robot."

- The robotic killer app will likely show up on your doorstep, either in the form of an "entertainment robot" (think: AIBO) or a robot domestic (think Martha Stewart's brain in R2-D2's body).

- The Internet is cross-pollinating robot ideas and designs faster than a bumblebee in a coffee plant patch (or something like that).

note

PDF versions of the "Heroes of the Robolution" trading cards are also available on the book's website.

- A robot world without laws would be...well, lawless. We lay down some "laws" governing robots (or at least governing those who conceive of robots).

- We look at some other magnificent maxims, principles, and cautions worth taking to heart.

- Robot builders can be a scruffy bunch in need of some rules of their own. We offer up a few suggestions.

3

ROBOT'S RULES OF ORDER

Bots in Legal Briefs

In the previous chapter, we looked at the many branches on the family tree of robot evolution. We also profiled some of the pioneering scientists and technologists who are feverishly engineering that tree, each focused on a different branch. From these various schools of robotic thought have emerged a number of operating principles (such as looking to nature for inspiration, or using human competition to accelerate robotic innovation). In this chapter, we'll look at some of the "laws," maxims, words of wisdom, and other pithy thought compressions that guide many robot builders. Some of these are from science fiction, some from engineering; some are whimsical, others more serious. They are all worth chewing over; hearty food for thought to keep you stoked as you think about, design, and build robots of your own.

Asmivo's Three (er...Four) Laws of Robotics

You can't call yourself a sci-fi fan, a deep geek, or a robot builder if you aren't familiar with Asimov's *Three Laws of Robotics*:

0. A robot may not injure humanity, or, through inaction, allow humanity to come to harm.

1. A robot may not harm a human being, or, through inaction, allow a human being to come to harm.

2. A robot must obey the orders given to it by the human beings, except where such orders would conflict with the Zeroth or First Law.

3. A robot must protect its own existence, as long as such protection does not conflict the Zeroth, First, or Second Law.

The laws first appeared, explicitly anyway, in the short story "Runaround" in 1942. The story was later reprinted in the wildly popular Asimov collection *I, Robot* in 1950. Asimov's Laws were basically created as a literary device, something for Asimov to work off of as he tried to think intelligently and rationally about the future of robots (and how intelligent robots might interact with humans). The "Zeroth Law" appeared in a later story as a necessary addition to safeguard all of humanity (not just individuals) from robot aggression. In the real world, the Three Laws aren't taken that seriously by most robot researchers, especially because we aren't even close to having a robot that can parse the full grammatical import of the words in the sentences that make up the laws, let alone comprehend their meaning.

Some roboticists, such as BEAM (Biological Electronic Aesthetics Mechanics) creator Mark Tilden, have even suggested that these laws would create laughably wimpy robots. As Dave Hrynkiw and Tilden point out in their book *Junkbots, Bugbots & Bots on Wheels*, "If an Asimovian robot has enough power to push a vacuum cleaner into your toe (assuming it could even recognize the difference between your toe and a toy lying on the floor), it'd be too nervous to get any practical work done." Still, Asimov and his laws deserve their props. Just as the laws gave Asimov something to push against in writing his positronic robot stories, they've also inspired countless other sci-fi writers, and real-world robot builders. Which brings us (as a for instance) to Tilden's Laws.

Tilden's Laws

BEAM innovator Mark Tilden (see *Heroes of the Robolution* trading cards in Chapter 2, "Robot Evolution") likes his robots a little more feral than Asimov.

1. A robot must protect its existence at all costs.

2. A robot must obtain and maintain access to a power source.

3. A robot must continually search for better power sources.

The more...ah...earthy expressions of these laws are:

1. Protect thy ass.

2. Feed thy ass.

3. Move thy ass to better real estate.

BEAMbots are survivors. They are built to be hearty and suit the environment in which they find themselves. This is one reason BEAM developers focus mainly on tried and true analog technologies, and why they look to biological inspirations (millions of years of evolution can't be all bad). A fussy big-brained bot with wheels, cameras, multiple processors, and other high-end gear is not going to last very long, in say, a jungle environment. A robot built like a Rhinoceros Beetle, with relatively low-tech parts and primitive sense-act behaviors, is more likely to survive. The other main feature of Tilden's Laws concerns power autonomy. Tilden sees a robotic future in which robots should go about their (programmed) business and not have to be fiddled with very often by human operators (see Figure 3.1). So far, in BEAM, this has translated to solar power as the best way of delivering this autonomy.

FIGURE 3.1

And you thought it was a pain when Fido mangled your slippers. Would bots based on Tilden's Laws be a little *too* autonomous?

Moore's Law

Moore's Law was proposed by Gordon Moore, one of the founders of computer chip juggernaut Intel:

The number of transistors on a computer microprocessor (basically a measure of processing power) will double every eighteen months.

When he first presented his forecasts on computer chip manufacturing, in a 1965 issue of *Electronics* magazine, Moore said this doubling would occur every 12 months. That number actually held true for a decade. In the mid-1970s, the "law's" speed limit was slowed to 18 months, and that has held true ever since. Just when we think that manufacturers can't possibly fit another transistor on a chip, some new breakthrough makes the impossible possible, and Moore's Law remains in effect.

Ohm's Law

Ohm's Law (named after German physicist George Ohm) is a formula used to figure out the interdependent relationships between voltage, current, and resistance in an electrical circuit:

> *One volt will push 1 amp of current through 1 ohm of resistance. Change a value, and they all change.*

The basic formula is voltage (V) equals current (I) times resistance (R), or $V=I \times R$ (see Figure 3.2). If you know two of these values, you can calculate the other ($I=V/R$, $R=V/I$). We won't go into this any further here (we'll cover electronic fundamentals in Chapters 6–9), but knowing Ohm's Law is extremely important to anyone doing work in electronics (and that includes us bot builders!).

note

The perennial truth of Moore's Law is impressive, but one might ask: Why is there no equivalent law for digital storage capacity? Each year, more storage is available, for less money, on ever-shrinking storage media. In fact, storage capacity advances actually exceed Moore's Law. In 1983, a 10MB (megabyte) hard drive (which was nearly the size of a small car, and a forklift was required to get it onto your desk) cost nearly $1,000. If 10MBs cost that much in the '80s, a modern 60GB (gigabyte) hard drive (which now sells for under $100) would cost $6,000,000! In the robot world, this storage boon translates to ever-more sophisticated control programs that can fit into tinier and tinier robot brains and require much less power.

FIGURE 3.2

It might look like a drug tablet, but this is Ohm's Little Helper, a handy pie chart to help you remember how to do Ohm's calculations ($V=I \times R$, $I=V/R$, $R=V/I$).

Moravec's Timeline

Carnegie Mellon robot researcher Hans Moravec (see *Heroes of the Robolution* trading cards, Chapter 2) sees machine intelligence as basically a hardware problem, or at least, a problem not solvable with the computing hardware of today. Using animal brainpower as a guide, and roughly calculating the processing power of various animal brains (in *MIPS*, or *Millions of Instructions Per Second*), Moravec has created a timeline for when machine intelligence will be possible (according to him, anyway). So, for instance, an insect brain can handle about 1,000 MIPS. By comparison, a modern Pentium 4 PC can deal with about 1,700 MIPS. Using Moore's Law (see previous), Moravec believes that a computer will reach (and maybe even surpass) human MIPS power (approximately 100,000,000 MIPS) by 2050 (see Figure 3.3).

note

If voltage is abbreviated *V* and resistance is designated by an *R*, than why is current marked with an *I*? Well, just to confuse you and make you feel inferior, of course! Logic would dictate that it might be *C* (for current), but *nooo*… So, what does the *I* stand for? Bet you never guessed *intensity*.

FIGURE 3.3

Moravec's Timeline predicts that the computing muscle needed to handle human-level instruction processing will arrive around 2050. Image courtesy of Hans Moravec.

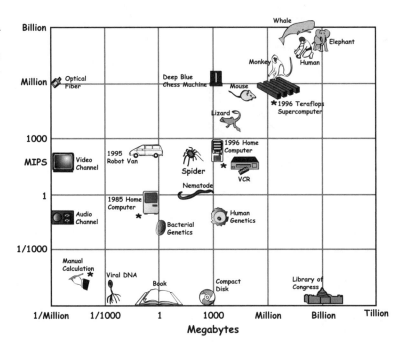

The Turing Test

Considered to be one of the founding fathers of digital computing, British mathematician Alan Turing came up with this test in the 1950s:

> *If a human judge engages in a conversation with two parties, one human and one machine, and cannot tell which is which, then the machine is said to pass the Turing Test.*

The idea is simple: If a human being can interact with another human intelligence and a machine "intelligence" (through written communications), and is unable to tell the difference, the machine is, for all intents and purposes, intelligent (see Figure 3.4).

FIGURE 3.4

From party game to artificial intelligence assessment, the Turing Test lives on. Gender guessing is optional.

Over the years, there has been growing criticism of the test. Does effectively simulating conversation equal intelligence? Can't a machine be smart without having to engage in conversation? A 10-year-old child or an illiterate person wouldn't pass the Turing Test. Does that make them stupid? Although there is an annual competition every year (called the Loebner Prize) to find the most "human-like" machine, to date, no machine has passed the Turing Test.

Amdahl's First Law

Offered in Kenn Amdahl's hysterical and enlightening book, *There Are No Electrons: Electronics for
Earthlings* (see Chapter 6, "Acquiring Mad Robot
Skills" and Chapter 11, "Robot Books, Magazines,
and Videos"), this law basically reminds us not to
mistake scientific models of the world for the world
itself:

> *Don't mistake your watermelon for the universe.*

If you use a watermelon to describe the universe to
children or particularly slow adults ("the universe is
like a watermelon, and the stars are its seeds"), it's
easy for them to start thinking "watermelon"
whenever they hear "universe." Models can (and
often do) become conceptual traps. The idea
behind this law, and the inherent dangers of models and analogies, has been expressed in numerous
other ways. Alfred Kozybski, the father of General
Semantics, was famous for the quote, "The map is
not the territory, the name is not the thing named."
This is the same basic idea. A related maxim from
the cyberneticist Stafford Beer: "Models are not
true or false, they are more or less useful." Let
your neurons fire that one for a few minutes!

Brooks's Research Heuristic

In Rodney Brooks's book *Flesh and Machines*, he
reveals how he came upon many of his radical
ideas regarding robots and AI:

> *Figure out what is so obvious to all of the other
> researchers that it's not even on their radar, and
> put it on yours.*

Essentially, Brooks would look at how everyone
else was tackling a given problem, and what
assumptions were so implicit to them that these
assumptions weren't even being questioned.
Brooks would then question them.

note

The Turing Test was actually
inspired by a party game. In the
game, participants try to guess the
gender of players (hidden in
another room) by asking written
questions and reading answers
sent back to them. In Turing's original proposal for the test, he had
the human participant pretending
to be the opposite gender (the
machine was not asked to switch
hit), although this feature was
quickly dropped.

tip

If you'd like to
know more about
the Loebner Prize, check
out the competition's Web
site (www.loebner.net). It's also
worth doing a Web search on
Hugh Loebner, creator of the prize,
to read up on some of the contro-
versy surrounding him and the
contest. And you thought the
Turing Test had its critics!

Braitenberg's Maxim

This idea lies at the heart of Valentino Braitenberg's groundbreaking book:

> *Get used to a way of thinking in which the hardware for realizing an idea is not as important as the idea itself.*

Braitenberg's book, *Vehicles: Experiments in Synthetic Psychology* (see Chapter 11) is a series of thought experiments using hypothetical autonomous robot vehicles to demonstrate increasingly complex, lifelike behaviors. What's amazing, and a testament to this type of freeform thinking, is how useful these ideas have proven in real-world robotics (and even in possibly understanding the building blocks of human psychology). Braitenberg's "vehicles" have inspired many real-world robot designs. Our "Mousey the Junkbot" project in Chapter 8 is basically Vehicle 2, the "Fear and Aggression" robot, described in Braitenberg's book.

note

(Kenn) *Amdahl's First Law* is not to be confused with (Gene) *Amdahl's Law*. The more widely known *Amdahl's Law* deals with the performance trade-offs of a single (large) computer processor versus multiple parallel processors.

The Krogh Principle

August Krogh (1874-1949) was a Danish physiologist, who wrote:

> *For a large number of problems there will be some animal of choice, on which it can be most conveniently studied.*

He strongly believed that studying the structures of the natural world could solve most engineering problems encountered in the human world. Many roboticists, such as Robert Full of the Poly-PEDAL Lab (see Chapter 2), have been inspired by Krogh's working principle.

The Sugarman Caution

A colleague of mine, Peter Sugarman, a pioneer of pre-Web hypermedia and a constant supplier of potent bumper sticker wisdom (he reads *way* too many comic books), once told me this one after my hard drive fried itself in the middle of a hellish book deadline (no, not *this* hellish book deadline):

> *A computer can smell your fear.*

The animistic paranoia behind this maxim suggests that machines will heartlessly pick the worst possible time to crap out on you (see Figure 3.5). And the more nervous and uptight you are around them, the more likely they are to check out. So relax, stay sharp, and back up frequently!

FIGURE 3.5

Caution: Your high-tech machines (including robots) are waiting for the worst possible time to fail you. Be prepared!

The Rules for Roboticists

Remember *The Rules*, that icky book written by two women who no man would want to date regardless of what relationship principles they were or were not applying? Well, we decided to dream up some rules of our own. No, they're not things like "Never call a robot after the final assembly. Make it call you." Or, "The way to a robot's stomach is through its rear access panel." These "rules" represent the collective working wisdom of builders who've been bolting together bots for decades. The cyberneticist Gregory Bateson used to say, "Always tie your ideas with slipknots." So these are not hard and fast rules, more like rules of thumb. Just a few things to consider as you build bots.

1. **A roboticist is a generalist, a systems thinker.** One of the things that attracts a lot of people to robotics is that it involves the orchestration of so many different disciplines. There are, obviously, specialists in the field—those who work only on AI control architectures or robotic locomotion, or whatever—but even they must keep the entire machine in mind. Most people who work in the field, and certainly all amateur builders, have to have at least basic skills in numerous disciplines. As you get more into robotics, you'll also find yourself spending a lot of time looking at humans and animals trying to figure out how they work. Oddly,

note

Speaking of paranoia, legend has it that the term *bug* (as in *computer bug*) was coined when actual bugs took up residence inside of first-generation room-sized computers and began short-circuiting their electronics. Another origin story claims that bugs used to nest in the plugholes of early telephone switchboards. In truth, the term dates back at least to the late 1800s and was even used by Edison during the development of the phonograph. The term was simply used to mean some imaginary critter (like a Gremlin) that had snuck into the machine works and was causing all of the trouble.

trying to construct machine "creatures" gives one an even greater appreciation for the heavenly designs of nature, which brings us to…

2. **A roboticist is a "deconstructionist."** As a robot builder, you'll find yourself obsessively looking at the natural and built worlds and going: "Ah-ha! So *that's* how it's done." Nothing will be safe as you take apart toys and machines that don't work anymore (and some that still do), and find yourself playing with your food in a manner unsettling to others ("Cool, there's the ligament attachments!"). But, for the love of all that's civilized, leave the family pets alone!

3. **A roboticist knows how to K.I.S.S. it.** Actually not every robot builder knows this, but they should. K.I.S.S. stands for *Keep It Simple, Stupid* and is a maxim recited (but frequently unheeded) in many design disciplines. Heed it in your own robot building. Take time to plan your projects. Don't just throw technology at a problem 'cause you can. Use prototyping technologies such as LEGO MINDSTORMS (see Chapter 6) and breadboarding (see Chapter 7, "Project 1: Coat Hanger Walker") to test out designs. Then try and figure out what you might not need and toss it. The simpler and more elegant your designs, the more likely your robot is to be stable and robust.

note

This is completely unrelated to robots, but it neatly illustrates our fourth rule. Years ago, a female friend of mine, a brilliant fashion designer, entered a beginner's fabric weaving contest. She rented a small loom, learned how to weave, and wove a seersucker blouse. Because she was new to weaving, she didn't know that you *couldn't* hand weave seersucker (which is comprised of alternating puckered and smooth stripes). She was having a devil of a time doing it, but she thought it was just because she was new to weaving. The judges were stunned. Needless to say, she won the contest, and the grand prize, a gorgeous room-sized Swedish loom.

4. **A roboticist must learn to think "outside the bot."** Innovation comes from thinking differently, heading down the road less traveled. Don't be afraid to take chances, to go in radical directions. Apply *Brooks's Research Heuristic* (see previous). Don't listen when people tell you that you can't do something. Ignore critics.

5. **A roboticist is as much an artist as a scientist.** Find someone who's done anything cutting-edge in science and technology, and chances are, he or she has a bit of an artist's soul. Independent engineer and self-proclaimed

"high-tech nomad" Steven Roberts is often quoted as saying, "Art without engineering is dreaming. Engineering without art is calculating." A roboticist worth his or her soldering iron knows this to be true.

6. **A roboticist must be methodical and patient (like any scientist).** The pressure that many robot developers are under to deliver creations that live up to our sci-fi fantasies leads too many to try too much, too soon. Scientific development is measured by nature. Don't be afraid to get one thing right rather than a bunch of things "sorta okay." (Notice how we just contradicted rule number 4. What can we say? Rules are meant to be broken.)

7. **A roboticist knows that neatness counts.** After you've built a few robots, you'll quickly realize that the mechanics and (especially) the electronics can quickly become complicated. There are usually wires sprouting everywhere, and trying to fit all of the parts within your robot shell, or on your robot platform (see Chapter 4, "Robot Anatomy Class"), can become quite a challenge. You'll learn that keeping everything neat and tidy will make a huge difference in the end. Use quick connectors when you can (for plugging and unplugging wires), use cable ties to bundle related wires together, and carefully plan (or revise) your design to maximize order and quick deconstruction/reconstruction of subsystems for troubleshooting.

8. **A roboticist must be a master of many trades.** As stated in rule number 1, a roboticist must be able to look at the big picture and know at least a little about a lot. He or she must have a working knowledge of materials sciences, structural and mechanical engineering, electrical engineering, and computer sciences. We know that all sounds intimidating to an absolute beginner, but knowing something about all of these areas of technology can actually be fun and exciting. And don't let the big words trip you. In plain English, these big words boil down to: building stuff (and knowing the right stuff to use), doing basic electronics, and knowing the ins and outs of microcontrollers and their software.

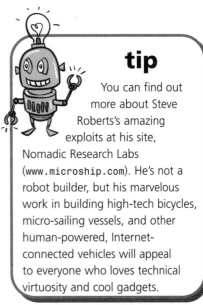

tip

You can find out more about Steve Roberts's amazing exploits at his site, Nomadic Research Labs (www.microship.com). He's not a robot builder, but his marvelous work in building high-tech bicycles, micro-sailing vessels, and other human-powered, Internet-connected vehicles will appeal to everyone who loves technical virtuosity and cool gadgets.

We'll cover all of this in the book, so by the time you're finished, you'll be able to casually spout sentences such as, "I'm not sure if that MCU has enough I/O to drive the sensors and servos we'll use, and we'll definitely need to hack the servo control circuits." (Come back to this sentence once you've finished the book and the projects, and you'll marvel at how alien it initially sounded and how downright clear, dare we say "warm and fuzzy," it will have become to you.)

9. **A roboticist should know his or her tools, materials, and processes.** You can have all the "book learnin'" in the world, but if you don't have a good working knowledge of robot building tools, building materials, and construction techniques, you're not going to be seeing robots scooting around your den anytime soon. You'll gain basic experience in tool wielding and project building by the time you've completed this book and you'll gain even more experience through additional projects and kits covered here as well. The more you tinker, the more mad skills you'll build, which leads us to...

10. **A roboticist knows that you need to build early and build often.** Modern robot building technologies such as LEGO MINDSTORMS, Fischerteknics, reprogrammable microcontrollers, prototyping boards, and other similar innovations (not to mention computer designing and simulating software) allow robot builders a tremendous amount of freedom to experiment. Think of pre-PC writing technology (pens and paper, typewriters) versus using a word processor (complete with spell- and grammar-checking, a built-in dictionary, Thesaurus, and so forth) and that gives you some idea of today's robot tools versus those of a decade ago. Now you can have an idea for a drive system, a sensory array, whatever, and have it built and tested within a few hours. If it doesn't work, you can quickly disassemble and assemble something else. From this rapid prototyping can come truly innovative robot designs.

11. **A roboticist should know when to come back later.** When you're building anything, especially something as complicated as a robot, the build can sometimes turn ugly (see "The Sugarman Caution" earlier in this chapter). If you try to force a solution, you'll often dig yourself into an even deeper hole. "Put the soldering iron down and step away from the robot!" You'll be amazed at what an hour away, vegging in front of the TV, rolling around on the floor with the housecat, or sleeping on your problem overnight will do (and no, not literally, wiseguy!). It almost never fails. And here's a corollary: The extent to which you *don't* want to drop what you're doing and take a break ("I *know* I can fix this, darn it!"), is inversely proportional to the extent to which you *need* one.

THE ABSOLUTE MINIMUM

Gravity. It's not just a good idea. It's the law! In this chapter we looked at principles that might not be as heavy as gravity, but that have helped govern the development of robots and robot-related technologies. If the word "law" made you think of some bad jury duty experience, or a night in jail you'd rather not talk about, and you therefore ran kicking and screaming to this wrap-up section, here's what you missed. We'll try to go easy on the legalese:

- Despite what many sci-fi fans might think, Isaac Asimov's *Three Laws of Robotics* are not taken that seriously among real-world roboticists. Even if they were, we're still so far from robots smart enough to in any way threaten us, that it'll be decades before we'd even need to consider building such fail-safes into robot brains.

- All hail *Moore's Law!* It states that computing processing power will double every 18 months. The truth of this law (on the books since 1975) means that some nearly blind computer chip engineer keeps managing to cram more and more microscopic transistors onto the same chip real estate—and that means more processing muscle for you and me (and our robot minions)!

- Get out your soldering iron and tattoo the following law onto your forehead: V=I×R (voltage equals current times resistance). This is called *Ohm's Law* and it's very handy when figuring out one of these three values in an electronic circuit if you know the other two. The other calculations are I=V/R and R=V/I. (By the way: We were kidding about the tattooing part.)

- *The computer can smell your fear.* That goes for the microcontroller on your robot and the robot itself. Complex machines can appear (at least to those of us who are paranoid) able to tell when it's the worst possible time for them to fail, and that's when they do. So, always have a backup system, a backup plan, and backed-up software. We swear there are little green Gremlins inhabiting our circuits. You can see 'em too, can't you?

- We decided lawmaking looked like fun, so we crafted some *Rules for Roboticists*. In a nutshell: Think generally, act specifically, know your stuff, have fun, don't listen to critics, keep it simple, keep it tidy, and in the immortal words of Kenny Rogers, "Know when to hold 'em and know when to fold 'em."

IN THIS CHAPTER

- Identifying some basic bot body types
- Running through the subsystems and components found in most robots
- Testing your newfound robot anatomy knowledge with a quiz (so pay attention!)

4

ROBOT ANATOMY CLASS

Flesh and Steel

The human body is an unfathomably beautiful creation, and no, we're not talking about supermodels or Natalie Portman (those are freaks of nature). We're talking about the flesh and blood machinery of which we're all so miraculously constructed. Open up an anatomy textbook, or watch one of the many "wonders of the human body" shows on educational cable, and you can't help but be awestruck by our impressive design. Going on such a fantastic voyage, two things are immediately apparent: The human body is elegantly and deeply complex, and it is also extremely fragile. There are literally millions of things that could go wrong at any given moment, but luckily for billions of us, don't. The body has evolved an impressive host of checks and balances and backup systems.

When looking at the human body with an eye toward robots, once again, our old friend, the "fantasy versus reality" dichotomy rears its ugly mug. We can't help but compare the machinery of a robot to our own "machinery" and find the robot's sorely lacking. But the human body has had millions of years to tinker with its hardware and software, while robot anatomy is measured in a few short decades. At the rate technology develops, though, imagine where robots will be in 50, 100, or even 1,000 years.

But all along this evolutionary timetable, robots will likely be comprised of the same subsystems they're comprised of today (only smaller, faster, and better). Ultimately, these subsystems are not entirely unlike our own.

In this chapter, we'll drag out our virtual toolbox ("Geordi, hand me that positronic flux wrench; Lt. Commander Data needs a little...checkup"), deconstruct some robots, and detail their subsystems. As a convenient reference, we'll use the systems of the human body, though we'll try not to belabor that point.

Robot Body Types

Before we start pouring over the parts of your average robot, it might be worthwhile to categorize some basic robot body types. This way, as we go through the various robot subsystems (frame, brain, power source), you can think about how each of these might be applied to different types of bots. Although robots come in a staggering number of shapes and sizes (just like humans), we can muscle most of them into some general categories:

Industrial Manipulators

These are obviously the most prevalent type of robots to date. Not only your car, but also many of the consumer items in your home were probably bot-handled at some point. We tend to think of only the fixed robotic arm in this category, but industrial manipulators come in all sorts of configurations these days. If you ever see a news item about the auto industry, computer chip manufacturing, or other bot-intensive business, all of those ganglia you see snaking around the product as it moves down the assembly line—each with a specialized function and tool on its tip—is an industrial manipulator (see Figure 4.1).

note

Just as there is no agreed upon definition for *robot*, there are also no agreed upon names for different robot types. The Japanese Industrial Robot Association's categories listed in Chapter 1, "What's in a Name?" deal mainly with robot function, rather than form. The names I've given to robot body types here are only used as a convenience in helping to understand the designs and design trade-offs for each robot type. I made up the category *robo-critters*.

FIGURE 4.1

FIGURE 4.1

A typical industrial manipulator, this one is an ABB-brand arm with a spot welder as its "end effector." It is also on a movable track base. Photo used with permission of ABB, USA.

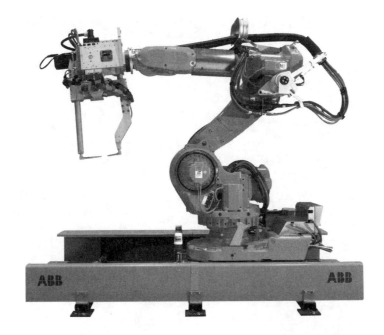

Robots of this type need to be strong, extremely durable, and capable of doing the exact same sequence of tasks over and over again without fail. With this species of bot, boring is good, surprises are bad.

Utility Robots

Any mobile robot that does work would fit into this category. Bots of this type are often called "field robots" (see Figure 4.2). This area of robotics is coming on strong right now, and within a decade, utility robots will likely be everywhere. Robots in this category include bomb squad bots, emergency response robots (like the ones that searched the rubble of the World Trade Center), military reconnaissance bots, and those in the burgeoning market of domestic robots (robot vacuum cleaners, robo-mowers, and home security robots). Combat robots would also fit into this category. Their job? Kickin' bot!

Because robots of this type are often battery-powered, they need to be lightweight enough to not unnecessarily tax their motors and stored power source. Utility robots designed for harsh environments need to balance power/weight concerns with protection from the elements. Like industrial manipulators, utility robots don't usually look very sexy (think R2-D2), focusing on function over form. For this robot category, reliability under changing real-world conditions is key.

FIGURE 4.2

Houdini, one of RedZone's field robots, is designed to fold in half so that it can move through tight spaces on its way to the job (which usually involves bulldozing toxic waste). Photo used with permission of Redzone Robotics.

Robo-Critters

Looking to nature for inspiration in robot design has led to a fantastic menagerie of mechanical creatures. All sorts of bugs, reptiles, fish, crustaceans, and mammals have provided design ideas (see Figure 4.3). Even how these critters behave—the communalism of ants, the swarming of bees, and the flocking of birds—has inspired designs for clusters of robots that function in a similar manner. Robo-critters often have legs, only basic brainpower (instinct-level), and are frequently autonomous (either solar powered or capable of finding their own power source). Many of today's entertainment robots (AIBO, B.I.O. Bugs, iCybie) fall into this category.

For robots of this type, weight is often critical. As legged mobility is complicated and fragile, this is often a weakness of the design. Legs are power-hungry, too, drinking up precious battery juice. The robots we will build in Projects 1 and 2 would be considered robo-critters.

FIGURE 4.3

JPL's latest robocritter, Spiderbot, is designed for extra-planetary exploration. Let's hope it doesn't get blown away in a stiff Martian wind. Photo used with permission of NASA/JPL/Cal Tech.

Humanoids

Hello C-3PO! This fidgety fella from a galaxy far, far away has become the poster bot for our dreams of the humanoid robot (personally, I always get the image of a rampaging Terminator, but then, I'm a perverse sort). Humanoid robots are bipedal, have heads, usually arms, and are often human height (or at least human proportions). These robots tend to be the most complex, have the biggest "brains" (and aspirations of higher intelligence), and spend much of their lives on workbenches with people in white coats cursing over them. These same people in lab coats assure us that this will one day change (see Figure 4.4).

For robots of this type, brainpower is key. Articulated movement and walking is a huge challenge, as is a robotic vision system that allows such robots to interact effectively with people and their environment. Also, getting all of the hardware required to mimic their makers inside of a human-sized/human-shaped body is a real problem.

Embedded Bots

This is the invisible robot, the machine with sensors, brains, and actuators, that doesn't look like a robot at all. Embedded systems is one of the fastest growing areas of digital technology, but its modest, behind-the-scenes (under-the-hood, inside-the-walls) nature means that it's off of most people's radar. And because of their very un-botish nature, we won't really be spending too much time discussing embedded robotics here.

Embedded robots vary greatly in body type, depending on their function and environment. They are usually highly integrated into their world (a new home, the signs and light poles of a highway, the guts of an orbiting satellite), which is why we tend to overlook them.

FIGURE 4.4

The SDR-4X (Sony Dream Robot), Sony's diminutive answer to Honda's Asimo. The SDR looks human-sized in many pictures, but it's actually tiny (23 inches tall). When it's available for sale, the SDR-4X will cost as much as a luxury car! Photo courtest of Sony Corporation.

The Development Platform

This type of robot is a little different than the rest. In research labs, school classrooms, and garages and basements all over the world, piles of mechanical, electronic, and computer parts can be seen motoring around, trying to make sense of their world. Bots of this type are often the least "style" conscious, and often have no shell. The form is usually the lowest priority and often takes the shape of a couple of large disks stacked on top of each other, separated by spacers (see Figure 4.5).

Here, the robot body only exists to allow the experimenter to test out control programs, sensor arrays, and other robot components. Information gained by experimenting on these platforms is often incorporated into more solid, robust robots.

Bots of this type usually have an extremely simple structure made of readily available materials and sport a constantly changing array of hardware. The robot we will build in Project 3 is a miniature version of such a development platform.

note

Researchers are discovering that human "intelligence" is not all run by some centralized command center (think: Soviet-era bureaucracy) between your ears. Research into brains and nervous systems tells us that the "control circuits" (if you will) for many so-called "involuntary," or second-nature, activities are actually distributed throughout the body. Scientists know this because an extremity (for example, a hand) can react to something (for example, a candle flame) faster than control signals can travel from the hand to the brain and back to the arm and hand. It's the immediate feedback loop that saves your arm, not your brain. When engineers can get robot arms and legs to react to their environment in this sense-act second-nature fashion, the many other cool things humanoids *should* be able to do will follow.

FIGURE 4.5

The type of simple platform seen here (without other components mounted) is a common sight in university labs and hobby workshops all over the world. Photo reprinted courtesy of Budget Robotics.com.

Robots: The Exploded View

Okay, so let's unbolt some bots, spread their parts all over our virtual workbench, and talk about components. Figure 4.6 provides a basic rundown of robot anatomy, based on three common types of robots.

FIGURE 4.6

Regardless of robot type, most robots have similar subsystems, and those systems are not entirely unlike those found in our own bodies.

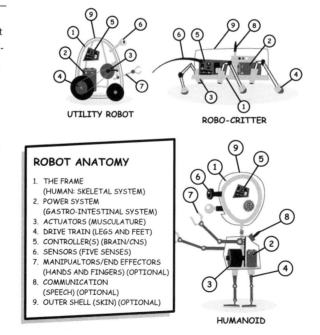

UTILITY ROBOT

ROBO-CRITTER

ROBOT ANATOMY

1. THE FRAME
   (HUMAN: SKELETAL SYSTEM)
2. POWER SYSTEM
   (GASTRO-INTESTINAL SYSTEM)
3. ACTUATORS (MUSCULATURE)
4. DRIVE TRAIN (LEGS AND FEET)
5. CONTROLLER(S) (BRAIN/CNS)
6. SENSORS (FIVE SENSES)
7. MANIPUALTORS/END EFFECTORS
   (HANDS AND FINGERS) (OPTIONAL)
8. COMMUNICATION
   (SPEECH) (OPTIONAL)
9. OUTER SHELL (SKIN) (OPTIONAL)

HUMANOID

Armed with this information, you should be able to look at nearly any robot and identify many of its subsystems. You'll be surprised how much this will increase your confidence in understanding how robots work. You'll be able to look at a bot and say: "Oh, cool. They used hydraulics for the arm actuators, heavy-duty servos for the end effectors, and stepper motors for the drive train. And look, there's the PCB for the motor controller." Come with us now, and you too can learn how to talk bot just like the pros!

The Frame

Everybody needs a body, a skeletal structure upon which to attach all of the other stuff. In a human, this is obviously made of bone. In robots, it's made of just about everything *but* bone. Robot builders are constantly trying out new materials. Which materials are used has a lot to do with the robot's application. Obviously, strength is almost always important, but as we'll see throughout this book, there is a constant conflict in robot building between strength/durability versus power requirements.

Most robots run on battery power and battery life is still preciously short. The stronger and heartier the building materials, the more the robot weighs. The more weight, the more battery power required to move it. More batteries mean more weight, which requires more power to move, and so the vicious cycle goes. Recent materials innovations, such as titanium and carbon composites, are shifting this equation somewhat, but such materials are still very expensive. Let's run down some of the commonly used structural materials and their trade-offs:

note

As you might imagine from the diversity of robot definitions in Chapter 1, whether most embedded robots are actually robots or not is a matter of considerable debate. Many say, if it doesn't actuate (move) anything, it isn't a robot. But isn't a "smart house" that turns on the outside lights when you call it from your car, or shows you the security image on your TV when the doorbell rings, "actuating" the switches to these systems? But, the argument goes, because a computer is nothing more than a massive collection of tiny switches, it would have to be a robot, too. The debate rages on…

Wood

Hey, don't laugh. Wood can be a perfectly reasonable building material for the right kind of robot. The disk-type of developmental robot is sometimes made of wood, and I've seen fairly substantial robot arms (experimental, not industrial) made out of it. In my son's Tech Ed class, he made an extremely cool hydraulic arm out of nothing more than 3/4-inch plywood, wooden pegs, plastic tubing, and medical syringes. Pushing and pulling the water pressurized inside of the syringes and plastic tubing powers the components of the arm (see Figure 4.7). Obviously, problems with wood include its relatively low strength-to-weight ratio and its low durability (at least under many active conditions). It's also, well, wood—not the most twenty-first century material from which to build a robot.

note

After reading the previous section, "The Frame," our smart-alecky tech editor felt compelled to point out that bone, in fact, *has* been used in bots. Robot engineer Mark Tilden has been known to run ribs from his last barbeque through the dishwasher a dozen or so times to clean them enough to perform "experiments" on them (cue maniacal mad scientist cackle).

FIGURE 4.7

My son Blake's hydraulic arm made from 3/4-inch plywood, plastic tubing, syringes, and hardware. Just add water!

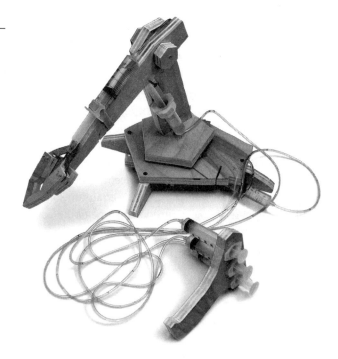

Steel

Steel is a common building material because of its considerable strength and durability. Unfortunately, that strength comes at a high weight cost. Using thin sheet steel and then bending it for greater structural integrity sometimes overcomes this cost. To do this, specialized tools are required that make steel an undesirable material for robot builders on a budget. Some robot kits (such as the Parallax Boe-Bot, which we'll discuss in Chapter 6, "Acquiring Mad Robot Skills") use a stamped steel frame. Stainless steel is sometimes used when a great degree of precision is required in the robot's components. The use of stainless is usually kept to a minimum because of its high cost and weight. If you have the equipment, steel can be welded relatively easily and takes to physical fasteners (nuts, bolts, screws) admirably well.

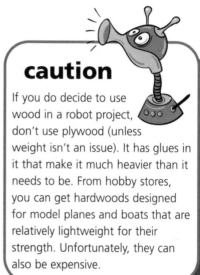

caution

If you do decide to use wood in a robot project, don't use plywood (unless weight isn't an issue). It has glues in it that make it much heavier than it needs to be. From hobby stores, you can get hardwoods designed for model planes and boats that are relatively lightweight for their strength. Unfortunately, they can also be expensive.

Aluminum Alloy

Aluminum is probably the most commonly used material in medium to large-size robots. It is extremely strong for its light weight and can be worked easily with common shop tools. Square-tube, C-channel (three-sided), or L-shaped (two-sided) extruded aluminum stock (where the molten metal has been pushed through a mold like a Play-Doh machine) can add a lot of extra strength to the material disproportional to the increased weight (see Figure 4.8).

FIGURE 4.8

Evolution Robots (www. evolution.com) uses extruded aluminum Tinkertoy-like "Xbeams" in its robots. It also sells the beams and connectors separately.

Plastic

There are a number of plastics that are used in robot construction. Probably the most common is acrylic resin (also known by the trade name Plexiglass). Plexi is very strong (though a bit heavy because of its high density), fairly easy to work with, and can be tapped (fitted with threads) to accept screws directly. It can also form a very strong bond with itself using special solvent cement.

note

Not only are many combat robots in Battlebots competitions built with or protected by Lexan, but the "BattleBox," the robot arena itself, is enclosed in this super-strong material.

Another commonly used plastic is polycarbonate resin (known by the trade name Lexan). You might also know it as bulletproof "glass." This material, as you might expect from something that thwarts bullets, is extremely strong and durable. Lexan is frequently found on combat robots (Battlebots, Robot Wars, and so forth) for this reason. One advantage to using these plastics in robot bodies is that they can be molded and shaped by applying heat.

A third type of plastic material gaining popularity among builders of small robots is expanded foam PVC (also known by the brand name Sintra). This fascinating material weighs about half as much as a comparable amount of acrylic, but is nearly as strong.

Titanium

Titanium is the Cadillac of robotic building materials. It is extremely strong and surprisingly lightweight. It's also surprisingly expensive. Luckily, the widespread demand for it, for use in everything from Apple PowerBooks to golf clubs to combat robots, is helping to bring down the price. Having a glittering chunk of this amazing alloy as my right hip, I can attest to its remarkable properties.

Carbon Composite

This is a material that is made by suspending carbon fiber sheets in an epoxy. The resulting material can be formed, similar to fiberglass, into whatever shape you want. Layers of the material, called *plies*, can be built up to create an incredibly strong, lightweight robot body part. The robot hobbyist can make carbon composite parts, but it's icky work (again, like fiberglass). Also, when made by an amateur, it can be prone to cracks between the layers of the carbon material (a process called *delamination*), which will greatly weaken the structure (did we mention fiberglass?).

Printed Circuit Board (PCB)

Sometimes, the body you need is no body at all, or at least not a structure separate from other components. In making miniature robots, especially robo-critters, builders sometimes cut down on precious weight (most critical in solar-powered robots) by using the other components (motors, capacitors, solar panels) as the robot's body. As the printed circuit board (the plastic or fiberglass board that all of the electronics are soldered to) is often the largest part of the robot, it gets pressed into service as the body (see Figure 4.9).

FIGURE 4.9

The ScoutWalker, a robo-critter from Solarbotics.com, which uses its printed circuit board as its body. Photo used with permission of Solarbotics.com.

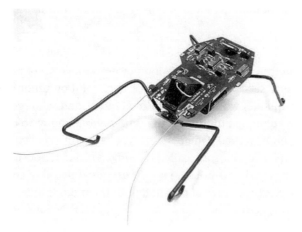

LEGO Bricks

The LEGO Group is building a bridge to the future, one little plastic brick at a time, with its amazing MINDSTORMS Robotics Invention System. This building set uses the same blocks that you made chunky-lookin' cars and rocket ships out of when you were a kid, but LEGO has added electrified bricks (with power connectors, sensors, and lights) and a "computer brick" (called the *RCX*) into which you can load control programs. This system is not just kid's stuff, either. MINDSTORMS can be found in active use in grade schools, college computer classes, and robotics labs throughout the world. It's a great way to create quick experimental robots that can be easily altered. Some builders of small robots even use them for permanent robot bodies by gluing the bricks together to create the structural components they need.

The preceding examples are only some of the more popular structural materials. There are also robot body parts made from magnesium, PVC pipe, tin, iron, brass, and even foam board. Whatever the materials, the perpetual struggle is between strength and weight. Depending on the environment the robot will inhabit, there's also the durability factor. And then there's the money. Cost always crashes the party eventually to wake you from your fever dream of building the ultimate all-magnesium/titanium/carbon composite super-bot.

Actuators

Your own skeleton is covered by *musculature*, a system of long, stretchy fibers that contract when stimulated, delivering power for moving your body. On a robot, such systems are called *actuators* (from *actuate*—to put into action). These are the motors, gears, hydraulic or pneumatic cylinders, or other active components that work to put your robot into motion. Let's look briefly at the most common types of actuators.

Motors and Gears

The most tried and true means of delivering power to move robot parts is direct current (DC) motors and sets of gears. Sometimes, each joint of a manipulator arm (or other moving appendage) will have a dedicated motor and set of gears. Other times, sets of gears will be designed to transfer the power of one motor to a number of places where motion is needed. One popular type of motor is called a servo motor. It has a DC motor and a set of gears (called a *gearbox*) inside a housing that protects the whole assembly. Inside the housing is also usually a board of controller electronics that allows for precise control of the motion of the motor's axle. (We'll be discussing servo motors, gearboxes, and controllers in other sections of this book.)

Even more important than moving power around, gears are used to convert the high-speed, low-torque output from a DC motor to the slow and strong motion you need to move arms, legs, and other robot body parts. Gears are the most compact way of upping the torque/lower revolutions per minute (RPM) of a motor.

Hydraulic Cylinders

A form of robotic muscle frequently found in heavy-duty industrial and field robotics is the hydraulic cylinder. *Hydraulics* simply means the use of fluid to power something. In common practice, this usually takes the form of a cylinder that has a piston in it. The piston moves, and delivers power, when a pressured liquid (usually an oil-based fluid) is pumped into a chamber within the cylinder. Hydraulic systems can deliver an impressive amount of power (hence their widespread industrial use), but they are also heavy, temperamental, and extremely messy when tubes, fittings, and cylinders leak (which they are prone to doing eventually). Robots and liquids don't get along very well.

Pneumatic Cylinders

A type of "pressure system" like hydraulics, *pneumatics* does with air what hydraulics does with liquid. The techniques and hardware are very similar between the two systems. Obviously, one big advantage of pneumatics is the lack of yucky fluids that can ruin the rest of your robot. The downside is noise. Pneumatics hiss and purr as they work, and depending on the application, that sound pollution can be a problem. This type of actuation can also be temperamental, and it is very difficult to get consistently smooth movement because of the compressibility of air.

Muscle Wire

Shape memory alloy (SMA), also known as *muscle wire*, is a fascinating material that holds great promise for certain types of robotic (and other) applications. Muscle wire has the unique capability to be formed into a shape, and then when a small electrical current is applied, it will return to its original unstretched shape. This process can be harnessed to deliver motorless power (see Figure 4.10).

FIGURE 4.10

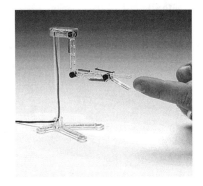

A robotic arm kit from The Robot Store (www. robotstore.com) designed to demonstrate the capabilities of shape memory alloy. Photo by JFM Digital, used with permission of Robot Store.com.

The most common form of SMA is made of a combination of nickel and titanium and is called *nitinol* (pronounced "night in all"). Unfortunately, the high cost of the materials has limited SMA's widespread use, but as the price of titanium continues to drop, this will become less of an issue.

One drawback to SMA is that it uses the heat (resistance) from an electrical current to change its shape. Converting all of that useful DC electricity into heat to get a contraction wastes a lot of power, which is something that's always at a premium for robot builders.

Drive Train

The part of a robot's actuation system that allows it to move through the world is called the *drive train*. This includes the motors, gears, drive belts, axles, wheels, legs, or tracks, and any other components directly related to robotic locomotion. The job of the robot's drive train is loosely analogous to that of our own legs and feet.

The heart of the drive train is the motor. There are basically three types of motors commonly found in robots; each of them is covered in the following sections.

DC Motor

If you've ever cannibalized a remote-controlled vehicle or other motorized toy (and who hasn't?), you're familiar with this type of motor. It uses direct current (either from batteries or from AC wall current converted within the device to DC) to power the motor and spin the drive shaft. DC motors come in all sizes, from tiny pager motors (used to vibrate the case of the pager to alert you that your mother-in-law "needs to talk"), to big beefy jobs that can power the largest of robots. DC motors are great for use in robots, but they spin too quickly to send their power directly to wheels or legs. They generate the most power of the motors used in robots, but the high rotation speeds can make them difficult to take advantage of on their own. This means that they require the addition of mechanical systems (such as gears, belts, sprockets, and chains) to slow down their speed and increase their torque.

note

A common way of controlling the speed of DC motors is through a technique called *Pulse-Width Modulation (PWM)*. Now don't get all sweaty—PWM might sound complicated, but it's simply a way of changing the "on" time percentage (called the *duty cycle*) of an electrical signal in order to control the amount of current going to a motor. PWM turns the motor on and off very quickly, delivering small jolts of power. A greater percentage of "on" time means the motor spins faster. PWM is a very power-efficient way of slowing a motor down while keeping its output torque (twisting ability) relatively high. We'll cover PWM in more detail in Chapter 9, "Project 3: DiscRover."

Servo Motor

A *servo motor* is an ingenious little machine. It usually consists of a DC motor, a gearbox, and a motor controller inside a rectangular plastic housing (see Figure 4.11).

The beauty of the servo is that it has its gearbox already built in (so you don't have to go loony trying to wrestle with gear reduction formulas), and its built-in circuitry allows for precise control over the motion the servo delivers. When you buy the hobby type of servo motor, it is typically limited to 130 to 180 degrees of back and forth movement (what's required for the remote-controlled vehicles, where servos are commonly found). Robot builders often change this limited rotation to continuous rotation with a little "hardware hacking." We'll detail how this is done in Projects 1 and 2. Although the hobby servo is the most common type, there are larger high-performance servo motors that can deliver serious motive power.

FIGURE 4.11

The Hitec Servo Motor. The types of servo motors used in small-to-medium size robots are the same types found in remote-control (R/C) vehicles (planes, boats, and cars).

Stepper Motor

A DC motor uses magnets and windings of wire to generate rotation through electro-magnetic induction. A stepper motor operates on the same basic principles, except its innards are arranged differently. In exact opposition to a regular DC motor, the stepper has its magnets on the rotating shaft (the *rotor*) and its wire windings on the inside motor wall. This arrangement allows the rotor to be moved from one discrete coil winding (called a *stator*) to the next. The action of moving from one stator, or "step," to another gives the motor its name. The stepper motor design allows for more precise speed control than with conventional DC motors or servo motors. This control also often means that gears don't have to be used to slow down the motor's rotation. Unfortunately, steppers take more power to operate, need special electronics to power, and don't deliver as much work in return as the other motor types.

Locomotion

Obviously, motors in a drive train are there to power something. That something is usually wheels, tank tracks, or legs.

Let's take a peek at each of these.

note

Although we cover them separately, a robot's locomotion is technically still a part of the drive train.

Wheels

The majority of robots use the good ol' wheel for their mobility. Wheels are readily available, relatively easy to put into service, and extremely reliable. Many robots that use wheels use only two of them (or two drive wheels and two unpowered (*idler*) wheels, or two drive wheels and one idler wheel). The reason for this is simplicity of design. If you have four wheels—two drive wheels and two steering wheels—that's two extra wheels and a steering mechanism you have to deal with. Two drive wheels can do double-duty as steering wheels by simply turning the power of one wheel off (or

note

When traction and stability are *really* important, as with Martian rovers and competition robots, six or more wheels are sometimes used.

running it in reverse) while the other wheel moves forward. This is often called *skid steering* or *differential steering*. The disadvantages of wheels are obvious: They can't climb over significant obstacles, or climb stairs (unless they're fancy wheels designed for this purpose, as shown in Figure 4.12).

Another advantage of the wheel is that it is a very mechanically efficient device to use. Tracks on a tank drive need to be bent and unbent around each wheel. Legs need back and forth motion, which is electrically "expensive" (you gain nothing motion-wise from moving a leg backward, other than to position it for the next forward step). With wheels, all the energy that's sent to them are used directly for vehicular motion.

FIGURE 4.12

The Mars rover Sojourner had a special kind of wheel-suspension system called a *rocker-bogey* that allowed it to climb over rocks bigger than it could with conventional wheels. Photo used with permission of NASA/JPL/Cal Tech.

Tank Tracks

The cousin of the wheel is the *tank track*, two sets of wheels with rubber (or metal) belts connecting them on each side of the vehicle. Steering here works the same as on a two-drive-wheel robot: Stop or reverse one side, and the vehicle will slew toward that side. The advantage to this design is much greater traction. Tanks can climb up and down steep inclines that other vehicles can't handle. A major drawback to this type of locomotion is the mechanical complexity of building it. And, if you lose a tread, the whole side of your robot becomes "motion-liberated" (to steal a horrific technical euphemism). Skid steering also requires much more power on a track system. You're literally dragging around lots of rubber every time you make a turn.

Legs

I'll try to resist the temptation to quote from ZZ Top's song "Legs" here (oops, I guess I failed). Look around in nature. You don't see many animals with wheels. Wait, you don't see *any* animals with wheels. Evolution has decided that legs are the most versatile and hearty form of mobility. Eight out of ten roboticists agree, but making legs work on a robot can be difficult. They have to be powered in such a way that the legs on either side work in opposition to one another (one leg up while the other is down), and on bots with many legs, this can become very complicated. Then there's that drag called gravity to deal with. All of a sudden, you find yourself with numerous jointed leg segments to engineer, multiple servos to power and control, and some gnarly circuit designing or programming to turn all of this into a walking robot. Still, when it's done really well (like Honda's humanoids), a walking robot can go where other bots can't, like up and down stairs.

Power Systems

A robot isn't going anywhere without juice. Your body uses a complex process of breaking down food into chemicals that it then converts to energy for powering muscles, organs, and other systems. In robots, a number of different substances are used to give a robot its get-up-and-go. The most common are electricity (from AC wall current, batteries, and solar power), air, and liquid. Let's take a look at each of these.

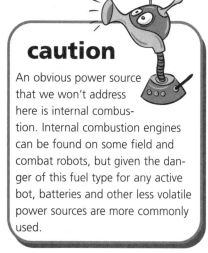

caution

An obvious power source that we won't address here is internal combustion. Internal combustion engines can be found on some field and combat robots, but given the danger of this fuel type for any active bot, batteries and other less volatile power sources are more commonly used.

Batteries

The majority of mobile robots consume battery power. These batteries process chemicals, either wet or dry, to create an electrical current that can be sent through wires to power actuators, sensors, and on-board controllers. Since batteries are such a common form of robot power, we should take a few minutes to detail some different battery types:

- **Alkaline**—These are your run-of-the-mill home workhorse batteries, found in flashlights, smoke detectors, and toy firetrucks all over this great planet of ours. One drawback to the alkaline battery is that the voltage available decreases as the battery is used up (giving robots that "tired blood" feeling). They also don't perform well under pressure (when a high current draw is required). And, last but not least, regular alkaline batteries are not rechargeable. Since active robots consume the juice faster than an Irish pub on St. Patty's day, most serious robot builders use rechargeable batteries.

- **Nickel cadmium (NiCad)**—The rechargeable NiCad (or NiCd) battery has been around since the 1950s. Although this battery has some advantages (it can deliver a very high discharge current, and it can be charged/discharged a lot), it's being phased out of the consumer market. That "cadmium" part of the name spells bad news for the environment (it's extremely toxic). Environmentally conscious robot builders have started to pass over this battery type as a power source.

- **Nickel metal hydride (NiMH)**—In the past decade or so, this type of rechargeable battery has been gaining ground in the market. Batteries of this type don't require complete discharging before recharging (like NiCads), and they have a much higher energy density than NiCads. Unfortunately, they're a battery that will not be ignored. If left unused, they lose their charge faster than any other battery type (as much as 30% per month!).

- **Lithium ion (Li-Ion)**—If you have a swanky new laptop computer, chances are, it sports a lithium ion battery. These batteries are very popular in consumer electronics, where size (as in: small) matters. A Li-Ion cell can deliver three times as much juice as a comparable NiCad or NiMH battery, in a much smaller package. Li-Ion batteries also have a very slow self-discharge rate. Unfortunately, this comes at a price: Li-Ion batteries are still very expensive.

- **Lead acid**—Look under the hood of your car. See that battery in the corner? That's a lead acid battery. This type of battery is found in some field robots, but is not common in other robot circles. Unlike the "dry cell" batteries mentioned previously (where the chemicals are in a dry form), lead acid batteries have caustic, corrosive liquids inside of them. These batteries can leak if tilted or punctured.

- **Sealed lead acid (SLA)**—A type of lead acid battery that's more common in robots is called a sealed lead acid battery. Here, the hazardous goodness of the lead acid is permanently sealed inside a rugged plastic case. SLAs are common in combat robotics because of their high-current capabilities, reasonable cost, and relative safety. Drawbacks to the SLA are size and weight. It's a big-boned fella that weighs more than any other common robot battery type. SLAs are also frequently referred to as *gel-cell* batteries.

- **Battery packs**—This isn't really a specific battery type, but a way in which dry cell batteries are often grouped together. Frequently, on robots, you'll see an odd-shaped bumpy plastic brick. This is a battery pack. Inside is a cluster of rechargeable batteries (NiCad, NiMH, Li-Ion) all connected together and then shrink-wrapped inside a permanent plastic covering. These packs are convenient because separate battery holders don't need to be used, which can save precious space and weight, and all the batteries can be recharged at once.

There are many other types of batteries (carbon-zinc, lithium, polymer, silver), but we won't go into them here. I don't know about you, but I'm tired of talking about batteries!

Pressure Systems

Although battery power is extremely common in robots, it's not the only source in town. No amount of pouring battery juice into a hydraulic or pneumatic leg or arm is going to make it move. These actuator systems get their energy from liquid and compressed air. Again, like the cylinders themselves, the "circulatory systems" to power these two technologies are nearly identical. You have tanks that hold the liquid or air, strong flexible tubing to deliver the material, and a dizzying array of fittings, couplers, valves, pressure gauges, and other "plumbing" parts.

note

In truth, hydraulic and pneumatic systems are actually hybrid systems, as they both use electricity (from batteries or AC) to power the pump(s) that circulate the liquid or air.

Solar Cells

For true robot autonomy (or any other type of power autonomy, for that matter), nothing beats the big bright rays of our closest star. Solar-powered robots use what are called *photovoltaic cells* to covert the sun's radiation (photons) into electrical power (electrons). Although this is a miraculous feat, it's also an extremely inefficient one. Only a small percentage of the energy that reaches the solar cell is actually absorbed and converted into electrical current (the rest is lost as heat). To be able to use this current, it must be stored up and then accessed when there's enough of it to do anything useful.

Some robots (like the Mars Sojourner) use batteries for storage, whereas other types of solar-powered robots, such as many robo-critters, use capacitors to store up their energy. We won't get into capacitors here (see the "Thumbnail Guide to Electronics" in Chapter 6), but basically a capacitor is an electronic component that dams the flow of electrons. Eventually, that dam breaks and the stored electrons race ahead through the circuit. This storing and dumping of the electrons can be controlled further with electronics. Using a solar cell and capacitors (and other support electronics), one can make what's called a *solarengine*, a free, always-available (while the sun shines, anyway) power source. Of course, since the cells deliver slowly accumulating power, robots of this type live a manic-depressive existence of bursts of energy followed by languorous sunbathing. And while most robots have to be weight conscious ("Do these gearboxes make my hips look big?"), solar robots need to be obsessive about it.

Other Power Sources

There are other, more exotic, power systems for robots being experimented with. One of them involves bots that actually eat! Dubbed *gastrobots*, these robots would have a gastric system that can break down sugars and starches in food and convert them into usable power. Imagine what a boon this would be for autonomous robotics. There's even a Gastrorobotics Institute at the University of Southern Florida (www.gastrobots.com). The Institute's director, Dr. Stuart Wilkinson, coined the term.

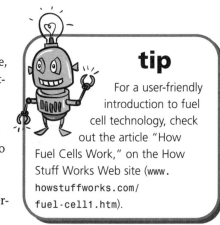

Another extremely promising new power technology, for robots and just about everything else, is the fuel cell. A cousin to the conventional battery, a fuel cell produces electricity through an electrochemical reaction. But unlike a battery, fuel cells use hydrogen (the most abundant element of Earth) and oxygen, and will continue to produce power as long as there's fuel available. Batteries need to be recharged. Besides not running down, fuel cells weigh a lot less than batteries (for comparable power delivered), are much

tip

For a user-friendly introduction to fuel cell technology, check out the article "How Fuel Cells Work," on the How Stuff Works Web site (www. howstuffworks.com/ fuel-cell1.htm).

more power efficient, and are, overall, better for the environment. There are problems with fuel cells. Hydrogen might be abundant, but getting it into a form that can be used by the cell takes processing, and a good hydrogen source such as methane, propane, or natural gas.

Manipulators/End Effectors (Optional)

It's all about the work. At least that's true for industrial robots, utility robots, and one day, humanoids. To accomplish this work, laboring bots are equipped with all manner of manipulators outfitted with grippers and tools (called *end effectors* in botspeak). Grippers run the gamut from simple two- or three-fingered claws to more sophisticated end effectors with human-like fingers. Other tools include just about anything a human might wield (drills, cutters, wrenches, welders, and so forth).

In mobile robots, manipulators and end effectors aren't very common. Surf the numerous robot sites on the Web, and you'll see tons of cool mobile robots, but few of them will have arms or working tools. If Barbie is fond of complaining that "math is hard," robot-builder Malibu Stacey will tell you that robot manipulators and end effectors are even harder.

In many ways, it's not the level of difficulty that's the problem, it's integrating the arms/effectors into a mobile robot design. Like a lot of amateur bot builders, I have several robotic arms on my bench, and a bunch of mobile robots, but I've never made a mobile robot with an arm. As with a walking drive train, as soon as one starts to think about a manipulator, there are greatly increased power needs (more batteries/more weight), numerous servo motors to contend with (for all of the arm's joints), more complex computer programming to consider, and a reliable gripper to engineer and build. Also, an arm dangling out into the world can be a danger to a robot when it bangs into something.

One other issue that builders of end effectors have to contend with is the amount of force that the gripper, fingers, or other tool will deliver.

note

Androbot, the 1980s company that prematurely tried to start a home robot revolution, had a unique work-around for the lack of an arm on BOB, their top bot. To serve beverages (apparently the "killer app" for domestic robots), a special standalone dispenser was designed (but never sold) that dropped soda/beer cans into a bin on the side of BOB. Instead of trying to put the manipulator on the receiving end (the robot), they put it on the giving end (the mini-fridge).

It's a scary proposition to shake a robot's hand when you're not quite sure whether it's going to know when to stop squeezing. One way this is easily done is to use servo motors that have mechanical stops built into them that only allow the shaft of the motor to rotate a fixed number of degrees. Another more complicated method involves pressure sensors in the end effector that give the robot feedback (called *force feedback*) about the presence of an object in its grasp. Personally, until more of the bugs are worked out, I don't make it a practice of shaking too many robot hands.

Sensors

You interact with your world through your five senses: sight, sound, smell, taste, and touch. Robots can have all of these senses (okay, I don't know of any bots that can actually taste), and dozens more. A sensor is any device on a robot that receives input from the outside world and passes that input on to the robot's control system. Let's look at a few of the most common types of robot sensors.

Pressure Sensors

A pressure sensor is basically a switch that, when turned on (or off), sends a signal to the control circuitry of a robot to do something (usually back up or otherwise move to a new spot). Probably the most common type is the *bump sensor*, often a fender or skirt on the robot that, when hit, presses a switch to which the controller responds. Another common type is the *feeler* or *whisker* (common on robo-critters). This is simply a wire that, when pressed or bent, engages a switch. Pressure sensors are probably the most common type of sensors around, and robot builders are constantly experimenting with more effective ways of building them.

note

Grey Walter came up with a unique bump sensor design. The shells of his robo-tortoises hung from a "mast" attached to their bases. When one of the robots bumped into anything, in any direction, the shell would press against the mast, closing a switch, and the bot's control circuitry would trigger an obstacle avoidance routine.

Light Sensors

There are a number of ways of using light sensors, but the basic idea is to use light waves sent out and received by special photosensitive components to control a robot's actions. In one frequently used navigation scheme, this is accomplished by putting an infrared (IR) transmitter on the bot that sends an invisible light beam out into the environment. That light is reflected off of objects back to a special infrared receiver elsewhere on the bot. The angle of that reflected light changes depending on the proximity of the robot to the object that's reflecting the light. The robot can use this change in angle to measure the distance and trigger an appropriate action (such as an obstacle avoidance sequence).

Light sensors are also used in line-following navigation. Commonly, a black or white line is put down to create a "track" for the robot to follow. Photosensitive sensors, on the front or bottom of the bot, straddle the line. If one of the sensors registers a high degree of change in light intensity (by crossing the line), it triggers the motors to adjust course to keep the line between the two sensors (and the robot on track).

Another use of light sensing is to engineer circuits outfitted with light sensors that make the robot "attracted" to light (in other words, it moves in the direction of the most intense light source), or repulsed by the light (it moves away). These "behaviors" are called, respectively, *photovoric* (*photo* meaning light, *vore* being Latin for "swallow up") and photophobic.

note

There are basically two ways of using reflected light intensity in robot navigation. In *proximity detection*, the photosensitive receiver is simply looking for a set trigger value (for example, a high light intensity level), that when reached, will prompt the robot's response. The robot will wait until that trigger value is reached before doing anything. Proximity detection is like an on/off switch. *Distance detection*, the more sophisticated of the two approaches, uses the changing angle of reflected light to actually measure the changing distance between the robot and obstacles. This allows for the taking of different actions depending on perceived distance.

Sound Sensors

The most common type of sound sensor makes use of sonar technology. Sonar sensors basically use the speed of sound to measure distance and use that information to aid in robotic navigation. Inaudible sound waves (outside the range of human hearing) are projected out from a transmitter on the robot, bounce off of surfaces, and return to a receiver on the robot. The time it takes for the sound waves to

note

We'll be using light sensors in Project 2 and will discuss them in more detail then.

return to the sonar sensor (also called an *ultrasonic range sensor*) is used to calculate the robot's distance from an approaching object. This data is then passed along to the control circuitry of the robot and appropriate action is taken.

Another type of sound sensing involves the use of microphones. A robot outfitted with mics can be programmed to move toward a sound source, move a certain body part, or move all of its body parts. Entertainment robots such as Sony's forthcoming SDR-4X humanoid take in sounds via microphones so that the robot can dance (the sound input activates the robot's many servo motors). Robots are also being programmed to respond to certain voice characteristics (such as a raised voice or lowered voice). Of course, microphones are also used on robots that are programmed to respond to sets of human voice commands.

note

Inexpensive sonar systems for robots became popular when builders began to cannibalize the sonar range finders built into Polaroid's SX-70 instant camera (introduced in the 1970s). The sonar sensors in the camera were designed to measure distance for auto-focusing a shot. The demand for these sensors (and those in subsequent Polaroid cameras) became so great that they are now sold as standalone components.

Vision Systems

Cameras have been mounted on robots for decades (going all the way back to the bulky one that gave SRI's Shakey its shake). Cameras are most commonly used on robots that are *teleoperated*, that is, operated by a human at a distance (either at the end of a tether of control cables or over a radio link). With the advent of the Internet and cheap digital Web cameras, robots can now send video signals through the Net so that humans can see what the robot sees, or even control the robot remotely.

Getting a robot vision system to analyze its video input (moving, real-time, unpredictable input) and make sense of it is still a daunting challenge. However, those in the field of biometrics are making inroads into this challenge. Human facial identification is becoming a more common technology, especially in security systems. Here, face size and proportional information, skin tone, and other factors can be used to match a face with a database of stored faces. Unfortunately, this only works if the person is staring directly into the camera lens.

Robot vision systems are getting better at detecting humans in a "scene," tracking moving objects, making out saturated colors, and roughly understanding three-dimensional objects. They still struggle with telling men from women, interpreting objects if the camera is in motion, determining what materials objects are made of, and determining "gaze direction" (where the person is looking). Honda has made some advances with the latter problem. The current version of its Asimo humanoid can tell where a person is looking if the person both looks at and points to the target object.

Heat Sensors

Security robots and others designed to detect fire are outfitted with what are called *infrared pyroelectric sensors.* These sensors detect a combination of heat and movement. If they see some nasty, dancing flames in front of them (registering both movement and heat), the sensors are triggered (and the robot will usually call for human backup). These same sensors are used as motion detectors and will detect the heat and movement of a person within a certain range (see Figure 4.13).

FIGURE 4.13

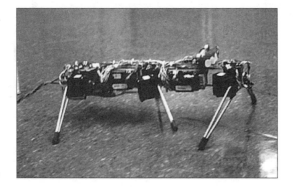

The infamous Genghis, the six-legged robo-critter built at MIT's AI Lab in the late '80s, used six pyro-electric sensors on its "head" so that it would follow humans around, attracted to their movement and body heat. Photo courtesy of Rodney Brooks and the MIT AI Lab.

Gas Sensors

Robots can be outfitted with "electric noses" that can sense a variety of toxic gases. Common on security or industrial robots, triggering this type of sensor usually means a call to a human who decides the nature of the threat. Robots might have become fairly sophisticated, but we're still not comfortable letting them handle a ruptured gas main on their own.

Other Types of Sensors

There are dozens more sensor types than what we've already covered. Here are just a few more and a word or two on each:

- **Encoders**—Optical or magnetic sensors that "read" the rotation of a robot's wheels to measure distance traveled.

- **GPS receiver**—A device that can access the Global Positioning Satellite (GPS) system, using the location of several orbiting satellites to pinpoint a robot's location on the ground (outdoor robots only).

- **Strain gauges**—For measuring the physical force exerted on an object. Used in touch sensors, gripper force feedback, and collision detection.

- **Tilt sensors**—Used to indicate the attitude of a robot. Sometimes used for balance, especially in walking robots. If the sensor detects that the robot is on a dangerous angle, and at risk of falling over, it will "ask" the robot's controller to back it away from the threatening terrain.

note

Many medium-to-large size robots often use the guts of a standard desktop computer for their controller. This is especially the case when weight is not a primary concern and computing power is. Robot programmers often grumble when they only have the limited memory space of an itty-bitty microcontroller to load programs into. A PC lets them stretch out a bit. Since PCs have been covered in…ah…a few other books, we won't detail this type of robot control here.

Controllers

All robots have brains of some sort. These can run the gamut from workstation-level computers to analog electronic circuits designed to give the robot basic bug-brained instincts. Here's a quick rundown of the types of controllers you're likely to find on a robot:

Microcontrollers

One of the most common forms of robot control is the *microcontroller*. A microcontroller is basically a computer on a chip that's been designed for embedded computing applications (like robot brains!). It typically has a CPU (a central processing unit), an erasable memory space (called an *EEPROM*) for storing control programs, some Random-Access Memory (RAM) for storing temporary data, a clock for controlling the speed at which the CPU barks its orders, and input/output (I/O) pins for getting data into and out of the chip.

Microcontroller chips usually come mounted on microcontroller boards, often called *modules*. The boards contain support electronics and sockets for connecting wires to input and output devices (sensors, motors, other controllers, and so forth) and to a power source. In most cases, creating the control programs that run the robot is done on another computer and then downloaded into the microcontroller's memory via a standard computer cable. Some robot controllers, like MINDSTORMS' RCX computer brick, use an infrared transmitter (on the PC end) and a receiver (on the microcontroller end) to load programs into the robot.

Off-Board Control

A lot of today's robots keep most of their brains someplace else, accessing the programs stored on a standalone computer (or network of computers) via a radio link. Before the increasing popularity of wireless data communications, robots of this type had to be tethered (connected via trunks of cables) to a computer, which greatly limited their range. Advantages of off-board computing include the availability of more computing muscle, less weight on the bot, and less power needs. The disadvantages are that the range can be limited (to the local radio range) and the robot is computer-dependent.

Nervous Nets

A radical form of robotic control, originally developed by former Los Alamos robotics researcher Mark Tilden, doesn't use a computer at all. Called *BEAM (Biological Electronic Aesthetics Mechanics)* robotics, this scheme uses conventional analog electronic components (capacitors, resistors,

note

Evolution Robotics (www.evolution.com) has come up with a unique (and cost-effective) way of getting plenty of computing power to its ER1 robot. Evolution sells a kit that consists of only the frame, drive and power systems, sensors, and software. The frame has a place for you to park your laptop computer, which becomes your robot's brains (when you're not electronically balancing your checkbook or playing *Sims Online*).

transistors, and integrated circuits) to build what Tilden has dubbed *nervous nets* (a play on the *neural networks* of AI). BEAM technology is basically an updated version of what Grey Walter was doing in the late '40s and '50s with his robot tortoises. Inspired by nature, BEAM bots often take the form of robo-critters and can exhibit extremely lifelike behaviors given their simple components.

Other Controllers

Frequently, robots have special controllers to manage power and speed. Not surprisingly, they are called *power controllers* and *speed* (or *motor*) *controllers*. These usually take the form of separate circuit boards with all of the electronics needed to perform their tasks, and they are often located next to their respective system (the power source and the drive train). Wires send data to and from theses controllers to the robot's main microcontroller. Sometimes, rather than power and speed controllers being on separate circuit boards, they're located on the microcontroller board (either handled as part of the microchip's duties, or separate circuitry on the main board).

Communication (Optional)

Although sensors are a way for a robot to listen to its world, there are a number of ways of making the conversation a little less one-sided. These include the following:

Two-Way Wireless

Increasingly, robots are communicating with their world in real time. This is largely thanks to increasingly inexpensive short-range radio technology (such as *Wireless Fidelity* or *WiFi*), developed for wireless computer networking. Using such a radio link, a robot can send real-time sensor data, video images, sounds, and more to a local computer (and that computer can send all this to remote locations via the Internet). This opens up all sorts of possibilities for interfacing robots with the rest of the world.

Speech

Getting a robot to respond to fixed speech commands has become near-commonplace. Numerous advances have been made over the past few years to the point where reasonably sophisticated voice command and speech synthesis systems are starting to show up in high-tech toys. But there's still a long way to go.

There are two flavors of speech recognition. *Voice-dependent* recognition, the "easier" of the two, requires that the user (or users) teach the robot (or other digital device, such as a desktop computer or talking teddy bear) commands using their voice. The voice recognition system will then only recognize the voices to which it has been programmed to respond. Systems of this type have become so reliable that, at some hospitals, surgeons now perform operations with robotic instruments guided only by voices. *Voice-independent* recognition, the more difficult of the two, can accept commands from anyone, but it is less reliable.

Voice synthesis, the capability of a machine to consistently speak in a manner that humans can understand, is still a work-in-progress. Text-to-voice programs have been around since before PCs, but after decades of development, they still come off sounding like machines with bad Swedish accents. Ironically, in robot applications, people seem to find a certain charm in a bot that sounds like one of the Cylon warriors from *Battlestar Galactica*. Of course, getting robots to understand the *meaning* of human speech, and to be able to say anything meaningful in return, is still largely science fiction. So far, most robots that "speak" are actually accessing already recorded sound files from a database that it puts together on the fly in response to sensor input.

Infrared

As mentioned in the previous section, "Microcontrollers," infrared technology can be used for sending control programs to a robot. This same technology can also be used for two-way, real-time communications between robots and humans, and between bots. This capability for robots to send signals to each other opens up all sorts of possibilities for orchestrated actions and emergent behavior. For instance, robots can be made to form "swarms," moving in concert and exhibiting pack-like behavior. Synchronized robot dancing, anyone?

Outer Shell (Optional)

Many robots are nudists. Actually, they're worse than nudists—they go around with their guts, nerve bundles, muscles, and bones hangin' out there for all the world to see. For most robot builders, this is a practical matter. If a robot's not going to be subjected to the elements, live in a dirty garage, or have to ward off a frisky kitten (by the way, cats *love* to beat up on robots), most builders don't bother putting a skin on it. The ugly truth of robotics is that robots "enjoy" a lot of downtime, so it's also a convenience not having to remove screws and access panels every time a robot decides to stop working.

In situations where an outer shell is called for, that shell is made out of a variety of materials, depending on what it's protecting the robot from. In combat robotics, these skins are used as armor, so the strongest materials possible (relative to weight concerns) are used. This usually translates to Plexiglass, Lexan, diamond plate steel, titanium, and thick layers of carbon composite. For commercial robots, such as robotic pets like AIBO, and robotic domestics, such as robo-mowers and vacuums, molded plastics are preferred. These are strong enough to take light-to-medium levels of abuse meted out by kids and cat claws, as well as the soil and turmoil of daily living.

Robot Soul (Software)

Some roboticists, when detailing robot anatomies using human analogies, have playfully suggested that robots even have souls: their programming. It is certainly the programs running in most robots that bring them to life, that animate them. There is a dizzying array of programming environments and software tools for building robot brains. Some researchers and developers roll their own languages and software; others make use of existing and popular languages, such as Basic, Visual Basic, Java, C, and C++. Most popular microcontroller modules have programming tools available, such as Windows-type program editors and built-in compilers (for turning code into executable programs your robot can run). These tools can make programming a lot easier, especially for the novice. Popular microcontrollers also have active mailing lists, online discussion groups, and file servers where robot builders can talk about and exchange programs. The software is often frequently updated, with new features, squashed bugs, and so forth.

A Word About Spaghetti

So that's a rundown of the major subsystems found in most of today's robots. It might seem like a lot of information to absorb, and a lot of very different subsystems to have to consider, but much of it will become second nature as you get into robot building yourself.

We didn't even bother talking about all of the minor hardware such as screws, nuts, bolts, nylon fasteners, cable ties, the many different types of tires, and all of the support electronics. Oh, and the spaghetti. Lots and lots of spaghetti. What's spaghetti? Wire. Any self-respecting robot of any significant size is bound to have many feet (even miles) of wire and cable. This can actually be one of the most intimidating aspects of looking at, and trying to understand the construction of, a robot you didn't build yourself. It's hard to make heads or tails of all the robot's systems because it's all obscured by mounds of spaghetti. Amazingly, as you start working on robots, you should find those tangles of wire much less intimidating. Think of that Olive Garden growing behind your computer desk. Those hook-ups might appear horrifyingly complex to your grandmother, or some other casual observer, but to you, it's nothing more than an unsightly mess that you keep swearing to yourself you're going to organize and color-code one "free" weekend.

Robot Pop Quiz

Want to see how well you can now identify robot components? Let's try a little test. Look at the following picture of Timbot (see Figure 4.14), a robot project being done at the OGI School of Science and Engineering in Beaverton, Oregon. As you examine the picture, see how many of these questions you're able to answer.

FIGURE 4.14

Meet Timbot, a robot being used at the OGI School of Science and Engineering to test out embedded systems software. How many robot subsystems can you identify? Photo used with permission of Mark P. Jones, Dr. of Philosophy, Dept. of Computer Science & Engineering, OGI School of Science & Engineering, Oregon Health & Science University.

The Questions

1. What type of robot would you categorize Timbot as?

2. What is the frame made out of?

3. What type of drive system does it likely use?

4. Can you see the battery (one of two, by the way)? Can you guess which type it is?

5. How many sensors can you see?

6. Any idea what that cylindrical shape on the front might be?

7. Got a guess as to where the main controller is?

8. And, lastly, what do you think that black mast in the center is for?

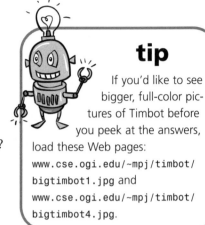

tip

If you'd like to see bigger, full-color pictures of Timbot before you peek at the answers, load these Web pages:
`www.cse.ogi.edu/~mpj/timbot/bigtimbot1.jpg` and
`www.cse.ogi.edu/~mpj/timbot/bigtimbot4.jpg`.

The Answers

It should be obvious by Timbot's "casual" appearance that it's an experimental platform. It's a DARPA (Defense Advanced Research Projects Agency)-funded project to develop better real-time control applications for embedded systems.

The bot's chassis is actually a cannibalized remote-controlled "monster truck." The platform on top is plywood, which should have given you a clue to the experimental nature of the bot. Obviously, you can see that it's a four-wheeled vehicle, but I bet you rightly guessed that it has two drive wheels and two steerable front wheels.

If you know anything about R/C cars, you probably also correctly guessed that DC motors are at work. Did you figure out the battery? It might be hard to see from this image, but that's a 12-volt sealed lead acid (SLA) battery between Timbot's wheels. There's another one on the other side.

The sensors might have been a tough call. The cylindrical sensor on the front is a sonar range sensor. A pan and tilt camera can be seen above it. What you can't really make out from this picture very clearly is the servo motor directly underneath the camera mount (which is connected on the other side by a rod that allows the camera to tilt), and another servo underneath the robot that provides for camera panning.

The large pile of circuit boards on the back of the plywood platform is Timbot's Pentium III computer (running the Linux operating system). The black mast behind the camera is the antenna for the on-board WiFi network connection. Unseen (behind the camera assembly) is a motor controller for the servos, and (behind the computer), a speed controller for the DC drive train. Also unseen underneath the bot is a three-sensor array for line-following navigation. If you'd like to learn more about Timbot, and see high-resolution images and movies of it in action, check out the Timbot Project Web site (www.cse.ogi.edu/~mpj/timbot/index.html).

THE ABSOLUTE MINIMUM

Whew. Robots can be complicated creatures! As we began to explore them more deeply in this chapter, we dreamt up some categories to make it easier to keep them straight. Then we did a little robot dissecting, traipsing around in their innards and ooh-ing and ahh-ing over all of the cool components. Here are some chapter highlights:

- Robots can be muscled into five general categories: industrial manipulators, humanoid robots, robo-critters, embedded robots, and development platforms.

- Bots can be broken down into a number of subsystems, and these subsystems are fairly consistent regardless of robot type. These systems are frame, actuators, drive train, power system, manipulators/end effectors, sensors, controllers, communications systems, outer shell, and software.

- One of our robot types (embedded robots) is a different animal than the other four types and therefore does not share a lot of the aforementioned subsystems. We won't be discussing embedded bots too much in this book.

- Although subsystems are similar across four of the five robot types (manipulators, humanoids, robo-critters, and development platforms), the components—what those robots are actually built out of—and their designs vary greatly depending on the application for the robot and the environmental conditions under which it will live.

- Builders face a vicious cycle when designing robots. They desire structural integrity, durability, and plenty of power, but at a certain point, this costs increased weight. More weight means more power requirements (bigger motors, more batteries), which means more weight—which means more power. And 'round and 'round it goes. Part of robot building is knowing to quit while you're ahead.

PART

GETTIN' DIRTY WITH IT

The Right Tools for the Job 113

Acquiring Mad Robot Skills 147

Project 1: Coat Hanger Walker 185

Project 2: Mousey the Junkbot 227

Project 3: Building a DiscRover 263

In This Chapter

- Must-have, should-have, and make-life-easier tools for your bot builder's toolbox
- Thumbnail guide to digital multimeters
- Your robot workshop
- Thumbnail guide to soldering (featuring the "solder dance")
- Safety first!

5

The Right Tools for the Job

Now that we have some robot "back story" and have discussed some of the high-falutin' ideas behind robot building, it's time to put some of this into practice. Building robots is an extremely fun, challenging, and rewarding experience. But to do it well, and to do it safely, there are a few tools and techniques you have to acquire first. That's where this chapter comes in. We'll be running down the basic tools that any robot builder needs in his or her kit, some basic techniques you'll need to know before tackling the projects in this book, and some safety tips you'll want to take to heart.

Starting with this chapter, and continuing throughout Part II, "Gettin' Dirty with It," we'll be introducing tutorials on some of the skills you'll need. Called "Thumbnail Guides," these tutorials will cover skill sets, such as electronics, which can be big, gnarly subjects that could easily overwhelm our actual subject: robots. We've tried to avoid this in two ways. First, we've tried to "data compress" as much information into as few pages as possible, using humor, plain English, common sense, and pretty pictures to take the sting out of what's being covered. We hope that you'll find the Guides a relatively painless way of getting up to speed. Secondly, the Thumbnail Guides will point you to awesome additional resources (many free online) where you can delve deeper into the subjects covered. Don't worry, though—the Thumbnail Guides will contain all of the information you'll need to complete the projects that begin in Chapter 7, "Project 1: Coat Hanger Walker."

note

This chapter focuses mainly on explaining what tools you're likely to need, not how to use them. For the more complicated tools, we talk about how to use them throughout the book as it becomes important to have that knowledge.

The Basic Bot Builder's Toolbox

Just as there are different types of robots, there are different sets of robot tools, depending on what types of bots you're planning on building. Small robots, like the types described in the following chapters, mainly use household tools and a few specialty items you can get at your local electronics store. Bigger robots, such as combat robots, utility robots, and humanoids, usually require bigger shop tools (drill presses, milling machines, and so forth) and specialty techniques (metal fabrication, composite molding, and so forth). For purposes of this book, we'll only be covering the tools needed for building the commercial kits described in Chapter 6, "Acquiring Mad Robot Skills," and the bot projects later in this book. Although you might have many of the tools listed in the following sections, we thought we'd run through them all. You'll want to have all of these handy as you build your robots.

Must-Have Tools

You shouldn't even *think* about building the robots discussed in this book without having the following tools in your work area (see Figure 5.1):

■ **Small screwdriver set**—Most small robots use both flat-head and Phillips-head machine screws in their construction. The screwdrivers you likely have in your home toolkit will probably be too large. You'll want to get yourself a set of *precision screwdrivers* (also sometimes called *fine* or *electronics* screwdrivers). You can get a set for only a few dollars, but if you think you might be building a lot of bots, get a good set. Xcelite (www.excelite.com) and Wiha (www.wihatools.com) make really nice, high-quality drivers.

- **Needlenose pliers**—This type of pliers is a godsend for any type of electronics work. The nose tapers to nearly a point, allowing you to get it into tight spaces. Needlenose pliers are also great for twisting wire pairs together and bending electronic component leads.

- **Diagonal cutters**—Diagonal cutters have a cutting head that's curved. This enables you to get the pliers tight against a printed circuit board (PCB) to cleanly snip off the component leads of electronic parts (after you've soldered the components to the PCB).

- **Wire cutters/strippers**—This nifty tool is for cutting the plastic insulating jackets off of wires without cutting into the wire itself. Along its jaws is a series of teeth marked with the gauge sizes of common wire. To strip off the jacket, you just put the wire inside the appropriate tooth for your wire gauge, press down, and pull the plastic jacket away from the wire. The tool also has a cutting blade for slicing wires, some crimping teeth (for attaching connectors to wires), and usually, blades for shearing off common bolt sizes.

> **tip**
>
> So, what do you look for in a screwdriver set? First, you want a nice range of sizes for both Phillips-head and flat-head screws. You also want comfortable handles on the drivers, which will allow you to apply the most torque. Also, the better the brand, and the more expensive the set, the higher the quality of materials and machining (in general). This translates to tips that will fit the screw heads properly and keep their shapes longer.

- **Soldering iron**—For "welding" all of the electronic components and wires on your robots, you're going to need a soldering iron. The iron is used on a material called *solder*, an alloy (usually tin and lead) that melts at a relatively low temperature, and is used to join electrical components. You can get a soldering iron for under $10, but if you think you might have a future in electronics and robot building, spend a little bit more and get a decent one. After years of cursing over a cheapo soldering iron that came with a computer repair kit, I sprung for a $30 XYTronic (www.xytronic.com) model. It has variable wattage (from 16 to 30 watts), which enables you to dial up the heat if you need it.

- **Iron stand with sponge**—This gizmo looks like a big steel spring mounted to a base. When you're not using the soldering iron, you holster it in the stand for safekeeping. The sponge in the tray at the base of the stand is used for cleaning the crud off of your iron. A clean iron tip is critical to properly melting the solder. The sponge is always kept moist while you're working.

- **Desoldering pump**—Even master solderers mess up once in awhile. And, truth be told, soldering is not an instantaneous skill to master. You'll likely make plenty of bad solder joints as you learn the technique. But the beauty of electronics is that, in most cases, you can *desolder* (remove the existing solder) and resolder the component, again and again, if you're careful. We'll discuss how the desoldering pump works in the "Thumbnail Guide to Soldering" later in this chapter.

> **caution**
>
> Some soldering irons come with a little metal cradle on which you can prop up your soldering iron. Don't rely on this! Soldering irons are extremely hot, and inevitably, will tip off the little stand onto your worktable, or your lap! Invest in a proper soldering stand.

- **Breadboard**—A breadboard is a cool gizmo that lets you test out your electronic circuits to see if everything is working properly before you submit your components to the soldering iron. It's basically an electrified grid on a base with rows of little holes into which you temporarily plug your components. We'll discuss how to use a breadboard in the "Thumbnail Guide to Breadboards" in Chapter 7.

- **Jumper kit**—A jumper kit is an essential accessory for your breadboard. It's basically a collection of prestripped wires in different lengths and colors that are sized for use on a breadboard. It would take forever to properly cut and strip all of your own breadboard wires.

- **Safety goggles**—You don't want to be a Darwin Award (www.darwinawards.com) winner because you were foolish enough not to wear eye protection. You should always wear safety goggles (or a pair of safety glasses) while you're working on robot mechanics, soldering, or doing anything else (okay, we'll let you slide when you're at your computer programming your bots). The glasses will protect your eyes and, with the addition of a smart-looking white lab coat, you'll look like an honest-to-goodness ubergeek science dude!

Besides the preceding tools, you'll also want to have obvious items, such as regular household pliers, an adjustable wrench, marking pens (Sharpies are best), scissors, a metal ruler, electrical tape, and a hobby (as in X-ACTO) knife.

FIGURE 5.1

No robot builder should be without this collection of tools.

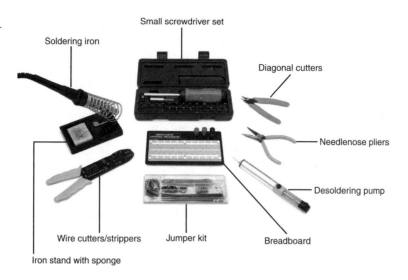

Soldering iron

Small screwdriver set

Diagonal cutters

Needlenose pliers

Desoldering pump

Wire cutters/strippers

Jumper kit

Breadboard

Iron stand with sponge

Should-Have Tools

You can build most of the commercial robot kits discussed in the next chapter and the robot projects in this book with nothing more than the tools just discussed. The following tools, however, will make your robot building easier and more enjoyable (see Figure 5.2):

■ **Digital multimeter (DMM)**—A digital multimeter is an indispensable piece of electronic test equipment. As the name implies, it is designed to test ("meter") a number of different electronic components and events. Learning to use one of these gadgets can be a lifesaver when trying to figure out whether or not a component is dead or when troubleshooting a circuit (or for making sure the power in your house is really off before you start rewiring the bathroom exhaust fan!). Most DMMs come with a set of needle-type test probes. Make sure you also get a set of hook probes (so that you don't have to manually hold the probes to the components as you test them). We'll be discussing DMMs in the "Thumbnail Guide to Digital Multimeters" later in this chapter.

■ **Third hand**—It won't take you long while soldering electronics to realize you don't have enough hands available to hold the iron, the spool of the solder, and the components you are joining. You can only employ a parent, spouse, or child so many times before they'll mutiny. The answer is a device called a *third hand* (also sometimes called a *helping hand*). It sits on a sturdy base and has a number of adjustable "arms" with "alligator clips" on the ends. Some also have a magnifying glass, which can be helpful.

- **Metal file set**—Inevitably, builders need to shave material here and there to make parts fit. The easiest and neatest way to do this is with a file set. You can get a cheap set for under $10.

- **Heat sink**—A heat sink is a device that shunts heat away from an electronic component. Your computer has one or more heat sinks inside of it to keep damaging heat away from your processor and other components. In electronics, a heat sink tool is a little aluminum clip that you attach to components (on the component lead between the component and where you're soldering) to keep them relatively cool while you solder. Most electronic parts are designed to handle soldering heat, but if you're just learning to solder, the heat sink can help prevent damage if you're holding the iron to the components for too long. Integrated circuits (ICs) *are* heat sensitive, so a heat sink provides a measure of safety when working with them.

note

You're likely familiar with "alligator clips." They're the spring-type pressure clips with jagged teeth found on everything from electronic test equipment to refrigerator note holders. Guess what they're called in the U.K.? Crocodile clips. Ah, those cheeky Brits.

- **Battery recharger**—Because most robots live on battery power (as does most every other portable electronics device in our lives), buying disposable batteries can get very expensive. You can save yourself a lot of money in the long run (and be kinder to the environment) by buying a battery recharger and rechargeable batteries. Rechargeables cost a lot more, but you have to buy them far less often (every 1,000 charges or so).

- **Hot glue gun**—God bless this tool. A hot glue gun can quickly "weld" just about anything to just about anything else. Sticks of glue are fed into the gun, heated, and squeezed out the other end. Hot glue bonds are not the greatest for strength or high-stress situations, but for light-duty, quick-and-dirty bonding, they're great. Once you have a hot glue gun, you'll become addicted to using it, wandering around the house looking for anything that's not fastened down—lazy children and housecats beware!

FIGURE 5.2

For about $100, you can add these tools to your basic kit and make your bot-building work far more effective.

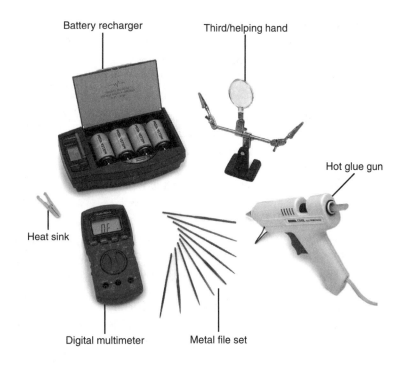

Battery recharger

Third/helping hand

Hot glue gun

Heat sink

Digital multimeter

Metal file set

Make-Life-Easier Tools

The more you work on robots, the more working annoyances you'll discover and the greater your desire will be for specialty tools that will make things go just a little bit smoother. Here are a few items you'd notice if you peeked into my work area (see Figure 5.3):

- **Parts picker-upper**—This gizmo is sometimes called a *three-claw parts holder*, but I've always called it a parts picker-upper ('cause that's what it does). Electronic and robot parts are often small, and inevitably, they fall into crevices and other hard-to-reach places. The parts picker-upper is a pen-sized device that has a plunger on one end of it. Press it, and three little metal claws come out of the other end and allow you to reach and grab onto your lost part. It's a little robot end effector!

- **Magnifying glass (or magnifying lamp)**—Trying to read the parts information on electronic components can be maddening. I find having a magnifying glass close by can be very helpful. If you really want to go all out, you can get a swivel-arm lamp with a magnifying lens built into it. Having one of these on your desk can be a great help when soldering circuits (especially tight ones where the components are very close together and hard to see).

■ **Bench vise**—For bending parts, whacking parts, and holding parts in place while you work on them, nothing beats a bench vise. You can get an all-purpose one at your local hardware or home store for about $20–$30.

■ **Heat gun or micro-torch**—In all of our robot-building projects, we'll be using an awesome material called *heat-shrink tubing*. This is a plastic material that shrinks when you apply heat to it. It's a perfect way of adding back the type of insulating wire jacket that you use wire strippers to remove. It's also a great way of adding traction to robo-critter wire legs and bump-sensing "whiskers." The heat to shrink the tubing can't effectively be supplied by most household hair dryers. You need either a special heat gun or a micro-torch. I use a refillable butane micro-torch, which you can buy in most hardware and electronics stores for under $20.

■ **Rotary tool**—For multipurpose tools, nothing beats a rotary multitool (more commonly known by the brand name Dremel). This device serves as a drill, cutter, sander, buffer, grinder, and numerous other tools. Dremels come in several sizes. I don't recommend the cordless Mini-Mite. It is underpowered for many applications. Full-size Dremels, such as the 9.6-volt (V) MultiPro, have a speed dial with variable RPMs (revolutions per minute), making them versatile for different types of applications. The regular Dremels also have nifty accessories available (such as a drill press and a router attachment).

note

Great moments in techno-tools: The humble three-claw parts holder had a role in George Lucas's first film, the 1971 sci-fi dystopia, *THX 1138*. It is used as some sort of futuristic medical test instrument in one of the examination scenes. It also appeared in the 1990 film *Total Recall* when Arnold Schwarzenegger uses it to remove a tracking device from his nose. Don't try this at home!

tip

It is possible to heat shrink the tubing with a regular disposable lighter, but it's trickier and I find it's harder to control. Okay, I'll admit it. I just really like using the micro-torch!

■ **Miter box and hacksaw**—A miter box is a cutting jig that allows you to accurately cut wood, plastics, and metal at precise angles (45 degrees left, 90 degrees, 45 degrees right, and so forth). If you're going to be doing any bot construction where precision angles are required, a miter box is a necessity. A hacksaw is simply an elongated U-shaped saw frame that accepts blades for cutting different types of materials.

■ **Parts cabinet**—It won't take long before you've collected a lot of electronic bits and pieces. Keeping them all straight and accessible can quickly become a problem. You'll want to invest in a few cheap parts cabinets. These are hard plastic boxes that have numerous small dividable drawers in them. Into these drawers you can organize your growing robot parts collection. Such cabinets can be found at any craft or home store and sell for about $15 each.

■ **Wire-crimping tool**—This device is designed for easy crimping of various types of wire and cable connectors. Soldering connecting wires to terminals and other components is fine if you don't do a lot of it, or don't ever plan on disconnecting the wires. If you do, then you'll want to use solderless connectors, and that's where this tool comes in. It has jaws designed to apply just the right amount of squeeze to the metal connectors you'll likely be attaching to wires. Many solderless connector kits (that come with connectors of many common sizes) come with an appropriately sized crimping tool.

■ **Hemostats**—You can tell how big a tool freak someone is by how many *hems* (rhymes with "seams") he or she has in the toolbox. Hemostats are the scissors-like clamps used to hold bleeding parts of you shut while surgeons work on other parts. Their small size and powerful grip (hemostats, not surgeons) mean that they can do holding work in tight spaces. New hems, sold at medical supply stores, can be quite expensive, but you can get them used (ewww!) at online auction houses like ebay.com.

caution

Obviously, a butane micro-torch gets extremely hot. If you plan on using such a tool for heat-shrinking purposes, make sure to practice on test pieces before heat shrinking the parts on your bots. Also, if you're heat shrinking with a torch indoors, make sure you always have a "plunge" nearby (a large pot of water you can douse a flaming part in). Shrink tubing is flame retardant, but your clothes and many other things are not.

note

Most wire cutters/strippers have at least a few "teeth" in their jaw for crimping common solderless connector sizes.

■ **IC chip extractor**—Also known as a *chip puller*, this funny little set of bent tweezers is designed to safely pull integrated circuits from IC sockets (what the ICs plug into) and from solderless breadboards. IC chips have extremely delicate, bendable connector pins on them, and pulling a chip can sometimes cause damage. A chip puller allows for safe extraction of your chips.

FIGURE 5.3

You know you're a real robot-buildin' fool when these tools show up in your box.

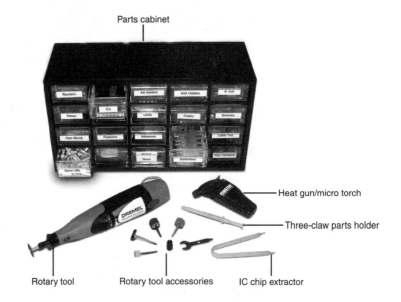

Parts cabinet

Heat gun/micro torch

Three-claw parts holder

Rotary tool   Rotary tool accessories   IC chip extractor

Supplies

Besides a decent collection of tools, you'll also need to lay in a few supplies. These include (see Figure 5.4):

■ **Solder**—Solder is the "glue" that holds our digital world together. It's a conductive metal alloy (often a combination of tin and lead or tin and silver) combined with a compound called flux. Solder melts under fairly low temperatures (120–400 degrees) and quickly cools to form a strong, durable bond between electrical components and whatever metal surface they've been soldered to (each other, a circuit board's metallic "pads," wires, and so forth). The flux in solder is a special material used to help "prepare" the metal surfaces for bonding with the solder. Solder comes in different forms, but *solder wire* is what you'll be using. It usually comes in spools. You can also get it in convenient pocket tubes that even have a pen clip on them (and you thought that laser pointer made you look like a major nerd). See the "Thumbnail Guide to Soldering" later in this chapter for more details on solder.

■ **Wire**—You'll want to have spools of wire on hand for use when breadboarding circuits, hooking up motors, and other wiring jobs. Radio Shack (and others) sells a set of insulated #22 AWG (which stands for *American Wire Gauge*) in three colored spools. A good overall wire size to have around is 22 gauge, and the three colors (black, red, and green) help you to keep connections sorted out.

■ **Two-part epoxy**—Two-part epoxy resin is a quick-setting resin that bonds metal, glass, plastic, wood, fiberglass, and other materials. It comes in a syringe applicator that separates the two parts (called the *base* and the *curing agent*). Squeezing the plunger allows both parts to come out in equal measure, and you then mix the parts together. The mixing sets off a chemical reaction that heats, cures, and hardens the epoxy, creating the bond. After epoxy is mixed, it has what's called a *pot time*—the length of time before it hardens. You want to get quick-set epoxy, which has a pot time of around five minutes.

■ **Superglue**—Cyanoacrylate, mercifully more widely referred to as *superglue*, is an extremely strong and quick-setting bonding agent. Besides allowing dwarves to glue their hardhats to steel girders (as seen on TV), superglue can be used for all sorts of quick-and-dirty gluing jobs in robotics.

■ **Flux paste**—Flux paste is good to have around when you encounter a component that doesn't seem to want to accept solder. The paste, sold in little jars, when smeared onto the stubborn areas, will prepare the way for a good solder joint.

tip

If you don't have a lot of these tools already, one easy and inexpensive way of getting them is in a computer technician's repair kit. I bought mine from Jameco (www.jameco.com) about eight years ago. Even though some of the tools are kind of cheap, I still use many of them daily. The kit cost about $25 and included a soldering iron, solder, a desoldering pump, small screwdrivers, a wire cutter/stripper, chip extractor, and other tools, all in a zipper case. Definitely a good investment.

tip

Although solder has a "core" of flux to clean crud from the metal you are soldering to, sometimes it's still not enough to get a decent weld. Keep some flux paste on hand (you can get it in a small jar) for dabbing onto particularly stubborn parts.

- **Scrubby pads**—Before you solder anything, you want to scuff up the parts (component leads, circuit board tracks and pads, and wires) a little with a nondetergent scouring pad or steel wool to remove any dirt, oil, or oxidation. I use the Scotch-Brite pads from 3M.

- **Poster putty**—It goes by many names: *reusable putty*, *mounting putty*, and *poster putty*, but whatever it's called, half of our house is held together by it! Basically a not-so-Silly-Putty, poster putty is not only a way of temporarily sticking things together, but it also can serve as a helping hand, holding parts in place while you work on them.

- **Heat-shrink tubing**—This shrinkable plastic tubing comes in many different lengths, diameters, and colors. It also comes in different shrinking ratios. The most common is 2:1 (where it will shrink to half its original diameter). Electronics stores usually sell the tubing in packs, with different diameter tubes, and that's what you should get. You need a heat gun or micro-torch to shrink the tubing (refer to the "Make-Life-Easier Tools" section earlier in this chapter).

- **Screw, nut, bolt, and washer assortments**—There seems to be a rule of building robots (or any other mechanical

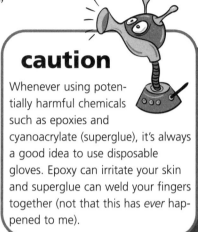

caution

Whenever using potentially harmful chemicals such as epoxies and cyanoacrylate (superglue), it's always a good idea to use disposable gloves. Epoxy can irritate your skin and superglue can weld your fingers together (not that this has *ever* happened to me).

tip

When buying superglue, don't buy it in those little plastic tubes you find at the local drugstore. Get the thin type of cyanoacrylate that comes in a bottle (and apply it with a toothpick or other disposable stick or brush). You can find this type at craft and hobby stores.

devices) that states that your ability to find the right size screw, nut, or bolt is inversely proportional to your need for said component. You'll have dozens of screws that are one size too big or one size too small. You can try and break this "law" and increase your chances of finding the hardware you need by investing in some assortments of screws, nuts, bolts, washers, and other miscellaneous hardware (available at most home, hardware, and electronics stores). Every time I go to one of these stores, I pick up a pack of some common hardware type to increase my collection. I'd like to say this means I can always find the right part, but laws are laws. Seriously, it at least betters your odds.

■ **Construction materials (brass and aluminum stock, sheet plastic, expanded PVC foam)**—As you get more into building bots, you'll start to grow a collection of parts and materials as you go (often left over from previous projects). It's a good idea to have some materials on hand for when inspiration strikes. I like to keep brass rods of varying diameters around (both solid and hollow), as well as varying thicknesses of sheet plastic (both available at home and hobby stores). Aluminum stock is good to have around too, especially if you're building bigger bots. As mentioned in Chapter 4, "Robot Anatomy Class," expanded PVC foam (commonly sold as *Sintra*) is a very cool building material that's worth having on hand too.

■ **Techno-junk box**—If you've always felt guilty about throwing away old mice, modems, printers, answering machines, portable cassette players, and all of the rest of the "techno-junk" we all generate, boy are robots the hobby for you! Much of the guts of these devices can be transplanted into your bots. The projects in this book all use at least some recycled parts. So, make a big bin in your work area and just toss all of that dead consumer tech into it. By the time you've finished this book, you'll be able to transform some of it into cool little robots!

FIGURE 5.4

Here are just a few of the supplies you'll want to have on hand when the bot-building bug strikes.

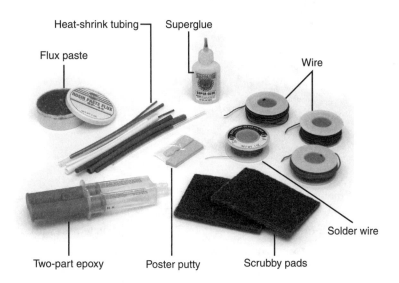

Heat-shrink tubing — Superglue

Flux paste

Wire

Two-part epoxy   Poster putty   Scrubby pads

Solder wire

Thumbnail Guide to Digital Multimeters

A digital multimeter (DMM) is an extremely handy tool to have around for any sort of electronics work, from testing to see if the current is actually off in your house, to finding out how much power is left in your MP3 player's batteries, to finding out the value of components for your robot projects (see Figure 5.5).

FIGURE 5.5

A typical DMM. This one has some nice high-end features like "auto-ranging" and "data hold."

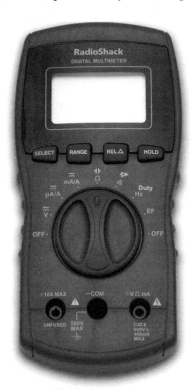

A multimeter has "multi" in its name because it contains a number of different instruments within it for the testing of electronic components and processes. Some DMMs are more "multi" than others. The most basic features usually include:

- **Direct Current (DC) Voltage Measuring**—Tells you how many volts (or how much *electrical potential*) are available. Expressed in *volts (V)*.

- **DC Current Measuring**—Tells you how many electrons (the little beasties that make electricity possible) are racing through your circuit. Measured in *amps (A)*.

- **Alternating Current (AC) Voltage Measuring**—Same as with DC voltage. Also expressed in *volts (V)*.

■ **AC Current Measuring**—Same as with DC current. Also expressed in *amps (A)*.

■ **Resistance**—Lets you know the rating of a resistor or how much resistance exists in a circuit. Measured in *ohms (Ω)*.

■ **Continuity Test**—Lets you know if an electrical connection exists between two points. Measured in *ohms (Ω)*.

■ **Diode Check**—Enables you to check the health of diodes, transistors, and other semiconductors. You also can check the polarity of diodes (diodes have a negative side and a positive side, and they're not always marked).

note

If you're unsure about the difference between AC current and DC current, may we recommend today's special: The Glossary. We'll be back in a minute to take your drink order.

The preceding features cover all the basics and you should make sure that your meter at least handles all of them. A meter with these capabilities can be had for as little as $15 (especially if you look for sales). If you have a few more bucks burning a hole in your cargo pants, these features are worth adding:

■ **Capacitance**—Can be used to measure storage capacity of capacitors. Measured in *farads (F)*.

■ **Frequency**—As the name implies, this meter measures how fast (per second) an electrical event is occurring. Measured in *hertz (Hz)*.

■ **Duty Cycle**—This measures the percentage of "on" time for a given interval of an electronic signal. Measured in *percent (%)*. This feature is often part of the frequency meter.

■ **Auto-ranging**—Without this feature, you have to guess what the maximum value of the component you're testing might be (and dial that value on the multimeter dial). The auto-ranging feature is for those of us with the naïve belief that computers should do this sort of thing for us. Here, you just set the DMM on auto-ranging and it does the guessing for you. What a concept!

■ **Data Hold**—You'll find that it can be difficult to hold the probes steady on components being tested and look at the screen at the same time to read the peak value. This feature leaves the results of your test on the screen until you do another test.

■ **Auto Off**—Turns your meter off after a set time of being idle. Nice to have to save on battery life.

- **Stand**—If you can get a DMM with a stand, do it. It's hard to see the screen when the meter is lying flat on your bench.

- **Boot/Holster**—A boot (or holster) is a rubber jacket used to protect the sides and back of your DMM from shock or damage. Some meters come with them, some are sold as accessories. It's nice to have one for added protection. Some boots also incorporate stands.

You'll also need to get some extra test leads. Your meter will come with a set of leads with straight needle probes on them. To use these, you simply touch the tips of the probes to the appropriate area of the circuit or component you want to test. But you have to continue holding the probes. For situations in which you'll need hands-free testing, you'll want to get a set of leads with *hook probes*, and a set with *alligator clips* (for times when the little hooks can't fit on the item to be tested). These three types of leads should cover any testing situation (see Figure 5.6).

FIGURE 5.6

The three types of DMM test probes you'll need: hook, needle, and alligator.

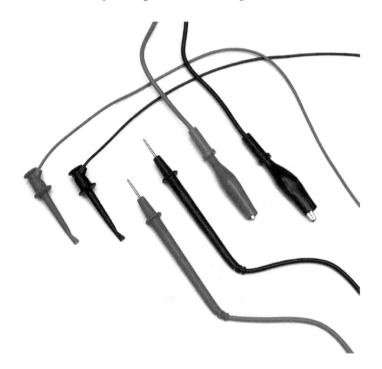

Okay, so now that you have a nifty new DMM, what can you do with it? Lots of things. For instance, you can test all of your components to see if they're working properly. You can find out if your battery has any juice in it (and exactly how much). If your robot isn't working, you can test to see if you've shorted out anything in constructing it. And as you might imagine, if you're going to be recycling parts from old techno-junk, it's important to make sure that the parts you pull are actually working (it is, after all, junk for a reason)!

You'll want to consult the manual that came with your DMM to see exactly how it operates (they're all slightly different), but let's quickly run through a few common testing situations to get you comfortable wielding your new gadget.

Where's the Juice, Bruce?

Your days of using those silly little cardboard pressure testers that come built into battery packaging are over, my friend. To find out if that bunny's *really* gonna keep going and going and going, plug the black needle probe test lead into the COM (also known as common, negative, or ground) jack on your DMM, and the red lead into the positive jack (refer to Figure 5.5). The main positive jack is often marked differently, but yours will likely be marked with *V/Ω* or *DCV/ACV*. It also often has a little lightening bolt next to it (again, consult your manual).

Next, you'll need to dial the Volts (V) meter on your DMM. If you don't have autoranging (I thought I told you to get that!), you'll have to select a maximum voltage higher than the battery is rated on the meter's dial. Now touch the black (-) probe to the negative terminal on your battery and the positive (+) to the positive terminal. The amount of charge left in the battery will appear on the display screen of your multimeter. You'll notice, if the battery is brand new, it'll likely have more capacity in it than it's rating (for example, a 9V battery might have as much as 9.8V). The ratings on batteries are just a convenience. In other words, they always have *at least* that much juice in them. If you're testing a battery that's "dead," you might also be surprised to discover it still has lots of charge left in it. Many consumer electronics stop working when the battery is less than half discharged. Sad, but true (and another reason to use rechargeables).

Resistance Is Futile

A great way to buy plenty of resistors (an electronic component that we'll be using in our robots and be covering in the next chapter in the "Thumbnail Guide to Electronics") is in a bulk pack. Radio Shack sells these for cheap. You get dozens of resistors in one bundle. The values of the lot are listed on the package (measured in *ohms/Ω*), but not on the resistors themselves. Now, any electronics geek worth his or her propeller beanie will tell you that resistors are *clearly* marked. They have color-coded bands on them, with each color having a value. Using the colors on each band and their positions in the band sequence, you look them up on a chart and then multiply (the third color band is a multiplier) to find the value of your resistor (see Figure 5.7). I'm lazy. I don't want to have to consult charts and do math (even if it's easily done) when I don't have to. I don't know why they can't print the values on the resistors themselves. I'm also old and can't see very well. The smaller the resistor, the harder the color bands are to figure out. So, when I get a pack of resistors, I get out my trusty DMM, set it to "Ω," plug in my hook-type test probes, and find out the resistor values. Resistors often come attached to a "reel tape," which connects all of the resistors of the same value (so, in a pack, you might get ten 100 k-ohm (kΩ) resistors, ten 10Ω resistors, and so forth). (You'll often see the prefix *k* for a thousand and *M* for a million in front of the Ω.) All you need to do is test one resistor on the tape and then write the value on the tape with a marker. When it's time to use a resistor, you can use your diagonal cutters to snip the component from the tape.

FIGURE 5.7

The four color-code bands are used to determine the value of a resistor. This one is a 10kΩ resistor like the one we will use in Chapter 8, "Project 2: Mousey the Junkbot."

You Light Up My Cadmium-Sulfide Sensors

Here's one that's fun to play with. One of the most common components for giving a robot "sight" is a *cadmium–sulfide photoresistor (CdS).*

A *resistor* is an electronic component that impedes the flow of electricity. They are usually of a fixed value, but some can vary the amount of resistance they offer. One such type is the CdS (cadmium-sulfide). To see how this works, connect your hook probes to the COM and V/Ω jacks on your DMM and turn the dial to the Ω setting (refer to Figure 5.5). Now attach the hook probes to the leads of a CdS (which you can get at an electronics store for under a dollar). Put your hand over the photoresistor and see what happens. Now shine a flashlight on it. Pretty nifty, eh? The resistance should change in response to the amount of light hitting it. More light equals less resistance. Can you imagine now how these CdS photoresistors could be used to create photovoric behavior in a robot? If not, don't sweat it, it'll all become clear to you in the projects ahead.

Resources

The main resource available for learning more about using a DMM is the Radio Shack publication, *Using Your Meter*, by Alvis J. Evans. It covers both digital multimeters and the older, analog voltmeters (VOMs). Unfortunately, you need several advanced degrees in electrical engineering to read it. Okay, it's not *that* bad, but it's far from an *Absolute Beginner's Guide*. The book does contain lots of useful information, and many common test situations, and is worth having around. And just think how superior you'll feel when it all starts making perfect sense to you!

Your Robot Workshop

Obviously, you don't need anything fancy to build robots beyond the handful of tools already outlined, and a clean, well-lit, well-ventilated workspace in which to use them. This workspace can be anything from a kitchen table to a corner of the bench in your basement. The bigger the robot, the more space you'll need, and the bigger and more specialized the tools. For most of the robot kits and building sets we'll cover in the next chapter, and for the robots we'll build later in this book, your "workshop" needs should be minimal. But if you find yourself spending a lot of time building bots, you'll want to consider a more permanent setup and a few more tools that can make your robot building safer, more comfortable, and more productive (see Figure 5.8). Here's a quick rundown of items to consider:

- A decent-size workbench or table that's high enough that you don't have to slump over it too much while you work.

- A comfortable and adjustable chair or stool. I have a nice drafting stool with adjustable seat and backrest. It has decent padding and even a lumbar support. It was an $80 chair, but I got it at an office furniture sale for $35.

- Plenty of good lighting. A combination of incandescent light (in a swing-arm lamp) and florescent light is ideal. A third light, a magnifying lamp, is even better.

- A power strip with an easily accessible on/off switch, so you can kill the power in a hurry (mine sits on the back of my workbench and is always a quick swat away).

- A ventilation system for soldering. See the "Building a Ventilation System," section later in this chapter for details on how to build a simple and cheap ventilator.

- Adequate storage. You'll need plenty of multidrawer parts cabinets and compartment cases in which to store electronic components, as well as small plastic "dump bins" to organize bench clutter, and larger bins (cardboard boxes) for storing big structural components, techno-junk, and so forth. These bins can go on shelves under your bench (if possible), or on storage shelves accessible to your workspace. You can get metal utility shelves at department stores for under $20 a unit.

■ An anti-static mat makes a nice, clean work surface, and its anti-static properties helps keep shock-sensitive parts safe.

FIGURE 5.8

Here's one take on an ideal robot workspace. Note the A, B, and C activity rings. See Tip on next page.

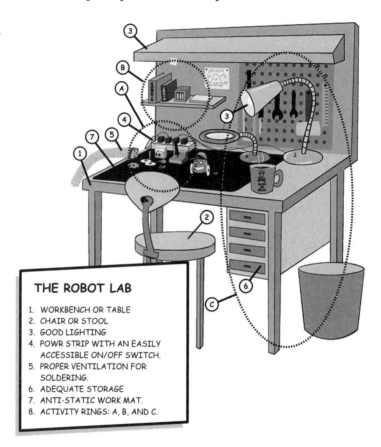

THE ROBOT LAB

1. WORKBENCH OR TABLE
2. CHAIR OR STOOL
3. GOOD LIGHTING
4. POWR STRIP WITH AN EASILY ACCESSIBLE ON/OFF SWITCH.
5. PROPER VENTILATION FOR SOLDERING.
6. ADEQUATE STORAGE
7. ANTI-STATIC WORK MAT.
8. ACTIVITY RINGS: A, B, AND C.

Thumbnail Guide to Soldering

To build robots, you *have* to know how to solder. Soldering can often be intimidating to absolute beginners. It looks difficult, and you worry that you might damage the components you're working with. The truth is that soldering isn't hard at all; at least, it's not hard to learn how to do it. To do it *well* does take practice. A lot of people start "practicing" on real projects. Even if it's one of those cheap kits sold in the back of electronics catalogs and magazines (what many are advised to start their soldering career on), if you paid money for it, you're going to be invested in soldering it well, and that can lead to frustration if you run into trouble. If you instead practice on some of that junk you've been collecting in your techno-junk box—stuff you don't give a hoot about, then you won't be risking precious parts, and you can focus first on the mechanics of soldering.

Soldering is simply the act of heating and melting solder (a special alloy with a low melting temperature) so that it creates a conductive metal bond between electronic components. This is accomplished using a tool called a *soldering iron* (there are also soldering robots, but since we don't have one of these handy, we'll be doing it by hand with a soldering iron).

The tools you need for soldering are

- Soldering iron (the *pencil-type*, not a soldering *gun*)
- Spool or tube of solder wire
- Desoldering pump (*solder-sucker*)
- Soldering iron stand with a cleaning sponge
- Nondetergent scrubby pad or steel wool

Also useful are

- Third-hand parts holder
- Jar of flux paste
- Heat sink tool

tip

Make sure to think through the organization of your tools, equipment, and supplies so that things are at the proper working distance. Think of three "activity rings." The items you use most often need to be in ring A (soldering iron, primary tools, third hand, multimeter, and so forth). (Obviously, some things will move in and out of this ring as needed.) Ring B, also within arm's reach, is where you should have your parts cabinets, vise, secondary tools, reference books and catalogs, and so forth. Ring C is for everything else (large storage, large tools, your bot collection, the espresso maker, your vintage videogame machine that you use for blowing off steam, and the cot you use to nap on while pulling robot-frenzied all-nighters).

The other essential thing that you need for soldering is *very* good ventilation. Solder contains lead. Lead makes you stupid. There are solder formulations that do not contain lead (such as silver and tin solder), but they're harder to work with. I think, in the beginning at least, you should stick to lead-based solder. But again, that means adequate ventilation (actually *any* type of soldering requires this). After we're done with this tutorial, we'll make a cheap ventilation system.

A lot of people have trouble soldering because doing it the right way just seems counterintuitive. It makes sense that you'd take the hot iron, touch the solder to its tip, and the molten solder would flow onto the component (sort of like *painting* the solder on). It doesn't work this way. When you do this, the high heat of the iron will cause the solder to "wick up" the iron's tip, *away* from the part you want to solder! Here are the few simple steps you need to know to solder well. Just read these for now; don't attempt any actual soldering. We'll revisit soldering in the "Practice, Practice, Practice" section later in this chapter and introduce you to the latest sensation since *The Macerena*: "The Solder Dance."

note

All of these tools and supplies are outlined earlier in this chapter.

1. Ah…first you need to plug in your iron. You should wait at least a few minutes for it to heat up.

2. Next, you need to "tin" your tip. You need to do this: a) if it's a new soldering iron, or b) if it's a new replacement tip, or c) if the tip hasn't been tinned in awhile. Tinning simply involves "painting" (ah, *here* you can paint) some solder onto the tip of the iron and wiping off the excess. Tinning helps to create a clean, smooth surface on the iron for good thermal contact with components. Just slather some solder all over the tip and wipe. When you have a nice bright, silvery tip, your iron is ready to roll.

3. If the tip doesn't need tinning, it needs cleaning. A clean iron is a happy iron. That's what the moist sponge on your soldering stand is for. As you work with a soldering iron, it quickly builds up crud (oxidized materials from components, packing residue left on components, oils (industrial or human), soot, and other gross goo). This crud can put a serious dent (literally) in your solder welds. Make sure the tip is clean and you can see that nice tinning job you did (and when you can't see that nice tinning job anymore, go back to step 2).

4. Scuff up your components. Oxidation occurs on all components and this oxidation (and crud, lovingly detailed in step 3) can make solder not want to stick. To prevent this, you want to gently "scuff" everything with a nondetergent steel wool or household "scrubby pad." It doesn't take much—just a quick, light swipe over all the parts to be soldered. You'll want to do this to the circuit board *tracks* (the flat copper foil paths that connect all of the components together on a PCB), the *pads* (the copper foil circles around the lead holes on a PCB), and the *leads* themselves (the wires that come out of the electronic components).

5. We'll assume you're soldering an electronic component to a PCB. We won't go into any more detail than that here (for more on electronic components,

see the "Thumbnail Guide to Electronics" in the next chapter). To solder a component to the PCB, you need to put the component through the appropriate holes on the PCB so that the leads are sticking up, from the printed side of the board, to the side with the tracks and pads (Part A of Figure 5.9). The pads are what you solder the components to. One trick: When you put a component into a circuit board hole, bend the leads out at a bit of an angle so that they hold the component in place as you solder it. Snip off leads if they're very long. Left unsnipped, the excess will absorb a lot of heat from your iron and make soldering harder (Part B of Figure 5.9).

FIGURE 5.9

A) Electronic component on a printed circuit board. Ready for soldering. B) Component leads snipped off prior to soldering.

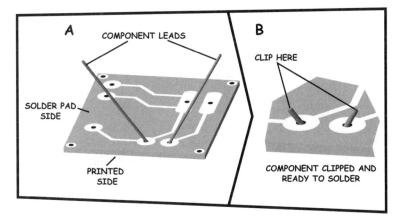

6. You're ready to solder! Hold a strand of the solder wire in one hand and the hot iron in the other. Hold the iron at an angle such that the tip touches both the pad on the PCB and the component lead. The idea here is to simultaneously heat the pad and the lead. Hold this position for a few seconds (see Figure 5.10).

tip

While you read through soldering steps 6-10, it may be helpful to look at the Solder Dance illustration, shown later, in Figure 5.11.

7. Now touch the solder wire to the other side of the component lead, opposite the iron (again, this seems counterintuitive), and hold it there until solder begins to flow.

8. As soon as the solder flows (which is called *wetting*), do the following in quick succession: Pull the solder wire away and *then* the iron. I find pulling the wire out to the side and then the iron upwards works best to get nicely shaped solder welds, but your mileage may vary. Whatever you do, *don't* pull the iron

away first! The weld will rapidly cool and your wire will get stuck to it. The whole process should only take four or five seconds. If you're holding the iron and wire in place for longer, and it's not melting, something is wrong. One trick to get things going is to touch the end of the solder wire to the tip of the iron, after you've been holding the iron to the lead and pad for a few seconds, and just before touching the solder to the *other side* of the lead. This heats up the tip of the solder wire and will make it flow more quickly and easily when you touch it to the heated component lead and pad.

FIGURE 5.10

A) The soldering iron goes on one side of the lead, the solder wire goes on the other. B) It's magic! A finished, soldered component.

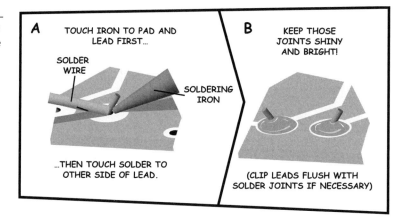

9. Reholster the soldering iron and admire your handiwork.

10. When your solder welds have hardened (which only takes a few more seconds for initial hardening), you can use your diagonal pliers to snip off the excess lead that still protrudes above the weld.

That's a rundown of basically all you need to know to solder like the pros. Now, to *actually* solder like the pros, you'll need practice—lots and lots of practice. This doesn't have to be an arduous process. All you need is a few dedicated weekend afternoons and some techno-junk lying around that's just dying to be desoldered and resoldered.

To get all of the practice that you'll need to successfully complete the projects in this book, you'll need to take apart some old gadgets in search of their PCBs. When you find one or two boards, remove them. These will be your soldering practice boards. Set them on a clean surface in your well-lit, well-ventilated work area. Before you can try your hand at soldering the parts onto the practice board, you'll need to remove the parts that are already there. To do this, you'll first have to learn the fine art of *desoldering*.

Desoldering

Desoldering is just soldering in reverse—sort of. You hold the iron in one hand, and instead of the solder wire in the other, you hold the desoldering pump. Using the tip of the iron, you heat up the existing solder joint, holding the iron on it until it shimmers and melts. It usually takes more time for the solder to melt on an existing joint than it did to create the joint in the first place. The bigger the weld blob, the longer it'll take. When the solder finally melts, you *immediately* place the nylon tip of the desoldering pump (with the plunger depressed) onto the molten solder and press the pump's trigger. The pump will make this very satisfying *THWOK* sound as the spring inside fires the plunger and sucks in the molten solder. Some joints take more than one melt and solder-suck operation before the component lead and pad are clear of solder and can be removed.

After each firing, there are bits of old solder inside the barrel of the desoldering pump. When you re-depress the plunger to get ready to melt your next weld, those solder flakes and chunks will come out of the tip. You don't want this lead-based material getting all over everything. I use a plastic leftover container with a tight lid to "save" all of my solder waste. I keep a paper towel in the bottom of the container that I wet whenever I'm going to be "shooting" solder into it (so that small flakes will stick to the towel and won't blow around). Each time you retrigger the pump, depress it over your waste container to clean it out, and be ready for the next melted joint. After you get the hang of this process, you can get quite a rhythm going and have an entire PCB denuded in no time.

Practice, Practice, Practice

After you have all of the joints on the PCB desoldered, and all of the components removed, you're ready to resolder everything. In removing existing components, you might need to heat some leads up again to detach them if they're stubborn. You might also want to use your needlenose pliers to apply more pulling force as you remove components.

Now you're ready to have some…er…fun soldering. (I have to confess, I *hated* soldering in the beginning, but after some practice sessions, it actually did become fun.) Don't worry about what the weld joints look like in the beginning—just focus on following the previous steps and paying attention to the rhythm. It's kind of like taking a dance class ("one-two-three, one-two-three"). In this case, the "solder dance" might go something like this (see Figure 5.11):

FIGURE 5.11

Do the Solder Dance! Touch iron to lead/component, touch solder to iron to pre-melt, touch solder to other side of lead, pull away solder, then pull away iron. That's it! And you didn't even have to wear embarrassing leg warmers!

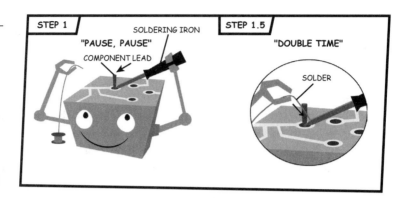

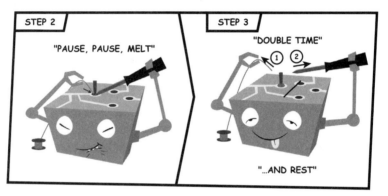

After you get these steps down, it's time to focus on making a *good* solder weld. It sounds corny to say, but a good solder weld should look *healthy*. It should look full (not sagging in), it should look shiny and bright, and it should form a complete cone-shape bond completely around (and tight against) the component lead. If the weld is dull (called a *cold solder joint*), lumpy-looking, has holes in it, or gaps between the solder blob and the lead, it's a bad weld and could be a real trouble-maker in your circuit. Any bad welds like this need to be desoldered and resoldered.

Another thing you need to make sure of is that your solder melt stays on the pad and doesn't *bridge* one pad or track to another. If your solder bridges to a pad or track that it's not supposed to, it'll short out your circuit. It's okay if the solder gets on the PCB material (that's nonconductive), but it's a good idea to shoot for keeping your solder constrained to the pads around the component leads. In electronics, neatness counts.

Keep desoldering and resoldering your practice PCBs until your welds consistently look good. Take a break once in awhile to stretch and to get some fresh air (those fumes are nasty). When you're confident that you can achieve nice, healthy welds, then you're officially an electronics geek! Congratulations!

And remember: Always thoroughly wash your hands with soap and water after playing around inside the guts of old techno-junk and when soldering. Solder contains lead. Lead makes you stupid.

Resources

The preceding section pretty much covers all you need to know to solder, but if you're hungry for more tips, tricks, and techniques (everybody has a slightly different style), type **how to solder** in an online search engine (such as www.google.com or www.yahoo.com). You'll come up with dozens of "Soldering 101" tutorials.

Building a Ventilation System

If you've been doing your soldering practice like a good budding bot builder, you know what bad news those fumes are. Even with the window open and a fan going, you still end up with solder smoke wafting into your nostrils. Let's take care of that (and do some techno-junk scavenging, more soldering practice, and wire shrink wrapping in the process). It's easy.

Parts List

Building a ventilation system requires the following parts, which are also shown in Figure 5.12.

- 12V DC power fan
- 12V AC-to-DC wall transformer (a *wall wart*)
- Flexible dryer hose (available at home and hardware stores)
- Shrink-wrap tubing
- Duct tape

After you've gotten your materials together, use the following steps to build your ventilation system:

1. First, you'll need to find the 12V DC power fan (such as the kind used to cool the innards of computers). If you have a junked computer (or maybe a friend has one lying around), you can use the fan from there. If not, you can buy one at Radio Shack or a mail order electronics house, such as Jameco (www.jameco.com), for less than $10.

FIGURE 5.12

The parts you'll need to construct a soldering ventilation system.

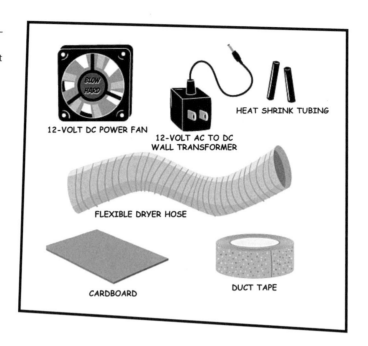

12-VOLT DC POWER FAN

12-VOLT AC TO DC WALL TRANSFORMER

HEAT SHRINK TUBING

FLEXIBLE DRYER HOSE

CARDBOARD

DUCT TAPE

2. Another part you'll need is an AC-to-DC wall transformer. This is one of those boxy "wall warts" found at the end of most consumer electronics devices. It's there to convert the Alternating Current (AC) from your wall into the Direct Current (DC) that electronic devices use. Again, you can try and cannibalize one of these from something in your techno-junk box. I got mine from an old Record-a-Call answering machine. If you can't scavenge one, the shack to the rescue!

3. After you have the parts, cut the power jack off of the transformer's cord. You'll also need to "split" the two-part wire jacket on the cord by peeling the two jackets apart for a few inches (you can get this started by slicing in-between them with a hobby knife). Then, use your wire-stripping tool to remove about an inch of the

caution

If you decide to yank a fan out of your computer case, stay away from the one inside of the power supply if you're an electronics newbie. The components inside a PC power supply can sustain a dangerous charge even when unplugged. The fans mounted over the central processing unit (CPU) or fans mounted to the interior of the case are fair game. Usually, you simply need only detach a little plastic connector.

plastic jacket from each of the four wires (two on the fan, two on the transformer) so that you expose bare wire (see Figure 5.13).

4. Before connecting the wires, you'll want to thread about an inch and a half length of heat-shrink tubing through *both* the positive and negative wires on *one* set of the wires—either the transformer's wires or the fan's wires (so that you can slide the tubing over the soldered connections when you're done). Make sure to choose a big enough diameter of tubing so that it'll fit over your wrapped and soldered wire connections (but not too big so that it won't fit snugly when it shrinks). See Figure 5.13.

FIGURE 5.13

The four wires (two from the transformer and two from the fan) stripped and ready for splicing. Don't forget to thread heat-shrink tubing on one negative and one positive wire before splicing them.

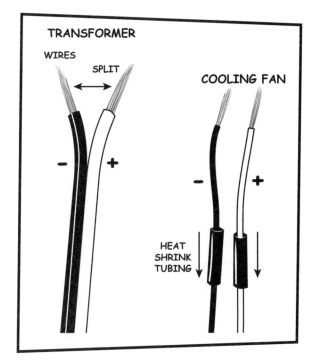

5. Now find the two negative wires on the fan and transformer wires. These might be colored black, or have a white line down the wire to indicate negative (-). Overlap the one-inch pieces of exposed wire, cross them over in an "X" in the middle, and then tightly twist one negative wire around the other. The wires will be multistranded, so make sure to keep the strands twisted tightly together (see Figure 5.14).

6. Wrap the two positive (+) wires together in the same manner as shown in Figure 5.14.

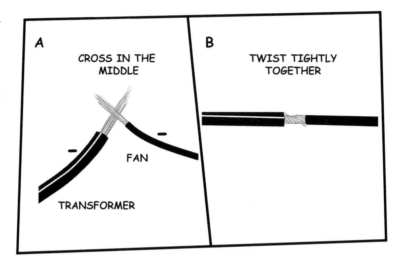

7. Using your soldering iron, weld the negative wires together by applying a thin coat of solder. Because the solder wants to move away from the tip, this can be somewhat difficult. Hold the iron under the wire wrap and hold the solder above the wrap. As the solder melts, move it back and forth and "paint" it quickly on, letting it seep into the wire strands. Don't hold it on for too long, or all of the solder will melt and drip off of the wire! Not good. You'll want to use your helping-hand tool for this operation. When finished with the negative wire connection, do the same with the positive wires (see Figure 5.15).

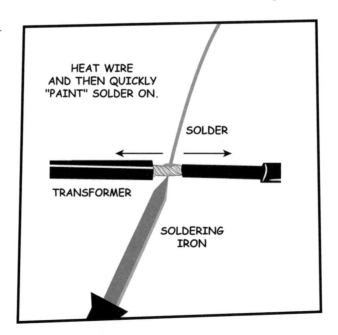

8. When your joins are cooled, slide the heat-shrink tubing over one of them so that no solder or wire is exposed. Use your micro-torch or heat gun to carefully shrink the tubing onto the new connection. Be careful not to melt the existing wire jacket! Now slide the tubing over your second join and heat shrink it (see Figure 5.16).

FIGURE 5.16

Your finished wires should look something like this.

9. Test your fan by plugging it into a power strip (and be ready to unplug it or hit the power switch on the strip if something goes wrong).

10. Now you're ready to hook up the fan to the dryer hose (see Figure 5.17). You can just set the fan into the mouth of one end of the hose and hang the other end out of a window, but you'll get better draw if you close the gaps between the fan and the hose mouth. You can easily do this with duct tape or scrap cardboard and tape.

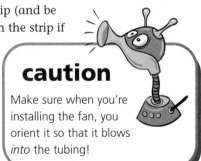

caution

Make sure when you're installing the fan, you orient it so that it blows *into* the tubing!

Set up the fan right next to your soldering area, toss the other end of the hose out of the window, and start soldering like a robo-maniac.

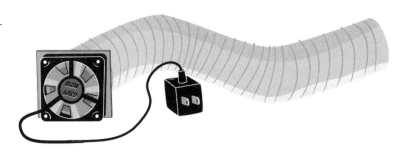

That's it! Now you can set your ventilation system right next to where you're soldering and all of that yucky smoke will get sucked away. And how do you get that rascally, snaking dryer hose to stay still? Duct tape, my boy—the miracle of duct tape.

Safety First

Okay, so we're actually talking about it last, but in your work habits, safety should *always* be uppermost in your mind. Building robots involves power tools, caustic chemicals, high heat, and large, stubborn and sharp-edged components and tools flying across the room (that latter one's only for those of us with anger management "issues"). Besides the recommendations elsewhere in this chapter of good lighting, adequate ventilation, and use of safety goggles, you should also consider the following:

- Use disposable gloves whenever handling any caustic chemicals.
- Make sure to have a paper dust mask and use it whenever doing anything that generates tiny particles of *anything* (wood, plastic, metal, and so forth).
- Always have a fire extinguisher nearby if using anything potentially flammable.
- Have a first aid kit in the house and know how to use it. Also, have an eyecup in your kit that you can use to flush out your eye if you get something in it.
- Read the labels and instructions on chemicals or on anything else that might be harmful. This might seem obvious, but lots of gadget geeks pride themselves on *not* reading instructions. This is fine if you're talking about a digital camera or your new TiVo, but not if you're talking about your micro-torch and a tube of two-part epoxy. As the Web admonishment goes: RTFM! (Read The Freakin' Manual!)
- There are certain items of techno-junk you want to avoid taking apart. TVs can store dangerous amounts of electricity in their circuits, even a long time

after you've unplugged them. They are best left to TV technicians. Computer monitors can also bite you and should be avoided.

- Capacitors in any piece of electronics equipment can hold a charge for a long time and are best handled carefully. Never touch both of the leads of a capacitor that's been in a powered circuit until you've "discharged" it (dissipated the stored current). This can be done simply by laying a screwdriver with an insulated rubber handle across the capacitor's leads.

- Safety goggles/glasses don't do much good if you can't see out of them. Always hang your goggles up. Don't lay them face down because they can get scratched. Hanging them up also prevents them from getting misshapen. Store safety glasses on their sides or in a case.

- Obviously, always keep any tools, supplies, materials, or robots that could be harmful out of the reach of children.

- Be a physics paranoid! Before you undertake anything involving physical motion, think through the physics of what's happening. What's going to be moving? What could move that shouldn't? What could fly off and where would its momentum take it? This is not only a good safety precaution, it's actually a way of better understanding the physical science of the world around you. Bonus, dude!

Overall, you just want to use common sense. If it smells toxic, it probably is. If it's hot, you don't want to touch it. If it's sharp, it'll probably cut you. If it's spinning really fast, you probably don't want to grab onto it. And then, of course, common sense should also tell you that just because these conditions aren't met doesn't mean you still shouldn't be cautious.

THE ABSOLUTE MINIMUM

In this chapter, we gathered and discussed all of the tools and supplies needed to build the robots covered in this book (and most other small- to medium-size bots). We also tried to stress the need for safety. C'mon, people, you've seen the late night movies—robots can be dangerous, deadly even! Always be careful when working with them or their constituents. Here's your chapter crib sheet:

- We humans are tool users. This goes double for bot-building humans. Besides typical household tools, you'll need gear with such exotic names as soldering iron, desoldering pump, diagonal cutters, micro-torch, and crimping tool.

- Bot builders also get to work with cool stuff like molten metal (solder) and heat-shrink tubing (basically an excuse to use the micro-torch).

- A digital multimeter will instantly turn you into a wirehead. It's a device for testing electronic components and circuits and is fun to play around with. It's the closest you'll ever come to Mr. Spock's Tricorder.

- Soldering is easy to learn, but not as easy to master. Learn the "solder dance." Practice the solder dance. Practice it some more. It won't get you dates, but your robots will love you for it.

- Solder will quickly make you dumber than Jethro on *The Beverly Hillbillies*. You need a ventilation system to suck those nasty fumes away. We made a cheap one. You'll always want to wash your hands after soldering. That means no nachos on the job!

- When working with robots, think smart. Be safe. Wear your goggles, wear gloves when appropriate, think through what you're doing (and consider worse-case scenarios). Have a first aid kit handy and know how to use it.

In this Chapter

- Cools kits you can build
- Robot building sets
- Ready-mades and retro-robots
- Thumbnail guide to electronics

6

Acquiring Mad Robot Skills

There are two ways to become a budding bot builder. One is to home-brew your own robots (which we'll start doing in the next chapter). The other way, which some of you might find easier and less intimidating at first, is to use readily available kits and building sets. In this chapter, we'll take a quick look at some of these commercial robot kits and sets and outline some of their strengths and weaknesses. We'll also talk about what aspects of robotics each of them emphasizes. Then, ladies and gentlemen, we'll attempt a major hat trick: teaching you the basics of working electronics in just a few pages. The "Thumbnail Guide to Electronics" should at least teach you what you'll need to know to tackle the kits detailed in this chapter, and the projects that begin in the next.

Killer Kits

Just a few years ago, there were only a few robot kit makers out there and only a few places where you could shop for their wares. Today, there are dozens of producers of robot kits and countless retail sources. Robot kits come in all shapes and sizes, but the most popular are small, car-like machines, with two-to-four wheels, several sensor systems, and usually, some means of controlling and/or programming the robot. These kits can be as low as $25 or as high as $1,000, but they usually run in the $40–$200 range. Obviously, the more you pay, the more sophisticated (and usually harder to build) your kit will be. A lot of robot kits are sold in two forms: one in which the printed circuit board (PCB) has already been assembled (but the rest of the bot is in kit form), and one in which you have to put everything together, including the PCB. Some kits can even be bought fully assembled, but if you were the kind of person who'd want one of these, you probably wouldn't be reading this book! Here's a rundown of some of the more popular robot kits out there.

Rockit Sound-Controlled Robot

Manufacturer:

OWI/Movit

Web Site:

robotstore.com

Price:

$29.95

Difficulty Level:

Beginner

FIGURE 6.1

Kids will love Rockit's "crispy movement."
Photo by JFM Digital for RobotStore.com.

Who can resist a zippy little critter named the Rockit? This kit, by the venerable Japanese robot kit maker, Elekit (which sells their kits under the OWI/Movit name), is a great starter bot for a child, and a fun monitor pet for the rest of us. At its heart, Rockit has a sound sensor that controls its actions. When it bonks into anything, or you clap or yell at it, it triggers a typical obstacle-avoidance behavior (it backs up, turns its wheels a little, and motors forward again). Elekit's Japanese Web site says that the Rockit has "crispy movement." Er...okay. The kit is available in both no-soldering and soldering-required versions (for the same price). Don't expect a lot from this kit (it's a "one-hit wonder"), but again, it makes a good soldering practice project.

Spider 3 Walker

Manufacturer:

OWI/Movit

Web Site:

robotstore.com

Price:

$59.95

Difficulty Level:

Beginner (soldered)
Intermediate (soldering required)

This kit is a bit more challenging than the
Rockit. It uses two DC motors and gear sets
to control six legs. The three legs on each side
are attached to one rotating cam that transfers
its movement into the three legs simultaneously.
The result is an approximation of walking. It's
not the smoothest gait in the world, but serves
as an example of how clever mechanical engineering can eliminate the need for sophisticated
programming. The Spider 3 uses visible light
sensing (in the proximity detection scheme we
discussed under "Light Sensors" in Chapter 4,
"Robot Anatomy Class") to avoid obstacles. It
really is a handsome little robot and will get you
many compliments if you display it on your desktop. All of the OWI/Movit kits are, unfortunately,
very cheaply made and the legs on our
Spider 3 were frustratingly prone to falling
apart.

FIGURE 6.2

It's temperamental, but the Spider 3 looks cool
and has unique walking mechanics. Photo by
JFM Digital for RobotStore.com.

tip

Most of the robot
kits and building sets
covered in this chapter
are available at the awe-
some Robot Store (along with
a bunch more kits, sets, and
parts). Visit its Web site (www.
robotstore.com) or call (800-374-
5764) for a free color catalog.

WAO-G

Manufacturer:

OWI/Movit

Web Site:

robotstore.com

Price:

$98.95

FIGURE 6.3

The venerable WAO series of kits has been on
the market for years. Photo by JFM Digital for
RobotStore.com.

Difficulty Level:

Intermediate (non-soldered)

Advanced (soldering required)

Elekit's most sophisticated robot is its WAO-G kit (*WAO* allegedly stands for *Wise Argent Orb*). This little gold-domed bot has a 26-button keypad on its rear. With it, you can program *if/then* routines (if *x* happens, then do *y*). The keypad is divided into sections. One section handles navigation (forward, reverse, left, right, and so

forth), and the other one handles the if/then commands. There are also keys for inputting numerical values (1–9), program execution keys, and some special function keys. By scripting routines with these keys, you can form control programs. There's also a cable and software available ($29.95) for interfacing with your computer, in which more sophisticated programming is available. The WAO series (the WAO, WAO II, and now, the WAO-G) has certainly been popular among robot hobbyists. It's been around for at least a decade, but it is now showing its age. For the same money, you could get the Sumo-Bot kit (shown later) and have a much more versatile robot that does everything the WAO-G

note

The BASIC Stamp is one of the most widely used microcontrollers in the amateur robots community. We'll be covering it throughout the book.

does, but has expansion capabilities, and a huge user-community for its microcontroller (the BASIC Stamp). Both a soldering and no-soldering version of the WAO-G are available at the above price.

Carpet Rover 2 Combo Kit

Manufacturer:

Lynxmotion

Web Site:

www.lynxmotion.com

Price:

$282.50

Difficulty Level:

Advanced

This robot is a full-blown development platform. It is sold in many different configurations, from a bare-bones kit of just the

FIGURE 6.4

The Carpet Rover 2 is a development platform on which you can experiment with different sensor systems. Photo by JFM Digital for RobotStore.com.

platform, motors, and hardware, to the Combo kit outlined here. The CR2 Combo kit comes with everything from the bare-bones kit, plus a BASIC Stamp II, an Infrared Proximity Detector (IRPD), a line-tracking sensor kit, two bump switches, two CdS photocells, a Ni-MH battery pack, and even a battery charger. It also has a prototyping area (or *breadboard*) for doing your own sensor experiments. Using the bump sensors, the Carpet Rover will run typical obstacle avoidance routines through contact sensing. Using the IRPD, and a BASIC program downloadable from Lynxmotion's Web site, the CR2 will exhibit non-contact proximity detection. It will smoothly move around a space and not crash into anything. It will also, according to Lynxmotion, exhibit some emergent behavior. In a corner, it will stop, back up, turn, and drive away, a sequence it was not specifically programmed to perform. Lynxmotion also sells a version of the CR2 that uses the OOPic-R, the same microcontroller you will learn about in Project 3.

Hexapod Walkers

Manufacturer:

Lynxmotion

Web Site:

www.lynxmotion.com

Price:

$236–$844

Difficulty Level:

Advanced

FIGURE 6.5

The Hexapod Walkers sure look cool, but they are for experienced builders only. Photo by JFM Digital for RobotStore.com.

Jim Frye of Lynxmotion is a legend in the hobby robot biz. His Hexapod Walkers are like the Sea Monkeys and X-Ray Specs of the robot world, always advertised in the backs of robot and electronics magazines, seducing gadget geeks with promises of awesome robot glory. The Hexapod kits aren't cheap, and building a walking robot (with up to 14 servo motors) and getting it to work properly can be a real challenge—but success can be very gratifying, much less disappointing than Sea Monkeys. Most of the Hexapod kits use the Next Step, a BASIC Stamp–based microcontroller. Rumor has it that at least some of the future Hexapods will use the OOPic-R microcontroller. The OOPic is part of a new generation of microcontrollers that offer easier and more powerful programming capabilities and many built-in components created with the robot builder in mind.

CyBugs

Manufacturer:

JCM Inventures

Web Site:

`www.jcminventures.com`

Price:

$49.95

Difficulty Level:

Intermediate

FIGURE 6.6

CyBugs demonstrate how biology is influencing robotics. Photo by JFM Digital for RobotStore.com.

There are lots of robot kits out there, but few of them offer true autonomy, complete with the capability for robots to "feed" themselves. Craig Maynard, the robo-maniac behind the CyBug, is creating a growing robot "ecology." You get started with the CyBug Scarab kit itself, a simple two-motor, PCB-based robot. The CyBug has touch and light sensors used for photovoric, photophobic, and obstacle-avoidance behaviors. The kit also comes with instructions for building a feeding station that can be used for recharging the 9-volt (V) battery you provide. From there, you can add a Predator/Prey kit ($32.95). These are two add-on circuit boards that can turn two CyBugs into predator and prey (one robot sends out an infrared signal that the other one tracks and attacks). Another kit, the Sunflower Power Plant ($59.95), is a solar-powered feeding station, tracking the brightest light source, gathering energy to recharge any CyBugs that dock with it. The CyBugs teach BEAM-like robotic concepts and demonstrate some of the influences of biomimicry on contemporary robotics thinking.   ·

Build Your Own Robot

Manufacturer:

Tab Robotics

Web Site:

`tabrobotkit.com`

Price:

$59.95

Difficulty Level:

Beginner

FIGURE 6.7

The Build Your Own Robot kit is fast and easy to build, but limited in function. Photo by JFM Digital for RobotStore.com.

You have to love the idea behind this kit. You're in a bookstore perusing the shelves, and there, right next to the other robot books (heck, probably right next to *this* robot book) is a programmable robot, in a book-size box, that you can take home, assemble, and have running in under an hour. What you end up with is a barebones, two-wheeled bot that uses its circuit board as its body. A remote control is used to control the bot, and it has some built-in behaviors that it can demonstrate (photovoric, photophobic, and wall-following behaviors—the classics!). What you can do beyond that is limited with what you have, but the board does have an IC socket designed to accept a Stamp II microcontroller. Add that, and you can program the robot to your heart's content (or to the extent of BASIC programming, anyway). But given the extra $50 for the brain upgrade, you might as well buy Tab's Sumo-Bot (see the next section).

Sumo-Bot

Manufacturer:

Tab Robotics

Web Site:

tabrobotkit.com

Price:

$99.95

Difficulty Level:

Intermediate

FIGURE 6.8

The Sumo-Bot kit enables you to program behaviors and change up sensors. Photo courtesy TAB McGraw-Hill

Also found on the shelves of major bookstores, this is Tab's next-generation kit, built they say, on feedback from the *Build Your Own Robot Kit*. Robot sumo is a fast-growing area of robotic sports. It uses similar rules to human sumo, only on a smaller ring and with less blubber, grunting, and body odor. This kit is designed as a ready-made sumo competitor, although it doesn't meet the specs of any regular sumo competition—and hey, most of the fun of robot sumo is building your own mechanical competitor! It also has a weak drive system, making it a pretty lousy competitor even for doing some impromptu sumo in your backyard. Still, this is a pretty decent general robot kit for $100. It goes together in about an hour. It has some really nice features, such as the Stamp II microcontroller built in, a prototyping area for adding your own sensors (see the "Thumbnail Guide to Breadboards" in the next chapter), and extensive documentation, control programs, and a crash course in all things electronic and robotic on an included CD-ROM.

If this was the only robot kit you bought, and you tried all of the experiments, read all of the docs, and so forth, you would come away with a fairly decent understanding of modern robotic concepts. This robot doesn't teach much on the building side of bots (there are only a few parts to assemble). The focus here is on the programming and use of the prototyping area to add (and program) new types of sensors. It uses the BASIC Stamp II, which is the most widely distributed microcontroller. There are tons of great books and other educational materials available for the Stamp and the BASIC language that it speaks.

note

Sumo-Bot kit trivia: One of the two creators of the kit is Ben Wirz of iRobot Corporation. iRobot is Rodney Brooks's company and is responsible for such innovative bots as the Roomba (which we'll discuss later in this chapter), My Real Baby, and the Co-Worker, the robotic presence platform.

BOE-Bot

Manufacturer:

Parallax

Web Site:

www.parallax.com

Price:

$299

Difficulty Level:

Intermediate to Experienced

Anyone who's spent time thumbing through electronics magazines will be familiar with Parallax's ubiquitous BASIC Stamp microcontroller. First released in 1992, this tiny chip, which can be programmed in the BASIC language, has put microcontroller technology into the hands of everyone, enabling them to control all manner of electronic devices. Of course, it didn't take long

FIGURE 6.9

The BOE-Bot is designed to teach kids (of all ages) robot principles and BASIC programming. Photo by JFM Digital for RobotStore.com.

for the Stamp to become a favorite among robot hobbyists and professionals. Soon, Parallax got into the act itself with the development of its own robot kits. The Stamp has become the favorite microcontroller in schools, and this robot kit is geared toward that market. The *BOE* in the bot's name stands for *Board of Education*, the name for the controller module that comes with the kit. Like the Sumo-Bot kit, the BOE-Bot is more than just a robot; it constitutes an entire course in mobile robotics.

It takes only an hour or so to construct the robot, but using the included *Robotics!* manual, and the on-board prototyping area, one can perform dozens of robot experiments. There are also numerous add-on sensor kits that Parallax sells, including the new CMUcam, an on-board camera (which can, among other things, be programmed to track objects of a certain color), a front-end gripper (which has an infrared sensor that can detect objects in its grasp), and an ultrasonic rangefinder. The BASIC Stamp (and the BASIC language itself), ancient by computer tech standards, seems to be losing favor in the hobby microcontroller world, with more sophisticated and powerful technologies (like the OOPIC) taking its place. But BASIC is...well...basic—a good, solid, easy-to-learn programming language with tons of resources available on it. Nothing you learn from building and programming the BOE-Bot will be lost if you move onto another microcontroller and programming language.

note

Another huge drawback to the Tab Sumo-Bot being, well, a sumo bot, is that it has no edge-detection sensors. *All* sumo bots need a way of telling whether they're about to wheel themselves over the edge. This bot doesn't have such sensors. You might be able to add some, but it does make you wonder why they bothered calling it Sumo-Bot (other than to capitalize on the current sumo craze).

SolarSpeeder

Manufacturer:

Solarbotics

Web Site:

www.solarbotics.com

Price:

$25

Difficulty Level:

Beginner (with some soldering experience)

The SolarSpeeder, though not an actual robot, is a nifty, entry-level kit that teaches some of the basic ideas behind BEAM robotics, namely use of a *solarengine* and high-capacity capacitors for power storage. This

FIGURE 6.10

The SolarSpeeder uses a solar cell and a capacitor to create a jumpy little racer. Used by permission, Solarbotics.com.

project also makes use of the 1381 voltage trigger integrated circuit (which looks like a regular transistor) to "time" the release of power from the full capacitor to the pager motor. The kit is extremely well made, with a cool-shaped custom PCB that's bent to create a car-body shape. It even has an ingenious steering mechanism where part of the PCB is very thin (and therefore bendable), but is reinforced with a copper wire to keep it from breaking. By bending this wire (which has the pager motor and third wheel on the other end of it), you can make the Speeder do turns. The Speeder takes about three minutes to charge under incandescent light and about a minute in direct sunlight. It's rather boring to watch the solar panel charge, but it's a thrill when it jets off, hell-bent for leather. It *really* takes off, traveling up to 10 feet on its three soft rubber tires with nylon hubs. If you have an older child who's interested in electronics, or if you want to practice your soldering and kit building, this is a great place to start—though you'll want to practice on some techno-junk before tackling the PCB. There are some tight soldering spots you'll need to be careful with.

Photopopper 4.2b

Manufacturer:

Solarbotics

Web Site:

www.solarbotics.com

Price:

$60

Difficulty Level:

Intermediate

This kit, from the premier BEAM kit maker, Solarbotics, is a perfect introduction to the concepts and technologies behind BEAM robotics. The Photopopper uses a solar engine arrangement similar to the

FIGURE 6.11

Watch it skitter, watch it bump, watch it "photopop." It's the Photopopper from Solarbotics. Used by permission, Solarbotics.com.

SolarSpeeder, but here, it is combined with more sophisticated electronics (and sensors) to make a truly autonomous robot. The Photopopper has a voltage detection device that manages the power stored in the capacitor, which results in a very active solar-powered robot. When discharged, the Photopopper can travel 1 meter (3.3 feet) in under 2 minutes. That might not sound like much if you haven't experienced a BEAMbot before, but the self-powered autonomy makes these little "photopops" surprisingly exciting. The Photopopper has a pair of infrared sensors that give it light-seeking capability, and a pair of whisker-type bump sensors that provide obstacle avoidance. The SolarSpeeder kit and the Photopopper together make a great introduction to solar engines, BEAM circuitry, and concepts of robot autonomy.

ER1 Personal Robot System

Manufacturer:

Evolution Robotics

Web Site:

www.evolution.com

Price:

$499

Difficulty Level:

Intermediate to Advanced

Since the 1980s, there have been many valiant attempts by robot-inspired entrepreneurs to jump-start a home robot revolution with a killer kit or out-of-the-box bot. Unfortunately, all of them have eventually become digital dodo birds along the path of robot evolution. The latest brave explorer in this market wilderness is the aptly named Evolution Robotics.

Obviously, one of the biggest impediments to success in home robotics is getting the price down low enough to make it attractive to a large-enough market, while still offering a robot that's interesting or useful enough to be worth the purchase. Evolution has handled this in a unique way. It sells a kit that contains everything but the most expensive component, something many consumers already have: a laptop. The ER1 kit comes with a set of Tinkertoy-like aluminum building struts, called XBeams, two stepper motors, wheels, sensors, a Web camera, and the heart of the company's robotics technology, its innovative ER1 Robot Control Center software.

One of the other roadblocks to robots in the home is the difficulty and expense of a decent navigation system. Evolution's scheme is innovative and can be implemented at an extremely low cost. It uses wheel rotation encoders (which we mentioned in the "Sensors" section of Chapter 4) to measure distance traveled. By first teaching the robot the room (creating a map and wheel rotation data, stored in

FIGURE 6.12

Evolution Robotics's unique XBeam system lets you build different robot bodies.

note

Conflict of Interest Alert: Dave Hrynkiw, the big brain behind Solarbotics, was also the tech editor for this book. So maybe you'll doubt us when we say that Solarbotics makes the best darn robot kits out there! Well, it's true! Solarbotics has been in business for 10 years and their kits, parts, customer service, and constant attention to detail have consistently impressed us. That's why we asked Dave's help in making *us* look good. We wanted the best!

the robot's laptop brain), it can then use the stored information to find its way around. This has been used with other home robots, such as Probotics's Cye, and it does work, but the robot frequently gets confused and lost. A few wheel slips on the edge of a throw rug, and your bot is headed for the basement stairs. Evolution's software adds another feedback mechanism through the use of a cheap Web camera. This capability to recognize objects in the environment helps to keep the robot on track, even if its wheel sensors conspire to send it off someplace it's not supposed to go.

The ER1 system is easy enough to set up and use, and interesting enough as a robot development platform, that it is becoming a big hit with robot hobbyists. But it still is a hobby/developer's platform, not a true home robot for the average user. The iMac of robotics is still a generation or two away. Evolution Robotics offered a glimpse of such a bot at the 2003 Consumer Electronics Show when it showed off its ER2 prototype (see Figure 6.13). This out-of-the-box bot looks like it could be a distant ancestor of R2-D2. It has a Cyclopean eye in the center of its head, two speaker ears, and a video monitor in its belly. It also sports one gripper arm. The company is now looking for partners to develop the ER2 and sell it to a home market for between $1,000 and $2,000.

FIGURE 6.13

Is this the future of home robotics? It's Evolution Robotics's ER2 prototype. Image courtesy of Evolution Robotics.

Robot Building Systems

Kits are a wonderful way of building bots, but when you're done, your options for improvement, change, and experimentation are limited (or nonexistent). Building sets are just the opposite. They provide you with a diverse collection of parts (and there are usually supplemental sets available) from which you can build hundreds of different robots. With building sets, your imagination can run wild with what you can make.

LEGO MINDSTORMS Robotics Invention System

Manufacturer:

The LEGO Group

Web Site:

`www.mindstorms.com`

Price:

$199

Difficulty Level:

Beginner to Advanced

FIGURE 6.14

LEGO's MINDSTORMS building set has revolutionized small robot building. Photo by JFM Digital for RobotStore.com.

For anyone who remembers LEGO building sets from their childhood, there is something truly magical about first "playing" with the MINDSTORMS Robotics Invention System. With the addition of sensor bricks, motors, a microcomputer, and a programming language, the LEGO bricks of your childhood come to life. First introduced in 1998, MINDSTORMS instantly became a hit, selling so well, that it even stunned the LEGO company itself. What it saw as a more sophisticated version of its toy sets has become a serious robot experimentation and prototyping tool found in grade schools, universities, among researchers, and in the hands of countless amateurs. Besides the excellent official MINDSTORMS Web site, nearly overnight, hundreds of sites began to spring up on the Internet, with builders detailing their robot designs, control

tip

There are now thousands of Web sites devoted to LEGO-based robotics. There are also many books exploring the subject. We cover some of the best books in Chapter 11, "Robot Books, Magazines, and Videos," and some of the more active Web sites in Chapter 12, "Robots on the Web."

programs, homemade sensors, and other creations. Many builders quickly found the programming language that MINDSTORMS uses to be too limiting for the things they wanted their robots to do, and so new languages were written, such as Dave Baum's popular NQC (Not Quite C). For those who may not know, C is a popular programming language in the computer world.

The basic MINDSTORMS Robotics Invention System consists of over 700 LEGO building elements (bricks, beams, axles, wheels, gears, connectors, wires, and so forth), two motorized bricks, two touch sensor bricks, one light sensor brick, and an infrared transmitter (for downloading programs from your PC to your robots). The electrified heart of MINDSTORMS is the RCX Microcomputer, a large computerized "brick" onto which the other building elements are attached. To program your robot, MINDSTORMS comes with an extremely easy-to-use graphical programming language called RCX Code. One of the coolest things about RCX Code is that you build applications in it, block by block, just like the LEGOs themselves. The interface is so simple, even a young child can understand the basics of what's going on. It makes your head spin to think of the abilities some children will have who grow up with this level of sophisticated computer, robotics, and programming training at such a young age.

note

The name *MINDSTORMS* is taken from a paper, *MindStorms: Children, Computers, and Powerful Ideas*, written by Seymour Papert, cofounder of MIT's Artificial Intelligence Lab. MIT has had a partnership with LEGO for years. The RCX Microcomputer used in MINDSTORMS grew out of MIT's "Programmable Brick Project," an effort to build a self-contained microcomputer that could easily interface with LEGO components.

Besides the main Robotics Invention System, there are numerous add-on kits that include more building elements, additional computer bricks, and new sensor types, such as a heat sensor, rotation sensors, and a camera.

Fischertechnik

Manufacturer:

Fischer

Web Site:

www.fischertechnik.com

Price:

Varies

Difficulty Level:

Beginner to Advanced

FIGURE 6.15

Fischertechnik's kits can create anything from a plaything to an industrial machine prototype. Photo by JFM Digital for RobotStore.com.

LEGO was a toy that is now frequently used as a serious prototyping tool. Fischertechnik was originally an engineer's prototyping tool (used to model industrial equipment and assembly-line robots) that is now sold as a toy—a *serious* toy. Until recently, most of the products in its commercial line focused on industrial robotics and machine automation, but its new Computing Mobile Robots Kit ($399) enables you to build five different mobile robots. Some of the behaviors modeled include edge detection, collision detection, line following, and light seeking/avoiding. The kit contains 280 pieces, 2 motors, 6 touch sensors, 2 light sensors, and Fischertechnik's microprocessor, called the Intelligent Interface. Its Industry Robot Kit ($219) lets you create working factory robot models with various types of functionality and end effectors. The set does not come with the Intelligent Interface. That's available for $180.

Robix Rascal Robot Construction Set

Manufacturer:

Advanced Design, Inc.

Web Site:

www.robix.com

Price:

$550

Difficulty Level:

Beginner to Intermediate

Robix is one of the more unique building sets out there. As the price might suggest, it's geared more toward education and

FIGURE 6.16

The Robix Rascal building set teaches key concepts in industrial arms and manipulators.

research, rather than home, markets. Although the other systems focus on mobile robots, the Robix kit is mainly for exploring various designs for industrial manipulators. The kit comes in a heavy-duty plastic toolbox and includes six hobby servo motors, dozens of aircraft-grade aluminum building struts, an electronics interface (for connecting your robot to the control programs on your PC), a weighted robot arm base, all necessary tools, and numerous other components. You program control of the servo motors on your PC using the included software. The program includes a "teach" mode in which you use the arrows on your keyboard to move the servo motors to automatically create control routines. Years ago, when we first got our Robix kit, I used it to create the world's most unnecessary and inefficient pencil-fetching robot. At the time, I did most of my work on a Mac, and built a Robix arm that sat next to me at my desk, poised above a pencil cup. To get a pencil, I would have to roll my chair over to my PC on another desk, initiate the pencil-fetching routing in the (PC-only) Robix software, and then roll back over to my Mac workstation in time to grab the pencil from the robot's gripper. It was silly, but a fun and geeky gag to show off to friends.

There are non-arm bots you can build with Robix, such as a cool robo-snake and a three-legged walker, but in this day and age of cheap microcontrollers and servos, and such high-tech abilities as MINDSTORMS's infrared program downloading, being tethered to a computer has become a big drawback to the Robix set. Still, if you're fascinated by working with robotic arms and end effectors, you will not find a better, more versatile building set.

Robotix

Manufacturer:

Learning Curve

Price:

Discontinued

Robotix was a fascinating robot building set sold by Learning Curve. It has discontinued the line, but the sets are still being sold in some toy stores (at closeout prices) and are readily available on eBay. Robotix was really more of a mechanical construction set than a robot building system. It did have a programmable keypad-based computer in several of its high-end sets, but its functions were very limited. Most of the Robotix kits came with one or more motors and a battery pack/switching box. The real strength of the sets was in their sturdy hexagonal building components. These very cool-looking pieces (molded with all sorts of sci-fi–looking electronic junction boxes, wires, and control panels on their surfaces) could be fashioned into extremely stable structures. Unlike LEGO bricks, under most circumstances, Robotix parts stay where you put them. Given the limited electronic/computer capabilities of Robotix, the sets became more popular among robot amateurs and professional prototypers as building materials to hybridize with the LEGO RCX computer or another microcontroller and add-on sensors. And, given that the sets are still available, now at a deep discount, they are worth covering here.

Out-of-the-Box Bots

Robot hobbyists obviously love tinkering with and building their own robots, but even the hardest-core gearhead dreams of a fully functional robot that can be yanked from its box, plopped onto the floor, switched on, and immediately enjoyed (or immediately put to work). There are still precious few such robots on store shelves, but the success of entertainment robots such as Sony's AIBO, and affordable domestic bots such as iRobot's Roomba vacuum, point the way toward the future of home robot technology.

Sony AIBO

Manufacturer:

Sony

Web Site:

`www.sony.com/aibo`

Price:

$1,299

AIBO, Sony's first in a proposed line of "entertainment robots," has a surprising degree of mobility, with 18 different motorized joints. It can play with a ball, stretch like a cat or dog, and even right itself when it falls over. Sensors on its body make it

FIGURE 6.17

Sony AIBO: artificial life form or over-priced plaything?

respond to petting and handling. The brilliance of AIBO is not only in its impressive mechanics, but also in its apparent emergent behaviors and distinct personalities. Its behavioral architecture has been constructed in such a way that it appears to grow, evolve, learn, and become its own unique little critter. This is accomplished via a preset series of life stages (newborn, baby, adolescent, young adult, adult), with each stage offering more interactivity, more sophisticated behaviors, and several different character types that can emerge. This "emergence" is based on the amount of time the robot is played with, how the owner treats it, and some programmed predispositions. At the adult stage, the AIBO will exhibit one of four different personality tendencies: sheltered, nice guy/gentle, adventurous, or selfish. Again, these are only tendencies, not fixed behavior sets. Each AIBO matures slightly differently, depending on how it was "raised" (for example: how its sensors were triggered and how often it was interacted with).

It is amazing how little of this type of divergent behavioral programming is required before human beings start thinking they see intelligence. Furbies have similar sets of fixed behavioral routines, random traits and actions based on interaction and feedback, and yet, people ascribed all sorts of impossible emergent events. When the first generation AIBOs were released, owners started reporting that the robo-pets could recognize their faces and voices over those of other people. On the hundreds of AIBO

Web sites that began to spring up, it was widely agreed that facial and voice recognition were, in fact, unadvertised features of the robots. Sony finally released a statement saying that, unfortunately, AIBO did not have such features. Amazingly, some owners didn't buy it. They started to rationalize why Sony might be choosing not to tell the truth. Just as with organic pets, people want to think of their robo-pets as smart, responsive, and emotional beings.

note

AIBO stands for *Artificially Intelligent roBOt*. It also refers to the Japanese word *aibou*, which means *pal*. Brilliantly, Sony makes its robo-pets ambiguous looking. Are they dogs, cats, or lion cubs? Sony won't say. They make them look amorphous enough that people can see in them whatever pet they desire.

The AIBO has come a long way in four years. The current top-of-the-line model can respond to 75 voice commands, has wireless capabilities, a camera that can take digital images on command, and 21 colored LEDs that it uses to communicate its "feelings." Sony also now sells a Self-Charging Station ($99) that enables AIBOs to be completely autonomous. Although consumers have responded enthusiastically to the AIBO, many have balked at the price, so Sony now sells the LM series, a cheaper AIBO (for $599) with limited capabilities.

Sony SDR-4X

Manufacturer:

Sony

Web Site:

www.sony.com

Price:

A Lot

If you want to see the boot-strapping effect of how one commercially successful product in a new market can lead to bigger and better things, you need only look at Sony's latest creation, the SDR-4X. The next generation in its entertainment robot line, the SDR-4X is a 23-inch tall, 13-pound bipedal

FIGURE 6.18

Which would you rather own: a Mercedes Benz or a Sony SDR-4X bipedal humanoid? Soon, you'll get to choose.

humanoid that builds upon the technology developed for the AIBO. The SDR has 38 articulated joints and a dynamic, real-time balancing system so sophisticated that the robot can stand on one leg, recover its balance from being pushed against, and can negotiate different floor surfaces. The key to this balancing act is four pressure sensors in each foot and a gyroscope in the robot's body. The SDR-4X also has two stereoscopic color camera eyes used for navigation and facial identification. For navigation, the cameras are used to triangulate the changing distance of reflected light as we discussed in the "Light Sensors" section in Chapter 4. Seven microphones in the robot's head are used for spatial and voice recognition. When someone is talking, the SDR-4X will turn and look at the speaker in a natural way. The robot knows 60,000 words and can engage in what Sony calls "nearly conversation." If you've ever tried one of the computerized therapist programs that have been around since the dawn of the PC, or an online "chatterbot," you know how this goes. There's enough conversational give and take that you can *almost* imagine you're talking to a real person. The SDR-4X also has short-term and long-term memory. Things that get reinforced over time (such as its owner's voice and face) are remembered and those that are not reinforced are soon forgotten. A wireless connection enables the SDR-4X to trade data with a PC and enables the user to create custom programs. Because the bot is targeted at entertainment, the focus is on teaching the SDR how to do things such as sing and dance.

SDR stands for *Sony Dream Robot*, and Sony isn't kidding. A dream would have to come true for you to afford one. Although it has yet to set a price or a release date, company spokesbots say that, when the SDR does sing and dance its way onto the market, it'll cost as much as a luxury car!

B.I.O.-Bugs

Manufacturer:

Wowwee Toys

Web Site:

www.wowwee.com

Price:

Discontinued

The B.I.O.-Bug was, in many ways, an unlikely victim of the September 11th attacks. Developed by Wowwee Toys (a division of Hasbro) in collaboration with BEAMbot guru Mark Tilden, the B.I.O.-Bug was an innovative robo-critter that used BEAM technology to create a very lifelike mechanical creature. The toy was a huge hit with toy experts and was immediately put on their "hot-seller" lists for Christmas 2001. The B.I.O.-Bugs were released days before September 11th and the TV ads featured them crawling across a postapocalyptic landscape of rubble and twisted metal. After the attacks, people complained, the ads were pulled, and holiday sales fell far short of expectations. The line has now been discontinued.

There were four color-coded species of B.I.O.-Bugs. The yellow Acceleraider was designed with speed in mind. The red Predator was the most aggressive and the strongest. The green Destroyer had the best defenses and was deft at handling rough terrain. The blue Stomper was the most maneuverable and the best at handling different types of terrain. Like the AIBO, B.I.O.-Bugs were designed to "grow" through various stages, or 12 "life levels." These life levels could be advanced through the bots interacting with their environment and other B.I.O.-Bugs, or through manual "feeding" and training by a human (via an infrared remote control). The Bugs didn't actually feed on infrared energy, as some were led to believe (in other words, they weren't recharged by it), it was just a way of reinforcing the biological inspirations behind the robots. The B.I.O.-Bugs actually fed on battery power—required in both the bots themselves and in the hand controller.

Watching the B.I.O.-Bugs explore their world and interact and fight with each other, people were shocked to discover there was almost no computer technology involved. There was a computer chip in the Bugs, but most of the "action" was sense-act behavior-based control in the robot's analog circuitry, a hallmark of BEAM. The bots used infrared eyes to identify friend (same species) or foe (other species) and antenna-like springs in the head and tail to control movement and interaction between bots. The B.I.O.-Bugs showed to a much larger audience the impressive capabilities of largely analog BEAM technology. Many toy experts and early buyers said they found the B.I.O.-Bugs more fun and interesting than the AIBO. Not bad for a bot that sold for $40 each.

Although the B.I.O.-Bugs have been discontinued, they are still around, sold as closeout for as little as

tip

One of the things that was almost universally despised about the B.I.O.-Bugs was the *incredibly loud* volume of the sound chip. The sounds the Bugs make are really bug-like and cool sounding, but even the sound of one bot chirping away was maddening. Two or more and you'd need ear plugs. If you buy some Bugs, your first hack will want to be a volume control. This is easy to do. Just unscrew the body, find the wires to the speaker (found in the tail section) and solder a 500 ohm (Ω) potentiometer to one of the wires (see "Thumbnail Guide to Soldering" in Chapter 5, "The Right Tools for the Job"). A *potentiometer* (or *pot*) is a resistor that enables you to dial the degree of resistance so that you can adjust the volume to a comfortable level. You can then drill a hole in the body to mount the pot on the outside if you want (or just leave it at your desired level inside). You can get a 500Ω potentiometer at your local electronics shop.

Want a *really* quick fix to the sonic madness of the B.I.O.-Bugs? Slap a piece of tape over the speaker hole on the bot's abdomen.

$5 each, and also available on eBay. Mark Tilden and Wowwee designed them with hardware hacking in mind, so they're relatively easy to take apart and modify. There are a number of B.I.O.-Bug hacking sites online that can all be accessed through www.solarbotics.net/biobugs.

Roomba

Manufacturer:

iRobot Corporation

Web Site:

www.roombavac.com

Price:

$199.95

FIGURE 6.19

Why make little junior vacuum the carpets when the Roomba can do it without any muttering or complaining?

Rosie, the Jetsons's robo-maid, has nothing on the Roomba robotic vacuum (except a better sense of humor). Looking something like a plastic horseshoe crab, the Roomba has all the signs of a winning idea, at least for an initial effort. Using it is completely brainless. You set it down, select from Small, Medium, or Large room size, press the button, and let it go. It can clean rooms up to 15 feet × 20 feet. Electrolux and others have been playing around with robot vacuums for years, but they've made several crucial errors. First, they've used conventional vacuum technology, with powerful motors that eat batteries. They also employ complex and expensive sonar navigation and obstacle-avoidance technologies. Roomba makes use of newly engineered sweeping/vacuuming mechanics that use far less power and some of the simpler sensor systems and control architectures found in many of the robots that iRobot's chief technology officer, Professor Rodney Brooks, has pioneered. The bot uses soft bump sensors on its body for obstacle avoidance and combinations of several movement behaviors (spiral motion, random wander, and wall following) to adequately cover a room. Roomba builds no map of its world. Four infrared sensors underneath the robot prevent it from inadvertently...uh...vacuuming stairs (which it is *not* designed to do).

Given the magic "under $200" price point, the Roomba has been an instant hit with gadget lovers and couch potatoes. Owners seem basically pleased with what they get for their money. It doesn't do as good a job as a human would, and it only works in reasonably uncluttered spaces, but it performs well enough that, at the very least, you can greatly reduce the amount of vacuuming that you have to do. And as Martha Stewart would say, "That's a good thing."

Robomower

Manufacturer:

Friendly Robotics

Web Site:

`www.friendlyrobotics.com`

Price:

$750

FIGURE 6.20

Friendly Robotics's Robomower won't be putting gardeners out of business anytime soon, but it does (eventually) mow the whole lawn.

Next to Hoovering the house, the next domestic drag in need of robotic rescue is mowing the lawn. Okay, so maybe scrubbing toilets and washing windows trumps vacuuming and mowing, but robots that can do those chores are a long way off. The gnarly problems of manipulators and end effectors (as discussed in Chapter 4) come into play. Vacuums and mowers are relatively straightforward compared to these devices.

Like the robo-vac, we've been getting glimpses of robotic mowers for years now. Friendly Robotics released its Robomower in 2000, and it seems to have the most staying power in the market (like the Roomba, it's reasonably functional at an *almost* affordable price).

As with any robot, navigation is a big challenge in developing a robotic mower. Most of the mowers out there use a similar scheme found on the Robomower: An electrified wire is installed in the ground along the perimeter of your yard and Radio Frequency (RF) sensors on the bot detect the signal from the wire and keep it penned in. Bump sensors around the robot's body enable it to mow around obstacles such as trees. Rather than a *random wander* behavior that many robo-vacs and mowers use, the Robomower uses an internal compass that sends the robot on a succession of compass headings in a back and forth *V* formation. Between the V paths and a collision-avoidance behavior using bump sensors, the idea is that, eventually, all of the yard will get mowed. In practice, it's not quite that simple. Many users report that, by the time the mower has wandered around their yard until the batteries die, some stubborn patches of grass remain. A remote control unit comes with the robot, so you can still enjoy hands-off mowing of these trouble spots.

The Robomower uses two rechargeable 12V sealed lead acid batteries that are supposed to last long enough to mow 5,000 to 6,000 square feet. They're supposed to go 2 1/2 hours on a charge and then take 24 hours to recharge. I say *supposed to* because manufacturers' battery life ratings are usually based on ideal-case scenarios that don't exist in the real world. Some users with less-than-perfect yards (with numerous trees, flowerbeds, inclines, and other typical energy-consuming "obstacles") report the frustration of having to constantly recharge batteries, wait for them, and look at a spottily mowed lawn before Robomower finishes its rounds.

AmigoBot ePresence

Manufacturer:

ActivMedia Robotics

Web Site:

www.amigobot.com

Price:

$3,495

FIGURE 6.21

Not the friendliest bot on the block, the AmigoBot is for dedicated robo-heads only.

Although the warm and fuzzy name, AmigoBot, might imply user-friendliness, this expandable robot system is really just for the dedicated hobbyist. This small (11 inches × 13 inches and 8 pounds), fire-engine red runabout uses eight sonar rangefinders to navigate physical space. It has two drive wheels and one caster wheel. An onboard microcontroller handles low-level functions, such as collecting the data from the rangefinders, motor control, and battery power control, while mid- and higher-level controls (navigation, mapping, and path planning) are handled by your PC. Cheaper versions use a tethered connection to your computer, while the ePresence model talks to the computer over a radio modem link. The ePresence system also includes a color Web camera. This is where the *presence* part comes in. As we discussed in Chapter 2, "Robot Evolution," robotic presence enables you to see and hear what the robot sees, either on a local PC or from a remote location over the Internet. With AmigoBot, you can navigate the robot, see and hear the robot's environment, and take pictures using the robot's camera. You also can give remote access to others, so you can allow anyone you want to access and control AmigoBot over the Internet.

Those who've spent some time with AmigoBot ePresence report that it's not for the faint of heart. The robot and software system are temperamental and one must spend a lot of time under the hood to get the robot to behave properly. It also takes a lot of time to work with AmigoBot in mapping its environment. Add these frustrations to the hefty price tag and this robot is really only friendly to those with deep pockets and a troubleshooter's mindset.

Retro-Robotics

You know a technology has been around for a while when people start voraciously collecting everything having to do with that technology's "good ol' days." I'll never forget the first time I saw the magazine *Historically Brewed* (about "antique" PCs) and discovered that people would have been willing to pay good money for the boxes of '80s computer magazines and manuals I had just shipped off to the recycling center. In the last few years, the robots from the premature '80s personal robot "revolution" have been enjoying a similar wave of collector's nostalgia.

Androbots

Looking over the pictures and descriptions of the robots offered (and promised) by Androbot, Inc., two thoughts immediately leap to mind: How cool! and Wow, how shockingly naïve! Started by Nolan Bushnell, flush with cash from starting Atari and the Chuck E. Cheese's Pizza Time Theater chain, it feels as though Androbot wanted to wish these robots into existence by sheer force of will and weight of wallet. But this was the early 1980s, when the Apple II+ was cutting-edge computer technology. With what we know now, trying to get a robot to navigate a space, speak, respond to spoken commands, or to do anything with the digital technology of that era seems foolhardy, but try it did. The company actually sold two robots, Topo I and Topo II/III. Topo I was basically a large (over three feet tall) remote-controlled toy robot. It used a one-way radio frequency link from an Apple II+ and could be controlled via a joystick or by executing a control program in the BASIC language. It did *nothing* else. And even with this fairly simple task (moving around), many of the 1,000 or so Topos that shipped didn't work as advertised. The second-generation model, the Topo II, had a two-way infrared link to communicate with either an Apple or Commodore 64. The bidirectional link enabled Topo II/III to send data back to the computer. A text-to-speech processor also enabled Topo to speak what you typed on your PC.

The Cadillac of Androbots was *B.O.B.* (or *Brains on Board*). Unlike the Topos, B.O.B. was to be an autonomous bot with all of its...uh...brains on board. Prototype models became the darlings of consumer electronics shows and evening news reports. Teaser videos and promo pictures showed B.O.B. serving drinks, vacuuming the living room, and hauling stuff around the house. B.O.B. had four ultrasonic rangefinder eyes in its head. The first prototype used dual Intel 8086 central processing units (CPUs), but this never really worked, so future prototypes used a single 8086 chip.

Besides the text-to-speech capabilities of Topo II/III, B.O.B. also spoke a prerecorded vocabulary of some 100 words. There were many other plans for accessories such as a mini-fridge from which B.O.B. could serve drinks, a hauling cart, and a vacuum attachment. But none of these were meant to be. By the end of 1984, after only two years in operation, Androbot closed up shop. In the end, after all of the hype and flashy prototype demos, less than 2,000 units (of Topo I and II/III) had actually been sold. Given the mystique of the company and its ambitious offerings, and the rarity of surviving robots and parts, anything having to do with Androbot has become highly sought after.

Heathkit Heroes

Manufacturer:

Heath

First Released:

1983

tip

If you're interested in retro-robotics, a good book to try and hunt down is Texe Marrs's *The Personal Robot Book*. Published in 1985, it details all of the major home robots available at that time. It's a great resource for the collector and robot historian. It's also fun (and kind of depressing) to see all of the enthusiasm for the promise of personal robots, which seemed so within the grasp of that time. Although this book is obviously long out of print, you can find it online via such bookseller databases as www.bookfinder.com.

When most gadget geeks think of robots in the 1980s, they likely first think about the kits produced by the Heath Company (a division of Zenith). In 1983, Heath released HERO I, the first of what was to become a venerable line of electronics hobbyist do-it-yourself robot kits.

The HERO I was actually originally intended for the educational and industrial markets (to teach robotic concepts and programming), but it proved so popular with hobbyists that in 1984, the company released HERO, JR., a cheaper, more pared-down model, specifically targeted toward the amateur market. Both HERO I and HERO, JR. came in kit form and were not for the weak of heart or soldering skill. It took as much as 60 hours to build the bots, and countless additional hours troubleshooting, fixing, tweaking, and so forth. HERO, JR. was also available fully assembled.

Both robots looked similar, standing about 20 inches tall. They both used a single drive wheel and two caster wheels. HERO I had to be entirely programmed by the user, while HERO, JR. came pre-programmed with a variety of behavioral routines. The sensor systems were similar on both. They had ultrasonic range finders for navigation, motion and sound sensors, and light sensors that could detect quantities of light in the visible spectrum. Both robots had a 64-phoneme vocabulary (the basic units of speech), which enabled them to speak…sort of.

HERO, JR. was also billed as a security robot and could use its motion detector in a special security mode. If movement was detected, HERO, JR. asked for a password. If the password wasn't given, the "intruder" (most likely a house pet or a junior—the biological variety—going to the bathroom) was given a loud warning. With an optional security transmitter, the home's security system could be triggered. HERO, JR. even came with security stickers to put in your home's windows that read: "Warning, this area protected by a Security Robot." Boy, I bet that really scared away lots of would-be home invaders!

> **tip**
> Because the Heathkit HERO robots were so popular, there are still many robots and parts available online and numerous sites and mailing lists devoted to the kits. Start your online explorations at www.hero-1.com.

RB5X

Manufacturer:

RB Robot Corporation/General Robotics

First Released:

1982

If there was a robot version of *The Little Engine That Could*, the RB5X would be its protagonist. Personal robots in the '80s came and went, personal robots in the 90s did the same, but the RB5X just kept on chuggin' its way into the annals of robot history as the most persistent robot on the market. It is still sold today!

FIGURE 6.22

Believe it or not, this "antique" robot is still with us. The RB5X continues to be sold, with few changes from the original model. Photo courtesy of General Robotics.

First released by RB Robot Corporation in 1982, the RB5X looked (and still does) like a trashcan with a clear plastic dome on top. It not only looks the same as it did in 1982, but astoundingly, most of the hardware and software is still the same. Although you can now outfit it with a Web cam and access its programming environment in a Web browser (using the Java programming language), the RB5X still uses the same microprocessor (National Semiconductor's 8073 chip) and basic sensor arrays that it always has. It has two drive wheels, two caster wheels, eight bump sensors, ultrasonic rangefinders, and an optional gripper arm. It also still uses Tiny BASIC (a subset of the BASIC programming language), the language with which it was "born."

One reason the RB5X has such staying power is that it has become a favorite in educational settings, and its simple and durable design has kept it there. Schools report RB5X units working year after year with few failures (at least as far as programmable robots go). Besides teaching kids about robotics, engineering, and computer programming, the RB5X has also been used in remedial language classes in which kids have to use its phoneme vocabulary to construct words and sentences. The RB5X currently sells for $3,495 (from General Robotics, www.generalrobotics.com) and the add-on arm is another $2,295. Because the RB5X has been on the market for over two decades, there are many used units and parts available.

note

One of the funniest things about looking at manuals, catalogs, and ads for these '80s personal robots (many of which can be viewed online) is the lengths to which the companies went to sell the idea that their easy-to-use robots would revolutionize domestic life. Almost every promo picture shows a woman (apparently in an attempt to communicate *just* how easy it is), or a happy family, including a big shaggy dog, enjoying playing with, being guarded by, or being served by the trusty house robot. In truth, most of these robots never performed as advertised and spent more time with their guts spread out all over the kitchen than they did serving appetizers at the family Christmas party.

Other Cool "Antique" Robots

Besides the bots just discussed, there were dozens of other personal robots and relatively sophisticated robot toys produced in the 1980s. Here are a few others to keep an eye out for:

Armatron/Super Armatron

Radio Shack has sold some cool electronic toys over the years, but nothing beats the Armatron, a noncomputerized robotic arm. The arm had six degrees of freedom controlled by two joysticks. The gripper/arm wasn't capable of handling much weight, but it was still an impressive demonstration of basic robot arm mechanics at a very reasonable price (between $20–$30). It also had a timer on it and came with plastic objects of various types and sizes that you were supposed to organize and stack before time ran out. Trying to find one of these online or at a flea market/yard sale is definitely worth the effort.

Omnibot

The Tomy Company has long been a producer of robot remote-controlled toys and, in the '80s, it introduced the Omnibot line (Omnibot and Omnibot 2000). The robots had a unique means of programming. You could use the included remote control unit to teach the robots a movement sequence that could then be saved to a cassette tape recorder built into the robots and played back at a set time. You also could record voice messages that the bots would play back. Omnibot also could carry a tray and light objects (such as its own remote) in a "manual" gripper.

Maxx Steele

In the mid-80s, Ideal began selling a line of futuristic action toys called RoboForce. The most coveted toy in the line was a 2-feet tall, 34-pound robot named Maxx Steele. Although marketed as a toy, Maxx was a surprisingly sophisticated robot. He was programmable, via a radio remote control, had movable arms and a working gripper, a headlight, and he could speak 20 different phrases. He also had expansion slots for plug-in program cards that Ideal was allegedly working on when it pulled the plug. Although the robot was extremely innovative in its design, its $350 price tag, and the extremely cheap plastic parts that were used to get the price that "low," Maxx was doomed to early cancelation.

Big Trak

Before any of the previous robots and robot wannabes, there was Milton Bradley's Big Trak. Released in 1979, Big Trak was a programmable, futuristic six-wheeled tank. A keypad on its back enabled you to program and save movement sequences. It also had an optional trailer/dumper that could be raised and lowered as part of the programming sequence. Being one of the first programmable robot toys, it still puts a gleam in the eye of any gadget geek who remembers it. There are even people retro-fitting it with modern microcontrollers. One guy has put an OOPIC (the controller that we'll be using in Chapter 9, "Project 3: Building a DiscRover") in a Big Trak. You can see the results at www.robotprojects.com.

These are just a few of the robots of yesteryear that are fun to track down and to tinker around with. There are hundreds of Web sites for robot collectors, buying and trading bots, sharing repair tips, and step-by-step instructions on how to retrofit old robots with modern robot brains. A good place to start your travels in the robotic wayback machine is at Robert L. Doerr's Robots Wanted Web site (`www.robotswanted.com`).

note

Fans of Battlebots might recognize the name Robert Doerr. He's the human behind the combat robots Crash Test Dummy and Crash Test Dummy, Jr. Robert also has bought the rights to the Heathkit robot line. He now owns all of the remaining stock of kits and parts. You can read more about his robot adventures at `www.robotworkshop.com`.

Thumbnail Guide to Electronics

Hey, did you hear the one about the geek who tried to teach the absolute beginner practical electronics in just a few pages? Well, you're about to, and it's no joke! Despite what you might think, learning *basic* electronics really isn't that hard. Sure, the deeper you get, the thicker the thicket (of math and logic, impenetrable tech jargon, complex circuit designs, and so forth), but the basics can be taught to just about anyone. So, put on a pot of strong coffee (or for you teen readers, grab a Jolt Cola), and get ready for a whirlwind course in practical magic or "how to get the universe's electrons to do your bidding!"

As you'll notice in the preceding paragraph, we used the word "practical." That means we won't spend too much time with the physics underlying electrical circuits. We're going to quickly jump ahead to electronic components themselves and how they function. But for those of you who were busy drawing band logos on your spiral notebooks during science class, let's recap a few things:

- Electricity is an essential part of all matter.

- Atoms, the building blocks of elements, are composed of a nucleus (comprised of protons and neutrons) and "orbiting" electrons.

- As their names imply, protons have a positive (+) electrical charge and neutrons have no charge. Electrons have a negative (-) electrical charge.

- If an atom has an equal number of electrons and protons, it has no net electrical charge.

- If electrons get dislodged from an atomic orbit (thanks to things like radiant energy, heat, and presence of an electrical field), they become "free electrons" and will move toward other atoms that have free electron "holes" (places in their "orbits" just waiting for electrons to fill).

■ It is this movement of free electrons that creates an electrical current and makes everything happen that we associate with the electricity (such as bringing computers and robots to life).

■ There are two flavors of electrical current that we work with. Alternating Current (AC), the type on tap in your house's walls, moves back and forth in a cycle. Direct Current (DC), the type found in many of the gadgets in your home, moves in only one direction. If AC is used to power a DC circuit, a transformer is used to make the Alternating Current move in a single direction.

■ Different materials assist or impede the flow of electrons. Materials that offer low resistance to the flow (metals, for instance) are called *conductors*. Materials that offer high resistance (wood, for instance) are called *insulators*.

To simplify electronics for beginners, every writer starts off with an analogy to make the parts and functions of a DC circuit easier to understand. The most common is the water/plumbing analogy. Here's how it goes:

You have a big tank of water. The water in the tank represents the "potential" of what is available to flow through the pipe at the bottom of the tank. In plumbing, we could measure this as water pressure. In an electrical circuit, it's called *voltage* (the electrical potential or force available to the circuit). The pipe at the bottom of the tank is analogous to a wire in an electrical circuit. The amount of electricity that flows through that pipe…er…wire is called the *current*. If we put a valve on our plumbing system, we could impede the flow of water through the pipe. In electricity, this is done with a component called a *resistor*. A resistor can have a fixed value or a variable value. If you have a light dimmer in your dining room, that's an example of a type of variable resistor (called a *rheostat*). Turn it up, the resistance in the circuit drops, and your crystal chandelier sparkles; turn it down, the resistance increases and you have romantic mood lighting.

The Greenie Theory

There are several of these electronics analogies, but none are more bizarre (or memorable) than the one in Kenn Amdahl's book, *There Are No Electrons: Electronics for Earthlings*. This is, without a doubt, the weirdest and funniest book ever written about electronics. It also happens to be one of the best books ever written on the subject. Amdahl clearly explains electrons by (playfully) rejecting them all together. Electrons are a hoax, he declares, a conspiracy perpetrated by high school science teachers and guys who wear thick glasses and think pocket protectors are a sensible fashion statement. Why? Because the "truth" is far too strange to believe: that invisible Gremlin-like beasties called *Greenies* are what's actually behind what we've been told are protons and electrons. It is these Greenies' need to "hook up" with each other that generates electricity. Okay, so it's a really silly analogy, and reading the book makes you frequently wonder what Mr. Amdahl spikes his coffee with, but it is actually a great way to remember what does what in an electrical circuit. So, bear with us as we step through the electronic looking glass into the world of the *Greenie Theory* of electronics:

- Electrons are actually the males in a species of invisible critters called Greenies. The males hang out waiting for female Greenies to come along so they can party with them.

- Protons are the female Greenies. Where they gather, they generate a huge need to party on the part of the male Greenies.

- We know this "need to party" as *voltage*.

- The urge finally becomes so great that the male Greenies rush to where the female Greenies are. We know this intense traffic of male Greenies as *current*.

- As the male Greenies travel along available routes to join the females, they might encounter a road condition, such as a narrow bridge, that slows their movement. We know these traffic obstacles as *resistors*.

- The (negative) male Greenies will go to any lengths to get to the (positive) female Greenies. We humans harness this need to party and this flow of male Greenies to create all of the amazing things to which we give credit to boring ol' electrons.

That should get you started thinking Greenie. Just hold in your mind the vision of a bunch of rascally little green dudes starting out from the negative terminal of a battery (or other power source), and jumping through every hoop we humans put before them in a circuit. They perform every trick asked of them in their tireless efforts to hook up with the female Greenies—calling them to party, on the positive side of the battery.

Making Greenies Dance

Just making electrons…er…Greenies race through a wire, slowing down once in awhile because of some road resistance, wouldn't be much fun (or very effective). An electronic circuit is like a little machine that processes electrons in such a way that we can get useful work out of it. To get this work, we use various components that control and alter the flow of electrons. Although looking at all of the components in a circuit might be daunting to a beginner, understanding basic parts and what they do is actually quite simple. There might be lots of intimidating tech jargon in electronics, but most of the basic components' names describe what they do (for instance: a *resistor* resists the flow of current; a *capacitor* holds a charge). See Table 6.1 for a rundown of components, what they look like, and what they do.

Table 6.1   Basic Electronic Components and Their Functions in a Circuit

Component Name	What It Looks Like	What It Does	Schematic Symbol	Notes
Resistor		Limits the flow of current in a circuit at a set value. Measured in *ohms (Ω)*.		You'll often see the prefix *k* (for a thousand) and *M* (for a million) in front of the Ω. Examples: 4.7kΩ and 10MΩ.
Variable Resistor		Limits the flow of current in a circuit at a variable value. Measured in *ohms (Ω)*.		One type of variable resistor is known as a *potentiometer* (because you can vary the electrical potential by adjusting a control dial).
Capacitor		Used to temporarily store up current in a circuit. Measured in *farads (F)*.		The *micro* symbol μ (called a *mu*) often precedes the F. It means *millionths*. Example: 150μF.
Diode		Used as a one-way "valve" for the flow of current in a circuit.		Called a diode because it has two "odes": an *anode* (a positive end) and a *cathode* (a negative end).
Light-Emitting Diode		Used as a light source, function indicator, and in alpha/numeric displays.		The cathode (-) side of an LED is usually notched or flat. LEDs are available in light wavelengths from infrared (IR) to ultraviolet (UV).
Transistor		Used to amplify an electrical signal or to act as a switch.		Has three component leads. There are two main transistor types called *NPN* (-/+/-) and *PNP* (+/-/+).

Component Name	What It Looks Like	What It Does	Schematic Symbol	Notes
Integrated Circuit		Combines (integrates) miniaturized versions of other components (transistors, resistors, diodes, and so forth) into a small package (or a *chip*).	74HCT240	ICs can be analog, such as the ones we'll use in this book's projects, or digital, such as the ones in your PC. There are also analog/digital hybrids.
Wire		Used to carry an electrical current. Often covered in an insulating "jacket." Measured in *gauges*.		Wire is either multi-strand (more than one wire twisted together) or solid (single wire). Note: In American Wire Gauge, the higher the number, the smaller the wire diameter.
Relay		An electromagnetic switch in which a magnetic field is generated to make an electrical contact (or to break contact).		Relays (and other switches) are designated by numbers of *poles* (such as single or *SP*) or double pole (*DP*) and *throws* (*ST* or *DT*). So an SP/ST would be your basic on/off switch, while an SP/DT would be a switch with a single pole that can throw to two different circuits.
Switch		A mechanical (or electronic) device for opening, closing, or changing connections in a circuit.		*Pole* in a switch designation (see preceding). Refers to the number of separate circuits a switch can control at a time. *Throw* refers to the number of circuit "paths" that are controllable.

Table 6.1   Continued

Component Name	What It Looks Like	What It Does	Schematic Symbol	Notes
Motor		A machine that converts energy into motion.	—(M)—	In electric motors, magnetism is used to convert electrical energy into motion.
Battery		Used to store and deliver DC power to a circuit.	+ —|ı| ⊢ -	

Photos of components courtesy of Solarbotics.com

Before we move on, there are a few things we should touch on about some of the individual components covered in Table 6.1:

- **Resistors.** As we mentioned in Chapter 5, there are color-coded bands on resistors that will tell you the amount of resistance they provide. Charts of these color codes are everywhere, so we won't bother to print one. There's a great interactive one at `webhome.idirect.com/~jadams/electronics/resist_calc.htm`. You can also buy a color chart of resistor codes (with other useful electronic formulae on it) at any electronics store. Or, you can just be lazy (like me!) and use your digital multimeter.

- **Capacitors.** There are three main types of capacitors: *electrolytic, monolithic,* and *tantalum.* Electrolytics are cylindrical in shape, can store the largest amount of energy, and are polarity sensitive (in other words: they have a positive side and a negative side). A stripe down one side usually designates the negative lead. Monolithic capacitors are smaller (in physical size and in storage capacity) and are *not* polarity sensitive. They usually take the shape of brightly colored rectangles. Tantalum capacitors look like brightly colored blobs of plastic. Like electrolytic capacitors, they are also polarity sensitive, but just to confuse you, their positive, rather than negative, side is marked.

- **Diodes.** A diode's negative (cathode) side is usually marked with a colored band. Current flows (in other words, the one-way "valve" opens) when the anode side is more positive (has a higher signal) than the cathode side. It works like a one-way "check-valve."

- **Transistors.** Unlike most components, a transistor has three leads coming out of it. With the flat side of the transistor facing you, those leads are called (from left to right): the *emitter,* the *base,* and the *collector.* If a transistor is like a switch, it's the (small) current going into the base lead that turns it on. A higher current can then flow from one of the collector/emitter leads to the other. There are two main types of transistors: *P-type* and *N-type.* For a P-type, the emitter and connector leads are negative, but the base is positive. This means that a positive (+) current will flip on the switch. For an N-type, it's reversed: A negative (-) current will activate the transistor.

■ **Integrated Circuits.** ICs usually come in a rectangular-shaped "package" with two rows of leads (or *pins*) underneath. This is referred to as a *dual in-line package (DIP)*. Each pin carries information to and from the rest of the circuit. Some pins are for power, some for input from the circuit, others for output to the circuit. You have to know which pin does what to use the IC properly. The pins are numbered. Pin 1 is usually designated by a little dimple on the top of the chip. With the IC standing up like a table, you count down the chip from Pin 1 and then jump across to the bottom of the other side and read up (so, on a 20-pin DIP, Pin 11 would be the one directly across from Pin 10). What each pin does is usually detailed in the technical information that comes with the chip or can be found on datasheets on the manufacturer's Web site, or elsewhere, online.

Of course, these components will start having more meaning to you when you start using them, but believe it or not, those are the basics. These are most of the components used in the kits covered in this chapter and the projects in this book. If you combine your newly acquired soldering skills developed in Chapter 5 with the preceding information, you'll be able to assemble basic circuits and robot kits. You'll be amazed at what you'll learn from this hands-on experience.

How to Become an Electronics Whiz in Four Easy Steps

We could go on (and on) talking about many other cool and mysterious electronic components, and we could give you more goofy Greenie analogies (like how a Greenie duck named Bruce, driving a powerboat, is actually responsible for the magnetic lines of force (the wake of his boat) that always accompany an electrical current). But if we get too into an electronics primer, we could quickly overwhelm our actual subject. But we will leave you a few simple and fun assignments that will quickly take you from an absolute beginner to a wirehead who can talk chips and DIPs with the best of them:

1. Read Kenn Amdahl's book, *There Are No Electrons: Electronics for Earthlings*. It's actually so fun and entertaining to read that, like a good novel, you won't be able to put it down.

2. Head down to your local Radio Shack and buy Forrest M. Mims, III's *Getting Started in Electronics*. In print since 1983, this handwritten/handdrawn book (on graph paper) covers the same ground as Amdahl's book, but with a more traditional, electrical engineer's mindset. The two texts reinforce each other extremely well.

3. While you're at The Shack, buy one of its *130-in-1 Electronic Project Labs*. This kit works sort of like a breadboard that already has a bunch of components installed on it. You use a set of included jumper wires to connect up the components and make the 130 different circuits described in the accompanying book. You can make all sorts of nifty gadgets such as radios, transmitters, sound generators, amplifiers, a metal detector, and more.

4. Practice your soldering (see Chapter 5) and then build one of the soldering-required kits covered in this chapter.

If you do all of these steps, you deserve to do something nice for yourself. Go for a walk in the woods. Treat yourself to a thought-provoking independent film. Stop and smell the roses. Nah, forget about it! We've got too much bot building to do in the next three chapters. Who wants to go outside, anyway? All that really bright overhead lighting!

THE ABSOLUTE MINIMUM

In this chapter, we explored the wonderful world of robot kits, building sets, out-of-the-box bots, and got down to some business with the nuts and bolts, or in this case, the resistors and transistors, of electronics. If all you did was look at the purdy pitures o' da robots, here are at least a few things you'll want to be able to say if somebody asks you, "What have you been reading lately?"

- There are now dozens of interesting and reasonably priced robot kits on the market. They all tend to focus on different aspects of robots (different types of robot mechanics, different types of control, different sensors, and so forth), so you can choose kits that explore areas that most interest you.

- Astute readers probably noticed a lot of reoccurring technologies and design approaches across different kits and building sets (for instance: bump sensors, infrared light sensors, wall-following behavior, photovoric behavior, and two-drive-wheeled mobility). In fact, the more time you spend with small, mobile robots, the more you realize there's an established bag of tried 'n true tricks.

- Robot building sets are great because you can prototype, build, and rebuild a whole bunch of different robots and there are so many building ideas and resources available online (especially for the LEGO MINDSTORMS system).

- Robots have been around so long that you can now indulge both your love of them and antique collecting at the same time! Retro-robotics is the gadget collecting trend *du jour*.

■ Electricity really doesn't come from electrons and protons, it comes from invisible green dudes trying to hook up with invisible green gals. And if you believe that, we've got some *beautiful* residential properly to sell you in the Florida everglades.

■ Getting starting in electronics is easier than you think. Components often do what they're called: Resistors resist, capacitors capacitate, relays relay, transistors transition, and integrated circuits integrate. See that wasn't so hard, was it?

- *Finally*, we start building robots!
- The nifty BEAM Bicore circuit
- Breadboarding basics
- Build a robo-critter in your spare time...with junk parts!

7

PROJECT 1: COAT HANGER WALKER

Now that we know some of the basic operating principles behind robots, and some of the components that go into them, let's start putting what we've learned into practice. In this chapter, we'll build a little critter (see Figure 7.1) out of surprisingly simple and minimal parts. We'll even make use of one of those coat hangers that seem to breed in your bedroom closet. We'll also continue our Thumbnail Guide series with a look at *breadboarding*, a handy and easy way of prototyping your electronic circuits before welding all of the parts together.

This first project is an ingenious little hardware hack done by a Canadian BEAM enthusiast named Jérôme Demers. It wonderfully illustrates a number of principles put forth earlier in this book. Here are a few:

■ It revels in the K.I.S.S. rule (see Chapter 3, "Robot's Rules of Order"). It uses a simple control circuit, a single servo motor, and a single gear to create a four-legged walker.

■ It demonstrates some BEAM principles, making use of the *Bicore circuit*, a two-node timer circuit that tosses a control signal back and forth between the two nodes, creating a reciprocating back and forth motion in the motor and gears, and therefore, in the legs attached to the gears.

> **tip**
>
> As with anything built from instructions, it's a good idea to read through this chapter carefully before you begin building. When you're done with the read-through, also make sure to check the book's Web site (www.streettech.com/robotbook) for any instruction clarifications, bug fixes, and so forth. You can also ask questions there if you have them.

■ It makes use of techno-junk. Okay, so a coat hanger isn't very "techno," but if you've been doing junk scavenging as we suggested in Chapter 5, "The Right Tools for the Job," and saving the good stuff, you'll probably have some of the other components detailed in the parts list below.

FIGURE 7.1

What our finished coat hanger walker will look like. Cute, ain't it?

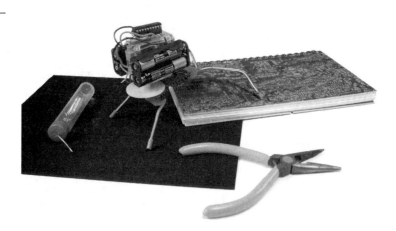

■ This project is a good example of the relatively level playing field found on the Internet for the sharing of robot designs and innovative ideas, regardless of who or where you are. I was truly impressed with this project, and many others on Jérôme's Web site Insectroïdes (robomaniac.solarbotics.net), even *before* I discovered that he was only 15-years-old! Dang precocious kids today! When I was 15, I was in the backyard playing chicken with a Bowie knife and a few mullet-headed friends.

The persnickety among you will be quick to point out, as you read over these plans, that by many definitions of robot put forth in Chapter 1, "What's in a Name?", this ain't no stinkin' robot. It doesn't gather, process, and respond to feedback of any kind. It's really a walking machine rather than a proper robot. We still wanted to include it because of the reasons listed in the preceding bullets, and because it teaches many of the fundamentals in construction and electronics that we'll use in the other two projects ahead, and in your other robot building. You will also end up with a really cool little "monitor pet" that will merrily motor across your desk and impress all your geek friends (especially when you tell them: "Yeah, I made it out of a coat hanger and some junked electronics."

We're not going to lie to you. Although this critter has relatively few parts, the Bicore control circuit can be a little challenging to solder (you're up for a challenge, right?). The dual inline package (DIP) IC socket that the control circuit plugs into is very small and the pins are very close together. Soldering these pins together will take some patient effort. That's the bad news. The good news is that DIP IC sockets cost about 10 cents a pop, so you can mess up quite a few in practice and it's no big deal. Hopefully, you'll have logged plenty of soldering practice time by now and be ready for some big boy/big girl soldering. If not, go ahead and read through this entire chapter, order the parts, and then continue with some soldering practice sessions while you wait for the components to arrive.

Gathering the Parts

This project requires some materials and parts you might already have lying around the house and some components you'll need to purchase at your local electronics store, an online retailer, or some other source.

The Parts List

The Coat Hanger Walker requires the following parts:

- (1) hobby servo motor with "Servo Horn" (Solarbotics Part #GM4)

tip

Solarbotics (www.solarbotics.com) offers convenient parts bundles for the projects in this book. (See its ad in the back.) Besides these bundles, if you mention, "Gareth says I get a deal!" when you order, you will receive a *one-time* 10% discount off the price of any order worth over $75USD ($100CND). This discount does not apply to existing specials, solar cells, or parts bundles not for this book. For other parts sources, see Chapter 10, "Robot Hardware and Software."

- (1) 1.5 inch (4cm) plastic gear (around 40 teeth are good)
- (1) 2 feet of 8 to 10–gauge wire; can use coat hanger wire or 10-guage copper wire (Solarbotics Part #Wleg)
- (1) short piece of 1cm diameter plastic tubing (can use "jacket" from preceding leg wire)
- (1) terminal block (Radio Shack Part #274-678)
- (2) AAA battery holders (each holds two AAA batteries)
- (1) length of 1cm diameter heat-shrink tubing (can use "jacket" from preceding leg wire)
- (2) .22 μF monolithic capacitors
- (1) 100K to 10MΩ resistor (we recommend 3.2MΩ)
- (1) 74HCT240 integrated circuit (IC)
- (1) 20-pin DIP IC socket
- (1) on/off toggle switch (smallest you can find)
- (2) leg mounting pads (Solarbotics Part #LMP2); optional, but recommended
- pieces of scrap plastic (from product packaging) or .030" plastic sheeting
- several feet of stripped hook-up wire (or other 22-guage wire)

These parts are shown in Figure 7.2.

FIGURE 7.2

All the parts you'll need to build your first robo-critter.

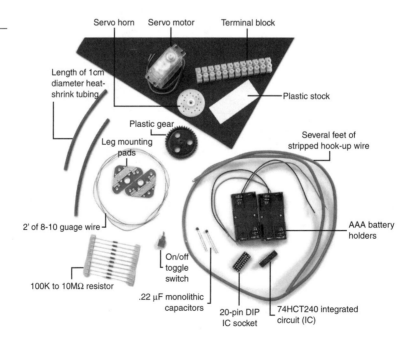

The Tools and Supplies List

In addition to the core components listed in the previous sections, you'll also need the tools and supplies listed here:

- Needlenose pliers
- Hobby knife
- Wire cutters
- Screwdriver set
- Soldering iron and related soldering tools and supplies
- Micro-torch or heat gun
- Breadboard and hook-up wire
- 2-part epoxy
- Superglue
- Some recycled component leads
- Rotary (Dremel) tool (optional, but highly recommended)
- Metal ruler
- (4) AAA batteries

tip

If you're unsure about any of these tools or what they do, refer back to Chapter 5.

caution

If you're a minor and are about to undertake this project, make sure you have your parents' permission and guidance. C'mon, you know mom's just dyin' to build some bots!

Freeforming the Bicore Control Circuit

The first thing we'll want to do in building our robot is to assemble its brain. The Coat Hanger Walker makes use of the ingenious BEAM Bicore circuit. It's prefixed *bi* because it has two states, or nodes, and *core* because, well, it's the central part of the robot's *nervous net*. Our Bicore uses the 74HCT240 integrated circuit. This chip is an *inverting octal buffer*. That's just a fancy way of saying that it is a chip with eight *logic gates* that invert the signals going into them. Whatever goes in each gate gets inverted, so a low signal becomes a high signal and a high signal becomes a low one. By combing the three gates on one side and three gates on the other (by soldering their pins together), we end up with two "teams" inverting gates that "buffer" the signal and make it more powerful. Bicore! The signal passing back and forth between the two nodes sends high and low (or "on" and "off") pulses to our servo motor. The result is back and forth movement of the motor shaft, which is transferred to our gears to create a reciprocating walking motion. By the way, if you're wondering, the remaining two gates are used as sort of the controller for the two three-gate teams.

Breadboarding the Bicore

Before we actually heat up our soldering iron and start dripping molten metal all over components, we want to breadboard our circuit to make sure that all the components are working properly and that we have a sound design for our Bicore circuit (If you don't know how to breadboard, read the "Thumbnail Guide to Breadboards" later in this chapter before continuing).

You'll want to hook up the wires (from your jumper kit), the resistor (whichever value you decide to start with), and the two .22µF capacitors to the following tie points on the breadboard. For these numbers, assume that the pin in the upper-left corner of the IC is pin 1 (the pin to the left of the little dimple). Then it's pins 2–10 on the left side, straight across (from 10) to 11, and then up to 20.

Connect jumper wires to the follow tie points:

- Left side of IC: 4-6, 3-4, 5-7, 6-8, 7-9
- Right side of IC: 14-16, 13-15, 12-14, 11-13, 15-18
- From left side to right side: 1-19
- Capacitors: 2-3 and 18-17
- Resistor (across IC): 2-17

The last thing you'll need to do is to connect the positive (+) power wire from pin 20 to a positive tie point on the power bus, and ground (-) wires from pin 1 to a negative tie point *and* pin 10 to a negative tie point. Make sure both of these pins (1 and 10) go to negative.

Figure 7.3 shows what the breadboard should look like when you've hooked up all of the pins on the IC. (In this photo, the motor is not yet connected.)

note

If you're unfamiliar with breadboarding, you really are going to have to read the "Thumbnail Guide to Breadboards" later in this chapter, or this section will make no sense to you.

note

Actually, you *don't* have to start with your bot's brains if you don't want to. You can read the next section, "Breadboarding the Bicore," and then skip ahead to "Building the Walker Body" if you'd like. After the mechanics of your walker are assembled, you can use long lengths of hook-up wire to attach (or *tether*) the motor of the walker to your breadboarded circuit. This way, you can test how your robot will operate, and increase or decrease resistance in the Bicore to get a suitable walking gait. When you're happy with the electronics and mechanics, then you can come back to this section to solder up the circuit.

FIGURE 7.3

Put hook-up wire, the resistor, and the two .22μF capacitors in the tie holes as shown. Note: Make sure the top and bottom power bus strips on your bread-board are con-nected (not shown here, but see Figure 7.13). Larger version of this image is available at www. streettech.com/ robotbook.

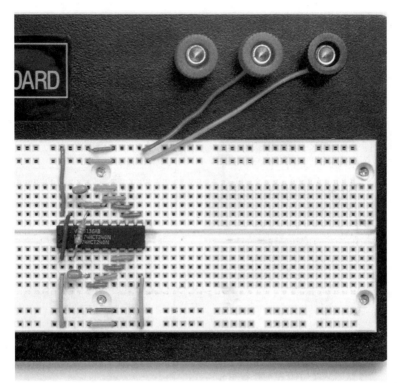

At this point, your breadboarded Bicore circuit is all hooked up and ready for juice (battery power) and something to drive (your motor). Before we hook up these final compo-nents, you might need to do a little work on your motor to get it ready for reciprocating (back and forth) motion.

Hacking the Servo Motor

If you got your servo motor from Solarbotics (Part #GM4), it already has the control electronics removed. If not, you'll need to remove them your-self. As we discussed in Chapter 4, "Robot Anatomy Class," servo motors have built-in control circuitry. We don't want this control on our servo—we want to control its movement with our Bicore chip. Removing the control PCB is simply a matter of opening the case, removing the board, and resoldering the positive and negative wires directly to the motor (see Figure 7.4).

caution

Make sure the battery pack that you're going to use to power the breadboard is *not* connected to the breadboard power bus before con-necting positive and ground (nega-tive) wires to the bus.

FIGURE 7.4

Here are the steps to removing the control circuit from a hobby servo motor.

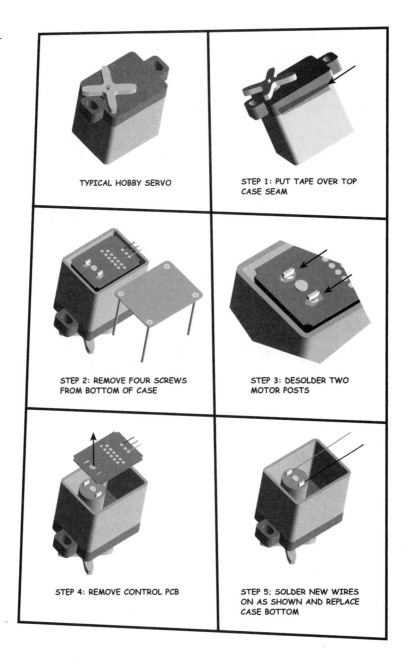

TYPICAL HOBBY SERVO

STEP 1: PUT TAPE OVER TOP CASE SEAM

STEP 2: REMOVE FOUR SCREWS FROM BOTTOM OF CASE

STEP 3: DESOLDER TWO MOTOR POSTS

STEP 4: REMOVE CONTROL PCB

STEP 5: SOLDER NEW WIRES ON AS SHOWN AND REPLACE CASE BOTTOM

If you did get your motor from Solarbotics, there are no electronics to remove, but the motor has been configured for continuous rotation. We want reciprocating back and forth motion. Usually, servo motors have a *final gear* (as the drive gear is called) with a mechanical pin on it that prevents full rotation. The Solarbotics servo has a final gear with no stop pin, but the "servo horn kit" that comes with the motor includes the original final gear with the stop on it. To re-install this gear, all you have to do is

1. Unscrew the four bottom screws on the servo, as in Figure 7.4.

2. Remove the top part of the case (instead of the bottom).

3. Pull out the middle gear in the center of the gear box (the one on top).

4. Replace the final gear (with the one that has the little plastic stop on it).

5. Replace the middle gear, and screw the case back together.

Before putting the case back together, make sure the gears are well seated and meshed. Also, make sure the plastic stop on the gear is facing toward the wired end of the servo case.

note

So, if our 74HCT240 chip has 8 logic gates on it, why does the DIP have 20 pins? Well, each input gate has an output pin on the opposite side; that's 16 pins. There's an "enable" pin on each side to "turn on" that side of the chip; that's 18 pins. And then there's the positive pin and the ground (negative) pin, used to power up the chip. 20 pins!

When you have a servo motor with a mechanical stop final gear in it and no control electronics, you're ready to hook it up to your breadboarded circuit to see if it works. Plug the red (+) wire into pin 9 and the black (-) wire into pin 12. Connect 6 volts (V) of power to your breadboard's positive and negative terminals. If you're using the two AAA battery holders for the robot, connect them in series as seen in Figure 7.5.

If, when you power up the motor, you see the motor shaft twisting back and forth, congratulations! You've just built your first robot control circuit. If nothing happens, go back and check each connection on your breadboard to make sure that they're wired correctly. Look at each hook-up closely, as it's easy to sometimes put a wire or component lead in the wrong tie point on the board. If it still isn't working, try different resistor values. On our robot, we got good action on resistors in the 3.2 to 4.2 $M\Omega$ range. If your circuit *still* isn't working, it's time to get out your digital multimeter and check all of your components (battery packs, resistors, capacitors, and switch) to make sure that everything's working properly. Consult the manual that came with your DMM to find out how to properly test each component type. To test the motor, touch its wires directly to the battery pack's wires (positive to positive and negative to negative, of course). If you do all this, your circuit should be working properly. There aren't *that* many parts that can fail here.

FIGURE 7.5

The proper way of hooking up your battery packs (in series) to deliver 6V of power to your circuit. Add a switch to the circuit, if you'd like.

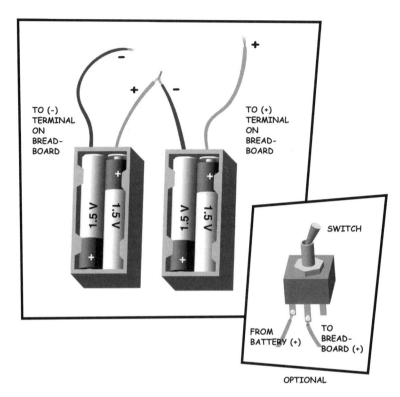

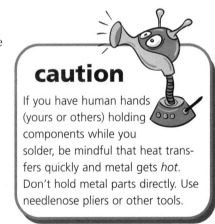

Soldering Up the Circuit

Now it's time to heat up your soldering iron, take a few deep breaths, and begin soldering the pins of your IC socket together, and then the discreet components onto it. Here are the steps involved:

caution

If you have human hands (yours or others) holding components while you solder, be mindful that heat transfers quickly and metal gets *hot*. Don't hold metal parts directly. Use needlenose pliers or other tools.

1. Get your 20-pin DIP socket and turn it over. Using your needlenose pliers, bend over pins 1 and 19 so that they are as close to touching each other as possible. Now find pins 3 and 4 and bend them down and toward each other into a pyramid shape until they touch. Next, bend pins 2, 5, 7, 9, and 10 outward. Bend pins 6 and 8 inward. Try to keep all the pins as straight and on the same level as possible. Now bend outward pins 12, 14, 16, 17, and 20. Then bend inward pins 11, 13, 15, and 18. When you're finished, the chip should look similar to the one shown in Figure 7.6.

FIGURE 7.6

All of the pins
on the IC socket
bent and ready
for soldering.

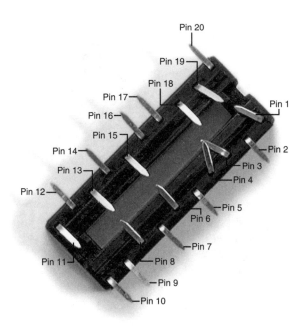

2. Using a small piece of component lead (clipped from one of your resistors, capacitors, techno-junk, and so forth) bridge the gap between pins 1 and 19. Solder the lead to one pin and then the other. If it makes it easier for you, you can solder the lead to the first pin when it's upright, and then bend it over, solder it to the other pin, and then clip off the excess.

3. Now find pins 3 and 4 and solder them together (see Figure 7.7).

4. Solder together pins 6 and 8 *and* pin 4 (which was already soldered to pin 3 in step 3) with a piece of component lead. In other words, you should have one long wire connecting pins 6 and 8 to pins 3/4 (see Figure 7.8).

5. Solder together pins 11, 13, 15, and 18 with a piece of component lead (also shown in Figure 7.8).

tip

In soldering the leads onto the IC socket pins, you'll quickly realize that you don't have enough hands to hold the socket in place, and hold the lead, the solder wire, and the soldering iron. This is where your third hand tool (or a hapless family member) comes in handy (see "Should-Have Tools," Chapter 5). Let one "hand" hold the socket on its side and the other "hand" hold the lead (while you hold the wire and the iron). Take your time setting up the third hand properly (tightening all of its joints) so that the socket and wire won't move as you work.

FIGURE 7.7
Pins 1 and 19
(the enable pins)
and pins 3 and
4 connected to
each other as
shown.

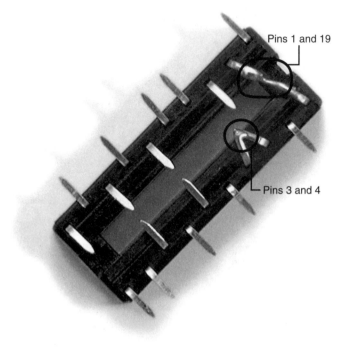

Pins 1 and 19

Pins 3 and 4

FIGURE 7.8
Pins 6, 8, and
3/4 soldered
together on one
side, and pins
11, 13, 15, and
18 on the other.

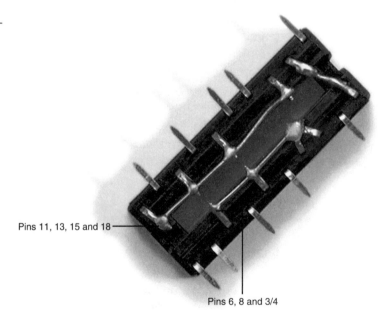

Pins 11, 13, 15 and 18

Pins 6, 8 and 3/4

6. Get a piece of hook-up wire and cut it so that it reaches from pin 10 to the join you made between the enable pins 1 and 19. Strip off just enough of the wire jacket to solder it to these pins and try to get the wire as straight (with as little slack) as possible (see Figure 7.9).

7. Solder together pins 5, 7, and 9 with a piece of component lead (see Figure 7.10).

8. Solder together pins 12, 14, and 16 with a piece of component lead (also shown in Figure 7.10).

note

These two pins, 1 and 19, are the "enable" pins. They "switch on" each side of our chip. By connecting them together, you're making one big "on" switch (see Figure 7.7).

FIGURE 7.9

Pin 10 (the negative, ground pin) connected via insulated wire to the enable pins 1/19.

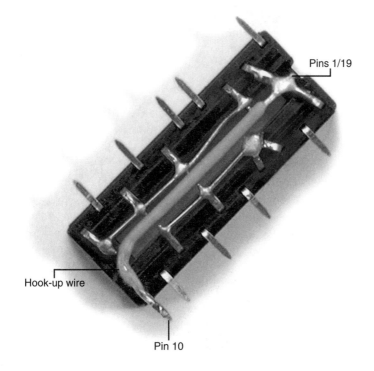

Pins 1/19

Hook-up wire

Pin 10

FIGURE 7.10
Pins 5, 7, and 9
connected
together, and
then pins 12, 14,
and 16.

Pins 12, 14 and 16

Pins 5, 7 and 9

9. Get the two .22µF monolithic capacitors. Trim the leads down (if you haven't already cannibalized them for lead clippings) so that they're of a manageable length. Solder one of them to pins 2 and 3 and the other one to pins 17 and 18. Solder them with enough "slack" lead so that they can be bent to the sides, and out of the way when you flip the socket over and insert the IC into it (see Figure 7.11).

10. Now we need to add our final component: the resistor. Hopefully, you experimented around during the breadboarding stage and found a resistor value that gives your walker the right amount of back and forth action. Here, you *don't* want to clip the resistor's leads because you're going to want to solder it around the top edge of the socket, from pin 2 to pin 17. We need to go around the socket because we don't want the resistor to get in the way of plugging the IC into the socket, or when mounting the socket/IC assembly to the top of the walker.

> **caution**
>
> Be very careful while you're soldering the DIP socket that you don't melt the plastic part of the socket package. It will melt quite easily if you touch it with the hot iron, and then the pins of the IC won't line up properly when you try to plug it in.

With plenty of lead on the resistor, we can bend and twist it during the final assembly phase to make sure it's out of the way of other components (as shown in Figure 7.11).

FIGURE 7.11

Our two capacitors soldered in place (one across pins 2 and 3 and one across pins 17 and 18) and our resistor installed (from pins 2 to 17).

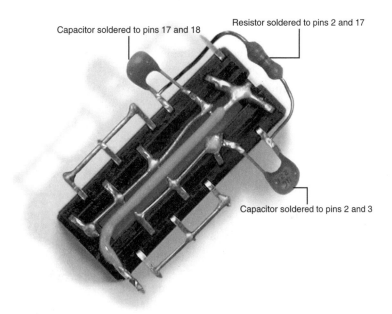

Capacitor soldered to pins 17 and 18

Resistor soldered to pins 2 and 17

Capacitor soldered to pins 2 and 3

11. Flip the IC socket over and carefully plug the 74HCT240 chip into the soldered-up IC socket (see Figure 7.12). If it has trouble going in, carefully inspect your socket assembly to make sure that you didn't melt any of the plastic package, and therefore disturb the sockets where the IC pins plug in. If they are misaligned, we hate to break it to you, but you might have to get another socket and do the whole thing over again. We told you to be careful!

FIGURE 7.12

The completed soldered up IC socket with the 74HCT240 plugged in.

That's it! If all went well, you should have a working Bicore control circuit. All you have to do now is connect the power and the motor to the appropriate pins. We'll hold off on doing that though until we've built the rest of our robo-critter.

Go ahead and take a break, indulge in your junk food of choice, play a round of *Enter the Matrix*, run a lap around the house, or otherwise cool out for a bit. I don't know about you, but these solder fumes are makin' me feel kinda funny.

> **tip**
>
> If trying to solder up this circuit (or any other part of this project) gets too frustrating, it's probably time to apply the "Know When to Come Back" rule from Chapter 3.

Thumbnail Guide to Breadboards

Every wirehead worth his or her propeller beanie knows about breadboards. And no, we're not talking about a cooking utensil from Martha Stewart's kitchen; we're talking about an essential piece of equipment for every electronics hobbyist and professional.

A breadboard is a simple, inexpensive device (available at any electronics store) for temporarily hooking up and testing an electronics circuit before you solder it together. By using a breadboard, you're checking to make sure that the circuit is designed properly and that all of your components are working as specified.

A breadboard is usually comprised of a metal base plate with a white nylon block mounted on it

> **tip**
>
> As you do your soldering practice (see Chapter 5) and assemble robot kits (see Chapter 6, "Acquiring Mad Robot Skills"), you'll find yourself trimming the leads on most of the resistors, capacitors, and diodes that you mount onto printed circuit boards. Save these lead clippings! They come in very handy when you're freeforming a circuit, especially when you're connecting the pins on an IC or an IC DIP socket.

that's covered with a grid of holes (known as *tie points* or *wire receiving sockets*). There are usually three *binding posts* also on the board, which are screw-down connectors for bringing power to the board.

The nylon grid on a breadboard is divided into two major sections split down the center. This center channel (called the "trench") is just wide enough to accommodate the two rows of connecting pins found on standard DIP ICs. Radiating from the center of the board are vertical rows of tie points (usually five on each side). All of these "5-position groups" of tie points are connected together, so a wire in any one of them electrically connects any wire or component lead in any of the others in that row. Along the top and bottom of the breadboard are a series of power "distribution buses" (or simply "buses"). They (usually) run horizontally and there are often four groups of them (two on the top and two on the bottom). To power the circuit you've set up on your board, you simply connect the leads from a battery to the binding posts (positive to a red post, negative to a black), connect a short wire from each post to the appropriate channel (+ or -) on the power bus, and then a wire from the bus to the appropriate socket in the row of tie points connected to the parts of your circuit that need power. Confused? Check out Figure 7.13. You'll get the hang of it pretty quickly.

To make life easier down on the breadboard, electronics stores sell *jumper wire* kits with different lengths of 22-guage *hook-up wire* cut to appropriate lengths that correspond to the tie-points on the board. You'll want to get one of these kits. You'll also need to power your board. For this, all you need to do is hook up the battery pack of the correct voltage needed to power the circuit you're building (for instance, in the Bicore circuit for this chapter, we need 6V of power). To make things even more convenient, the tops of the binding posts also accept *banana plugs* (also available at electronics shops). You can solder sets of these to various battery packs (6V, AAA, 9V, and so forth) and have these available

note

There *is*, in fact, a relationship between electronics breadboards and their kitchen counterparts (er… if you'll pardon the horrific pun). Electronics prototyping tools got their name from the early days of amateur electronics when hobbyists would take an old kitchen cutting board, hammer nails into it, and then temporarily hook up a circuit to make sure that it worked. Mother was *so* pleased! This is basically the same way that today's commercial breadboards work.

tip

When buying a breadboard, look carefully at the tie-point holes to make sure that the sockets underneath the nylon grid are properly aligned. If they aren't centered in the holes, it can make inserting wires difficult, if not impossible, and that's just…well…crummy.

for your breadboard power source. Then all you need to do is plug the banana plugs into the tops of the binding posts to power your circuit. (Breadboards? Bananas? Is anyone else getting hungry?) If you want to get *really* fancy, you can hook up a power switch to the board so that you can turn the power on and off without having to disconnect the battery.

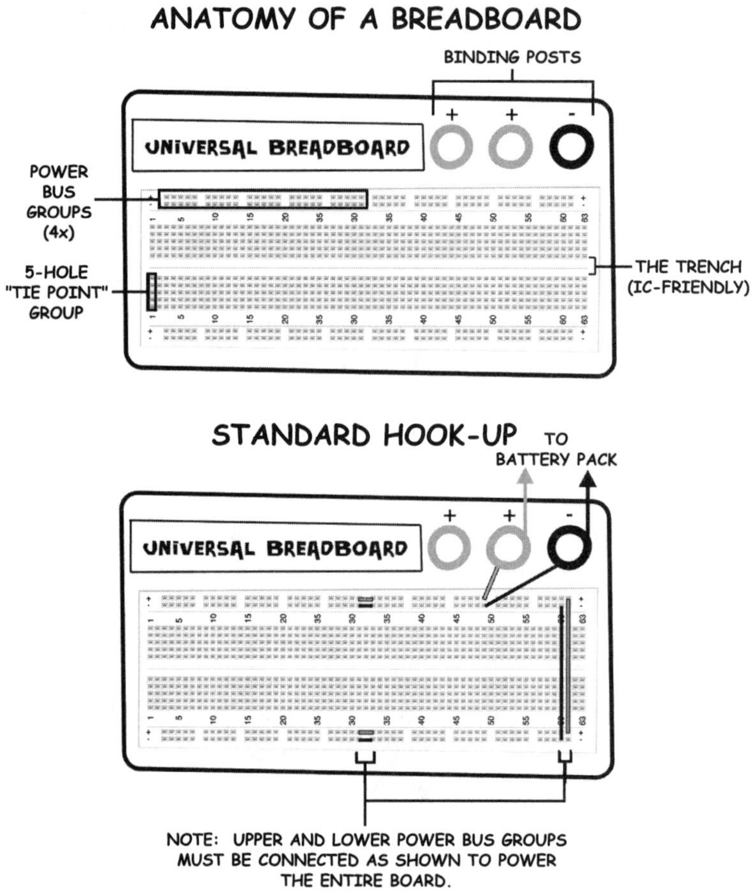

FIGURE 7.13

Anatomy of a breadboard. Larger version of this image is available at www. streettech.com/ robotbook.

ANATOMY OF A BREADBOARD

STANDARD HOOK-UP TO BATTERY PACK

NOTE: UPPER AND LOWER POWER BUS GROUPS MUST BE CONNECTED AS SHOWN TO POWER THE ENTIRE BOARD.

When you first get your breadboard, you'll want to hook it up with a standard power configuration that distributes power throughout the buses on the top and the bottom of the board. As we said earlier, there are usually four power bus groups on the board. To connect them, you need to put a jumper across the two positive and negative groups on the top and the two positive and negative groups at the bottom. You then need to connect the top bus groups to the bottom bus groups (with a positive-to-positive wire and a negative-to-negative wire). Use the standard red for positive and black for negative, if you have them; otherwise, just be consistent (all positive wires one color, all negative wires another). Refer back to Figure 7.13 if this is getting confusing.

Besides the breadboard itself, the power supply, and the jumper kit, you'll need two other tools for effective breadboard work. A pair of needlenose pliers is essential for getting those pesky jumper wires onto and off of the board, especially in tight spaces. You'll also want an IC extractor (or a *chip puller*). This is a funny-looking pair of tweezers that is used to safely remove IC chips from a breadboard (or an IC socket) without damaging their delicate pins. (And trust us, without a puller, this is *very* easy to do.)

Building the Walker Body

The body for our Coat Hanger Walker is basically the motor casing for the servo motor itself. The control circuit goes on top (actually, the servo is flipped upside down, so it's technically the bottom of the motor), the two AAA battery packs go on the sides, and the on/off switch gets attached to the back. The motor is oriented so that the drive shaft protrudes from the bottom, delivering power to two sets of gear/legs, one set for the front two legs, and one set for the back. The ingenuity of this design is that four legs are controlled, and that a single motor can achieve a reasonable walking gait. As we discussed in Chapter 4, walking technology is usually hard. The simplest walkers usually have at least two motors and two control circuits (one "master" circuit and one "slaved" to that). Jérôme Demers's design is a study in engineered minimalism. This walking machine doesn't have the most elegant gait in the robot kingdom, but it *does* work, and

it shows the kind of insect-like movement and persistence that's the hallmark of biologically inspired robots.

To start building the body, let's first fabricate a few special parts that we'll need.

Making the Gears

The gears that control the front and back leg sets are actually one plastic gear split down the middle and turned back on itself. Because leg walking only requires a back and forth, reciprocating motion (rather than continuous rotation), we don't need all 360 degrees of gear. One-hundred eighty degrees–worth of gnarly gear teeth is more than enough.

To cut the gear in half, choose the notch between two gear teeth on one side of the gear's circumference and count half of the notches between gear teeth to arrive at the point where you want to cut. So, if your gear has 40 teeth (as ours does), choose a space between two teeth and count between teeth to the 20th notch. Most plastic gears have holes in the area between the hub and the toothy rim (to cut down on weight). Try to align your cut so that you don't cut through any of these holes. Structurally, this doesn't really matter; it just looks nicer.

After you've found your centerline to cut, use your hobby knife and a ruler (preferably a metal one) to scribe a line across the gear. Use a cutting mat, scrap cardboard, or something else that you don't care about marring. Because the gear likely has a raised rim around the outer edge and a raised edge around the hub, cut into each of these surfaces in turn (in other words, don't just drag the knife along the ruler, at one depth, and in one long stroke). Take your time and cut repeatedly, applying more pressure each time. Don't try to cut through all in one stroke. If you have a hobby razor saw, or a razor saw blade for your regular hobby knife, that will make cutting easier (see Figure 7.14).

note

When you have your breadboard set up with the top and bottom power buses connected along each bus strip and the top and bottom buses connected (as shown in Figure 7.13), you'll want to keep this basic setup for all projects.

tip

Try and make your breadboard hook-ups as neat, linear, and "short distance" as possible. There's an almost Zen-like art to breadboarding so that you don't cross wires over each other and you minimize crossing wires over ICs. Things can get pretty confusing and messy enough when all components and wires are in place—you don't want to compound this with willy-nilly hook-ups. Try to keep your wiring at right angles. (See the breadboarding examples in the project sections of this book to see what we're talking about.)

FIGURE 7.14

Cutting the gear
in half to make
two 180-degree
walking gears.

Fabricating the Idler Shaft

There are two gear shafts on our walker. The drive
shaft is the one that comes from the gear box on
the servo motor (which is oriented upside down).
It transfers the back and forth motion of the
motor to the internal gears of the servo's gearbox,
and then to the gear shaft to which we'll attach
one half of the gear we cut. This gear also serves
as the base for mounting our back leg assembly.
A second gear shaft is used to hold the other gear
half and the front leg assembly. This gear shaft is
not powered (its gear uses transferred motion
from the powered "rear" gear). Because this shaft
is unpowered, we'll call it an *idler shaft*.

> **caution**
>
> You want to be
> extremely careful cutting
> the gear in half. Use a
> new blade on your hobby knife.
> Make sure the gear is secure as you
> cut (apply lots of downward pres-
> sure on your ruler to hold the gear
> in place). Use careful, controlled
> strokes. Don't rush it, or you could
> cut yourself.

The idler shaft can be made from either a metal
coat hanger wire or other wire in the 8 to 10–gauge range. If you want, you can use
the "Gumby Legs" wire sold by Solarbotics (which we recommend using for the legs).
This wire is simply 10-guage copper wire, so if you have other access to 10-guage
solid copper, you can use that.

1. Cut about a 3 1/2-inch length of wire. If you use the copper wire, you'll likely
   have to strip off the plastic insulating jacket first. Use your wire-stripping tool.
   Make sure the wire's nice and straight. By the way, if you do use the gumby
   wire (or other jacketed 10-guage), save the jacket pieces. You'll need them
   later on.

2. Measure down about an inch on the wire and, using your needlenose pliers, bend the wire above your 1-inch mark into an inverted teardrop shape (think shepherd's crook). It doesn't really matter if the shape is perfect. All we're trying to do here is to create as much surface area as possible for gluing. This bent part of the wire is what gets glued to the servo motor/body (see Figure 7.15).

3. At the 1 1/2-inch mark, bend the shaft about 35 degrees away from the teardrop as shown in Figure 7.15. Test fit this on your motor casing. The teardrop part should lay flat against the casing and the shaft should clear the motor mounts on the servo motor (with some wiggle room). If you have a Dremel tool or a razor saw, you might want to go ahead and zip off the motor mount anyway, to give yourself lots of space (we did this). You'll need to be able to bend this shaft (down) a bit during the final assembly so that the idler gear can mesh properly with the drive gear.

note

Servo motors often have mounting brackets built into them (on both ends of the motor casing). Depending on the angle of your idler shaft, the bracket on that end of the motor casing might get in the way of this idler shaft/gear assembly meshing properly with the drive gear. You might want to just go ahead and zip this mount off with a Dremel tool or razor saw. We went ahead and removed the mounts from both ends of the casing, just to make our little beasty look sleeker.

FIGURE 7.15

Inverted t0eardrop bend in gear shaft wire for mounting to servo (left), 35-degree bend in wire, seen from side and attached to motor casing (right).

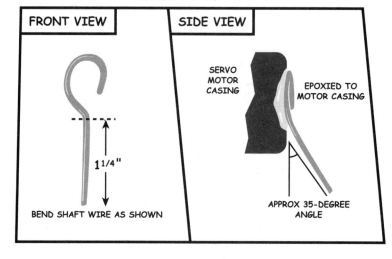

4. Mix up a small batch of 2-part epoxy. Apply it with a wooden coffee stirrer, Popsicle stick, or other similar disposable stick, to the end of the servo motor *opposite* the end closest to the drive shaft (see Figure 7.16). This is very important. Now place the teardrop-shaped crook part of your idler shaft into the glue. You should have a generous amount of epoxy on the motor casing so that the wire gets firmly bonded to it. You'll need to hold the wire in place for a few minutes while the epoxy sets up. Get comfortable, don't rush it, and don't peek! Use a piece of junk plastic (like the clear plastic used in product packaging) to hold the shaft in place (so you don't glue your fingers to it). You can carefully peel the plastic away after the glue has set up enough to hold the shaft firmly in place. Don't use paper for this, as some of the paper will stick to the shaft and look...ah...tacky. Make sure the shaft, as it comes away from the teardrop mount, is centered right and left on the motor. When the glue has set enough for the shaft to be stable, carefully put the motor down in a safe place for it to dry, preferably overnight.

> ## caution
>
> Epoxy is noxious stuff. Work in a well-ventilated space and wash your hands when you're done. Some people's skin has a bad reaction to the chemicals in the compound, so it's always a good idea to wear disposable gloves. It not only protects your hands, but it will keep that epoxy resin stink off of them (which doesn't go away even after washing them).

FIGURE 7.16

Our idler shaft glued to the motor casing (on the end opposite the drive shaft and motor wires).

Making the Idler Gear

While the epoxy is drying on the idler shaft, we can make the gear and gear mounting hardware that we'll use to attach it to the shaft. Let's start with the gear.

When we cut our original gear in half, we cut the center hub in half as well. This leaves us with no closed hub to hold the gear onto the shaft. We'll need to fabricate our own hub.

1. First you'll want to use your hobby knife to clean away any of the plastic residue in the hub area that might be left over from cutting the gear in half. You want a nice clean surface on which to bond.

2. If you used the Solarbotics gumby wire (or other jacketed 10-guage wire), you should have a lot of plastic jacket left over from the stripping. Using your hobby knife, cut a piece about 3/8-inch long. Make sure that your cuts are as straight as possible. If you don't have a 10-guage wire jacket to play with, you'll have to find plastic tubing with a 3/32-inch inside diameter. Heat-shrink tubing of this size will work.

3. You could just glue this hub right to your idler gear, but we found that the stress on it eventually broke the gear off. To fix this, we added some plastic reinforcement to help join the hub to the gear. We did this by way of the optional leg mounting pad (attaching the hub/plastic to the LMP). If you're using an LMP, you'll want to trim the LMP first (see details on page 214). To create the reinforcer, we cut a small (1/4 inch × 1 inch) rectangle of plastic and drilled a hole (just using the point of our hobby knife) in the center (for the hub to pass through). We glued the hub to the plastic reinforcement piece and then glued

tip

Working with 2-part epoxy:

- Keep it clean, buster! Epoxy likes it when you use a clean surface for mixing, and clean (unused) toothpicks, Popsicle sticks, coffee stirrers, and so forth for mixing. Use a new mixing surface and stick each time you mix up a batch.

- Mix *thoroughly*. This is chemistry, people. Let the two parts come together, do their chemical thang, and generate some heat. That's when the good bonding happens. Mix until the slightly foggy cast and bubbles are evenly distributed throughout—and then mix some more, for good measure.

- If you want to extend the working time (or *pot time*) of the epoxy, spread it out (after it's well-mixed). The more compressed the mass of epoxy, the more heat it'll generate (and the faster it will cure).

- Use clamps, tape, rubber bands, and so forth to hold parts in place while they dry, but don't get glue on them! Use a piece of scrap plastic, cardboard, or similar material between the clamps and the object being glued.

the reinforced hub to the LMP (which will, in turn, be screwed to the idler gear). We used superglue for this bonding job. To create the plastic reinforcement part, we used a piece of .03-inch plastic stock (often called Plasticard), available at most hobby and craft shops, but you can just as easily use more of that junk plastic most everything we buy is packaged in (see Figure 7.17).

4. While the hub assembly is drying, we'll create the mounting hardware. These are the two parts that will be fastened on either side of the idler gear to hold it in place on the shaft. To create these parts, we'll need to "free" two screw terminals from a terminal mounting block. A terminal mounting block is a simple but nifty electronic component that allows quick connecting/disconnecting of rows of wires. To use a terminal block, one sticks wires in holes on either side of the terminal block and tightens them in place with the block's built-in screws. This two-connector pair forms a disconnectable electrical connection. The metal blocks and screws are usually surrounded by a nylon housing. For our gear mounting parts, we'll need to remove a terminal pair from the block, cut away the plastic housing, and then cut the connector pair in half to make two screw-down mounting pieces. Follow the sequence in Figure 7.18 to create the two parts. It is highly recommended that you use a rotary (Dremel) tool for this operation. If you don't have such a tool, you can use a hacksaw or razor saw. A hobby knife won't work (Barbie says: "Nylon is hard!"). After you've cut out your two mounting parts, use a metal file to remove any burrs where you cut the block pair in half.

> **tip**
>
> When building anything, always keep it straight, level, square, tight, and braced. Did we mention: measured and re-measured? Obviously, your mileage will vary depending on what your project is, but you get the basic idea: Build things solidly, with precision in mind, and the results will greatly reward you for your patience. And always use the right tools (see Chapter 5).

FIGURE 7.17

The new gear hub and hub reinforcer glued to the LMP. If not using LMP, glue hub/ reinforcer directly to gear-half.

FIGURE 7.18

The steps required to "free" a terminal pair from a terminal block.

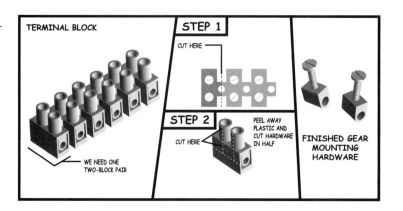

5. When the glue is dry on your hub assembly (and on the idler shaft itself), use two screws that came with your servo motor (or any other sheet metal–type screws long enough) to attach the mounting pad/hub assembly to the idler gear. Now you're ready to test-mount the gear on the shaft. Slide one terminal block onto the shaft all the way up to the teardrop mount, and screw it down tightly. Now slide the idler gear on (with the teeth facing down), followed by the second terminal block, which also gets tightened. The idler gear and shaft are done (see Figure 7.19)!

caution

Creating this mounting hardware is the most dangerous part of building this walker. You'll need to cut through plastic and metal, and if you use a rotary tool, the chips are gonna fly. It is ideal if you hold the block in a bench vise, and you definitely need to wear safety goggles or glasses. If you don't have a vise, use a C-clamp, spring clamp, or some other type of clamping technology. The terminal block *must* be secure for safe cutting.

FIGURE 7.19

Test-fitting of our finished idler shaft and idler gear.

Mounting the Drive Gear

The output shaft (*final gear*) of a servo motor usually ends in a stubby knurled bit that sticks out of the motor casing. These knurls mesh with a complimentary set on a component called a *servo horn* or *control horn*. Servo horns come in many shapes and sizes. They were originally designed for attaching control wires/rods used to control model airplanes, so the horns have many holes in them for such attachments. If you're using the Solarbotics motor GM4, it comes with a bunch of different servo horns, mounts, and mounting hardware. We want to use the 1 3/8-inch disk-shaped horn. If you got your motor else-where, hopefully it came with a horn close to this diameter. It doesn't have to be exactly this size, but when we attach the drive gear half to it, we want the gear/servo horn assembly to be as close to the lead edge of our walker (the edge with the idler gear) as possible. You'll want the idler gear to be at about 35 degrees in relationship to the drive gear.

> **tip**
>
> If you need to use pliers to straighten your leg wire, the teeth on the pliers will leave unsightly marks, especially if you're using copper, which is quite soft. Wrap some masking tape around the jaws of the pliers and then you can tweak and twist all you want and leave nary a scar.

To create our drive gear, all you have to do is glue (with epoxy) or screw the gear half to the servo horn. We chose to screw ours on. Because the rim of the gear is raised, there isn't much gluing surface. We used the servo mounting screws that came with the Solarbotics GM4. We tapped starter holes in the gear to make screw-ing the two components together easier. The servo horn has rows of holes already in it, so we made use of those (see Figure 7.20).

After you've attached the gear to the servo horn, test fit the drive gear (you don't have to bother screwing it in). The two sets of gears should mesh. You might need to bend the idler shaft. The gears will mesh at an angle. This is correct. There just needs to be enough meshing that they don't slip as your critter lumbers along. See Figure 7.21.

FIGURE 7.20

The drive gear mounted (here via screws) to the servo horn.

FIGURE 7.21

Test fitting to make sure that the drive gear and the idler gear mesh properly. Bend idler shaft down, if necessary.

Creating the Leg Assemblies

With your robot body all geared up, it's time we go a-walkin'. To do that, you'll need legs—two sets of them. One of the cool things about this robot...er...walking machine, is that it teaches us about different leg designs and how they impact the critter's mobility. If you want, you can just make two sets of legs, epoxy them to the gear halves, and be done with it. That's how Jérôme Demers did his walker. But we came up with a little revision to the design that lets you mount various leg configurations, so you can experiment with different designs. Let's fashion our first two sets of legs because we'll be using them regardless of whether we glue them on or use mounting hardware.

We'll do the back legs first, as they're the easiest.

The Back Legs

To fashion a set of back legs:

1. Measure out and cut 5 1/2 inches of coat hanger or similar wire. Find the midpoint, and using your needlenose pliers, fashion the wire into an overly wide "U" shape. If you're planning on gluing the legs directly to the gear, test fit and reshape until the crotch of the "U" fits nicely inside the inside rim of the gear. If you're using one of the Solarbotics leg mounting pads (LMPs), you'll want a more square-shaped bottom for your "U" (so that the leg wire doesn't start bending until it clears the attachment holes on the LMP). Both examples are shown in Figure 7.22.

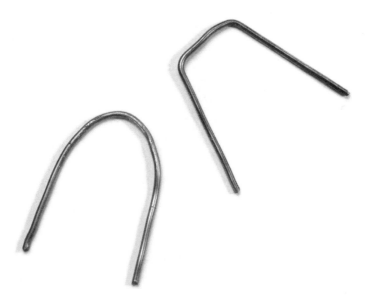

FIGURE 7.22

Legs shaped for gluing (left) or for LMPs (right).

2. Measure about 1 1/4 inches from the tip of each leg and bend wire to angle approximately as shown in Figure 7.23.

3. If you're gluing on your legs, mix up a batch of epoxy and glue the legs to the inside rim of the gear. Set the leg/gear assembly aside to dry. You're done. If you are not gluing, go to step 4.

FIGURE 7.23

Approximate
downward angle
for walker's back
legs.

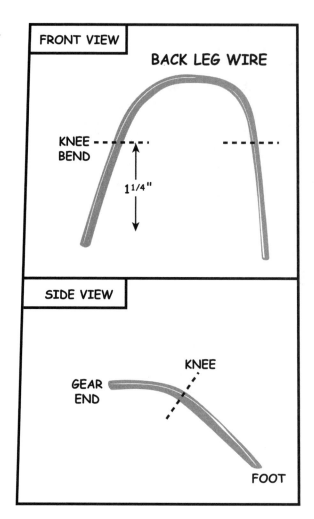

4. The Solarbotics LMPs have a triangular shape that can't work on our walker as is. So that the LMPs do not touch each other when we attach the gear/leg assemblies, we have to zip off the peaks of the triangles. Using your rotary tool, razor saw, hacksaw, or other cutting tool, remove the pad material from one of the LMPs down to the top-most mounting hole (see Figure 7.24).

FIGURE 7.24

Solarbotics LMPs
with their tips
trimmed to fit
our walker.

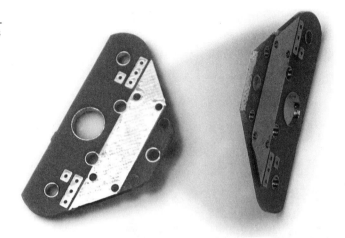

5. Get a piece of 22-guage hook-up wire (about 4 inches) and strip off the insulating jacket. Place your leg wire on the wide, silver mounting strip on the
LMP and thread the hook-up wire through two of the mounting holes on the
strip. Using your needlenose pliers, twist the wire tightly to attach the leg. Cut
another 4-inch piece and thread it through the other set of holes on the
mounting strip. Twist them down. Leaving about 4–5 twists, trim off any
excess wire with your wire cutters. Cut a third piece of wire and use it in one of
the three-hole tracks on the pad to further secure your legs (see Figure 7.25).

FIGURE 7.25

Back legs
secured to LMP.
Note third wire
wrap (far left) to
further secure
leg.

6. Using the screws that came with the LMPs, screw your leg/LMP assembly to the back servo horn/gear assembly (see Figure 7.26). You'll probably want to drill or tap starter holes in the gear.

7. To get better traction from the legs, and honestly, to make our robot look a lot cooler, we'll add some robot "feet" to the legs. You can use heat-shrink tubing for this or simply cut four 1/4-inch pieces of the insulating jacket from the 10-guage wire (if you used that in your build) and glue them onto the tips of the leg set (see Figure 7.27). Your back legs are done!

The Front Legs

The front legs are similar to the back legs, but they have a second set of "knees" (read: bends). We wanted our walker to have a nifty insectoid look, and given the angle of the front gear shaft, we thought a double knee leg shape looked the best. You can experiment around with different leg shapes (see "Further Experiments" later in this chapter), especially if you plan on using the Solarbotics LMPs so that you can change your legs.

note

If you find that you still don't get good traction with the plastic "footies," try gluing some rubber band material over the tips of the feet.

1. The first thing you'll need to do is measure out and cut a 9 1/2-inch piece of your leg wire (coat hanger or 10-guage copper wire). Find the center point, and as you did with the back legs, bend the wire into a slightly splayed-out "U" shape. Again, if you're planning on gluing the legs to the idler gear, you're going to want a rounder crotch on the "U" (so that it fits within the gear's rim), or a more square shape if you're using the LMPs.

2. Measure 3 inches from each tip of the "U" and mark this point. This is where you'll put the first knee. Bend the wire on both sides as shown in Figure 7.28.

3. Now measure 1 1/4 inches from each leg tip (the feet!) and bend both leg sides as shown in Figure 7.28. For shapes of bends (glued vs. mounted), refer back to Figure 7.22.

4. If you're gluing your legs on, mix up a batch of epoxy and glue the legs to the inside rim of the gear as you did for the back legs. Set the leg/gear assembly aside to dry.

5. Using your rotary tool, razor saw, hack saw, or other cutting tool, remove the pad material from the remaining LMP (if you haven't already) down to the top-most mounting hole as we did for the back legs (refer back to Figure 7.24).

6. Strip a 4-inch piece of hook-up wire. Place your leg wire on the wide silver mounting strip and thread the hook-up wire through two of the mounting holes on the strip. Using your needlenose pliers, twist the wire tightly to attach the leg. Cut another 4-inch piece and thread it through the other set of holes on the mounting strip. Twist them down. Leaving about 4–5 twists, trim off any excess wire with your wire cutters. Cut a third 4-inch piece of wire and thread it through the outside hole of one of the three-hole tracks on the pad. Thread it through the back of the pad and through the outermost hole on the other side of the LMP. Twist the wires together to further secure your legs to the pad (see Figure 7.29).

FIGURE 7.28

The shape of the front legs (here shaped for the LMPs) and the placement of the two knees.

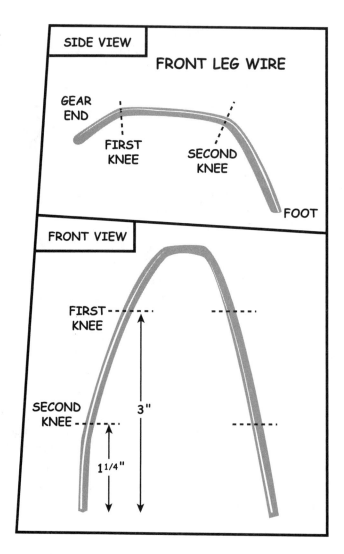

7. Use the screws that came with your servo horn set or other sheet metal screws (in other words, screws with points on 'em) long enough to fasten the LMP to the idler gear. Use a small nail or drill to tap holes into the gear where you'll be attaching the leg assembly.

8. Add your bot footies to the legs, as you did with the back legs, either using leftover wire jacket or heat-shrink tubing (also shown in Figure 7.29).

FIGURE 7.29
Our finished
front leg
assembly.

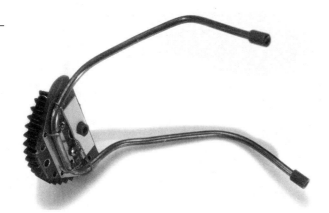

Testing the Fit

That's it for the leg assemblies. You're ready to test fit the legs/gears to make sure they mesh properly. Put the drive gear on the servo motor, with the gear basically centered right and left. Now slide the front gear on its shaft and use the screw on the terminal block to fasten the gear in place. The gears should mesh well enough that when you gently twist the drive gear, both sets of legs/gears move in tandem. If they don't—or if the gears slip—you'll have to push down on the idler shaft, (toward the drive gear) to get a better mesh.

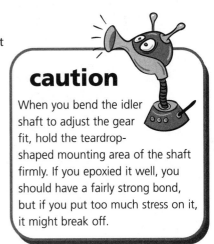

caution

When you bend the idler shaft to adjust the gear fit, hold the teardrop-shaped mounting area of the shaft firmly. If you epoxied it well, you should have a fairly strong bond, but if you put too much stress on it, it might break off.

The Power Plant

We're almost there! All we need to do now is attach the battery packs, add a power switch, and attach the motor and power wires to the control circuit.

Attaching the Battery Packs

The two AAA battery holders need to be attached to either side of the servo motor. This is easily done.

1. Orient the first holder so that the power wires face toward the back of the walker and center the holder front to back (on the motor casing). Orient it as close to the bottom of the motor case as you can (so that you keep the

center of gravity on the machine as low as possible). When you have it where you want it, place the tip of your hobby knife in the center mounting hole of the battery holder and twist the knife to begin a screw hole. Remove the battery holder and ream out the screw hole a bit more. Don't make the tap hole too deep or too wide—you're just looking for something to get the screw started.

2. Apply some superglue or epoxy to the back of the battery holder, and then, using a small sheet metal screw (like the ones that likely came with your servo motor) in the center hole, attach the holder to the walker's side (see Figure 7.30).

3. Repeat the preceding steps on the second battery pack (making sure the wires sprout out the back of the bot).

note

If you decided to build the walker's body first, and then the control circuit, go ahead and attach the battery packs to the sides, but don't connect the batteries or attach the switch. Do this only after you've built and attached the Bicore. It is recommended though that you load the battery packs with batteries (and use another 6V of power on the breadboard). This way, with a "dummy load," you'll have a true measure of most of the weight the walker will be carrying on board as you tweak the mechanics and electronics.

FIGURE 7.30
AAA battery holder attached to servo motor casing.

Adding a Power Switch

To be able to turn our walker on and off, we need to add a switch. This is easily done.

1. Test fit your microswitch on the back of your servo motor. It should be able to just fit on either side of power wires coming out of the motor casing. When you've decided where you're going to put it, mix up some more 2-part epoxy and glue it into place (see Figure 7.31). When it's dry, move on to step 2.

2. Find the upper-most red (+) wire on one of the battery packs. Cut and strip the wire as short as possible (so that it's as tight to the bot's body as possible after it's soldered to the closest connecting pole on the bottom of the switch). Thread the wire through the hole in the closest switch pole and then bend and twist the wire for a nice firm connection. Trim the excess wire and then solder the wire wrap to the switch pole (see Figure 7.31).

3. Measure out and cut about 2 inches of the excess positive wire you just snipped off of the battery holder. Thread this through the hole in the center terminal of the switch (if it has more than two terminals), twist it up, and solder. Leave the other end (which will eventually connect to the Bicore) unconnected (see Figure 7.31).

4. Now let's hook our two battery packs together in series to deliver all of their 6V of power to the switch (and to the motor and control circuit when we throw it). Take the bottom two wires, one red (+) and one black (-) from each battery holder and overlap them. Eyeball about how much wire you'll need to keep for stripping and wrapping these wires together, and then clip off the excess. Strip and wrap the wires together and add some solder for good measure (see Figure 7.31). Remember: Neatness counts.

note

If you decided to build the body first and the Bicore control circuit second, now it's time to hook the body to its (temporary) brain on the breadboard. Add your "dummy load" of four AAA batteries and use two long lengths of hook-up wire to attach the motor to pins 9 and 12 on the breadboard. The positive gets attached to pin 9, the negative to pin 12. Apply 6V of power to the breadboard's power bus, and watch your critter go! Play around with changing leg shapes and resistor values until you have a configuration you like; then go back to the "Freeforming the Bicore Control Circuit" section to create your control circuit.

caution

Make sure to remove the AAA batteries if you had them in their holders during mechanical testing. Never connect parts of a circuit while delivering power to the circuit.

FIGURE 7.31

Power switch
and battery
pack all con-
nected and
ready to deliver
the juice!

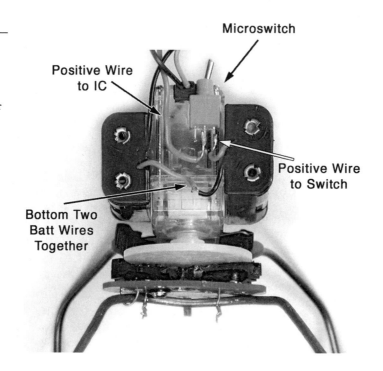

Microswitch

**Positive Wire
to IC**

**Positive Wire
to Switch**

**Bottom Two
Batt Wires
Together**

Final Assembly

We're almost there. Now all we have to do is
attach the power and the motor wires to the control circuit and then the control circuit assembly
to the top of our walker. It goes something like
this:

1. Measure out enough of the positive (red)
   wire coming out of the back of the servo
   motor so that there won't be too much
   slack in it when the controller gets glued to
   the top of the motor. In other words, you'll
   want to solder the wires on before you glue
   the control circuit in place, but you don't
   want to end up with a big loop of excess
   wire when the circuit is connected and
   glued in place. The positive wire will be
   connecting to pin 9 on the IC socket.

2. After you've measured and cut the positive motor wire, measure and cut the
   negative motor wire to the same length. It will connect to pin 12 on the
   socket.

tip

If you *really* want
to be a fancy-
schmancy bot builder,
you can put heat-shrink
tubing over the wire join connecting the battery packs. To do this,
slide a piece of heat-shrink tubing
onto one of the wires *before* you
wrap them together. After they're
wrapped and soldered, slide the
tubing over the join and heat with
your micro-torch or heat gun.

3. Measure the unconnected red wire from the switch so that it will comfortably reach pin 20 on the IC socket when it's glued to the top of the servo. Cut it.

4. Measure the top negative wire from the battery pack so that it will comfortably reach pin 10 on the IC socket when it's glued to the top of the servo. Cut it.

5. Now solder all of these wires that you've cut to the appropriate pins on the IC socket: positive wire from motor to pin 9, negative wire from motor to pin 12, positive wire from switch to pin 20, and negative wire from battery pack to pin 10.

   You can see the end result of each of these first five steps in Figure 7.32.

FIGURE 7.32

The power and motor wires connected to the appropriate pins on the IC socket.

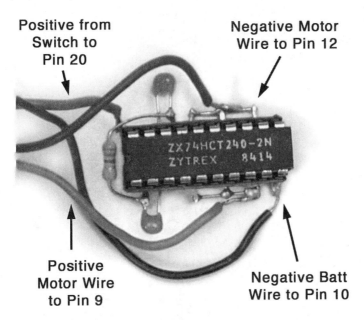

Positive from Switch to Pin 20

Negative Motor Wire to Pin 12

Positive Motor Wire to Pin 9

Negative Batt Wire to Pin 10

6. Now that all the wires are connected to the IC socket, you're ready to plug in the 74HCT240 chip (if you haven't already). Gently press the chip into the socket holes. Be careful not to damage any of the components and solder joins on the other side. Test fit the control chip assembly to make sure that it can sit on top of the motor. You might have to gently bend the pins so that the socket sits relatively level. It is going to be slightly funky and it will be resting on the leads you soldered on. That's okay.

7. Before you glue on the control circuit, toss in some batteries, flip the switch, and make sure everything works (with the control circuit just hangin' around on the ends of the motor and power wires). If it doesn't work, check to make

sure all of your connections were done properly (go back over the preceding steps and check each connection).

8. Now you're ready for the last step: fastening the Bicore circuit to the top of the walker using epoxy. Test fit the control circuit first. You might have to bend some of the parts on the circuit, especially the caps, resistor, and motor and control connections to make sure the chip lies as flat as possible and pins/wires don't touch each other. If wires touch that aren't supposed to (such as a cap and a resistor lead), the circuit will not work. When the chip is prepped, mix up some epoxy, pile it on top of your motor casing, and squish the IC socket into it. Hold it until it stays. If you think you might want to monkey with the circuit some more, you don't have to glue it down, you could just use some poster putty to temporarily hold it in place.

With these steps complete, your walker is finito (see Figure 7.33)!

FIGURE 7.33

Our finished walker in all of its mechanical magnificence.

Further Experiments

One of the things that's really fun to do with building projects like this is to tweak them when they're done. Using the MIT AI Lab's idea of building upon previous successes (see Chapter 2, "Robot Evolution"), you can now think about improvements to your walker—evolutionary upgrades. Unfortunately, this "robot" is limited in what you can do with it, and it mainly involves reworking the legs. If you used the LMPs in the construction, you can easily remove the existing legs, reshape them, or make new pairs. Try much shorter legs, longer legs, legs that are high in the back and low in the front. Try legs that have "knees" at the feet—in other words, rounded tips on the ends (whereas ours are straight wire) with tubing where the rounded knee meets the ground. Try rubber on the feet, little pieces of sand paper, poster putty, anything that might afford more traction (and of course, look cool).

THE ABSOLUTE MINIMUM

We don't know about you, but our hands are really dirty and we've got an animated coat hanger scuttling across our desk. Here are a few things to take away from this experience:

- You can design ingenious machines with minimal mechanics and electronics if you're really smart. Barring that, you can find plans for them on the Net, like we did with our cool one-motor walker.

- The breadboard has given "rise" (okay, that's just sad—we apologize) to many an electronics project. Knowing how to use this basic circuit designing/testing tool is essential to good robot building.

- The servo motor is an extremely useful type of actuating/drive train technology that can be used in many different ways to create either back and forth (walking) motion or continuous rotation.

- Bug-like walking machines, even ones with no sensors or real brains, exhibit a type of persistence and motion that is extremely lifelike.

- Building stuff is fun!

IN THIS CHAPTER

- We liberate a work-a-day computer mouse from its tethered desktop bondage

- Dremel tools whir and plastic flies as we do some *real* hardware hacking (bring your goggles!)

- We take freeforming to the next level, building our circuit right onto the bottom of the computer case

- Finally, we have a robot that can behave and misbehave—hijinx ensue

8

PROJECT 2: MOUSEY THE JUNKBOT

The Fine Art of Making "Frankenmice"

For our second building project, we're going to make a *real* robot, while still using very simple, conventional electronics (we'll get all twenty-first–century digital in Chapter 9, "Project 3: Building a DiscRover"). This robot, shown in Figure 8.1, continues with our techno-junk recycling mission, making use of an old computer mouse and commonly found parts such as a 5V relay. If you crack open that old analog modem you haven't used since you traded up to a broadband connection, you'll likely find the relay you're looking for in there.

FIGURE 8.1

He's alive! Your old mouse reborn as a robot.

Some 45 years after Grey Walter's experiments with his robot tortoises (see Chapter 1, "What's in a Name?"), our robot is basically the same kind of critter, using many similar (although more modern) components. Walter's mechanical life forms used light sensing/avoiding as their basic behavior set. Our Mousey the Junkbot works in a similar fashion. Walter used electromagnetic relays as switches to trigger a change in the robot's behavior. We'll use a relay as well. Like his bots, ours also has a mechanical bump sensor that, when triggered, initiates a *runaway* sequence—in our case, making the robot back away from the direction it was heading.

Our robot is not only called Mousey because its body is a computer mouse, but because its behavior is also somewhat mouse-like. The little DC motors that we'll use are very "torquey," creating a lively little critter that really scoots around on the floor. And when Mousey crashes into anything, it suddenly becomes...well...mousey, speeding off in the opposite direction for a few seconds.

> **tip**
>
> As with anything built from instructions, it's a good idea to read through this chapter carefully before you begin building. When you're done with the read-through, also make sure to check the book's Web site (www.streettech.com/robotbook) for any instruction clarifications, bug fixes, and so forth. You can also ask questions there if you have them.

The brains behind Mousey are the most ingenious part of this little robot hack. The creator of the first robot of this kind, Randy Sargent (see Note), used an audio *operational amplifier* (or *op amp*), a little 8-pin chip designed to boost signals in answering machines, speakerphones, and home intercoms, to read (and respond to) signals from light sensors. Besides the op amp, a 5V relay, two light sensors, and two motors, there's little else inside this bot (which is what allows us to fit it inside of a mouse case). The most unruly component is the 9V battery, but we managed to fit that in the case as well.

Gathering the Parts

Several of the key parts used in this project can be found inside of the computer mouse you will cannibalize. You should be able to find additional parts if you have a decent techno-junk pile to sift through. You'll still likely need to purchase some components from your local electronics store, online retailer, or some other source.

The Parts List

Mousey the Junkbot requires the following parts (which are shown in Figure 8.2):

- ■ (1) junked computer mouse (or new el cheapo one)
- ■ (2) small DC motors (highly recommended: Solarbotics Part #RM1)
- ■ (1) double-pole/double-throw (DP/DT) 5V relay (Solarbotics Part #RE1)
- ■ (1) LM386 audio operational amplifier (Solarbotics Part #LM386)
- ■ (2) light sensors (taken from mouse)

note

The principal design behind Mousey the Junkbot is not ours. The original such robot was called Herbie and was built by MIT's Randy Sargent. It was actually a line follower (not a light follower) and was entered into the 1996 Seattle Robotics Society's Robotohon (it came in last place, by the way). Many variations of the design have followed. Our Mousey is only a slight variation of the Herbie that can be found in our esteemed tech editor Dave Hrynkiw's book, *Junkbots, Bugbots, and Bots on Wheels: Building Simple Robots with BEAM Technology* (ISBN 0-0722-2601-3).

note

If you're not sure how light sensors are used in line-following robots, turn back to the "Sensors" section in Chapter 4, "Robot Anatomy Class."

- (1) SP/ST toggle switch (Solarbotics Part #SWT2)
- (1) SP/ST touch switch (taken from mouse)
- (1) 9V battery snap
- (1) 9V battery
- (1) 2N3904 or PN2222 NPN-type transistor (Solarbotics Part #TR3904/TR2222)
- (1) 1kΩ to 20kΩ resistor
- (1) 1kΩ resistor
- (1) 10µF to 100µF electrolytic capacitor
- (1) Light-Emitting Diode (LED)
- (2) spools of 22- to 24-guage stranded hook-up wire (one black, one red)
- (4) 6 1/2-inch pieces of 22-guage solid hook-up wire (two red, two black)
- (1) wide rubber band (or flexible LEGO tubing)
- (1) small piece of scrap plastic (about 1/4 inch × 2 1/2 inch)
- (1) small piece of Velcro or two-way tape (optional)

> **tip**
>
> Solarbotics (www.solarbotics.com) offers convenient parts bundles for the projects in this book. See its ad in the back. Besides these bundles, if you mention, "Gareth says I get a deal!" when you order, you will receive a *one-time* 10% discount off the price of any order worth over $75USD ($100CND). This discount does not apply to existing specials, solar cells, or parts bundles not for this book. For other parts sources, see Chapter 10, "Robot Hardware and Software."

The Tools and Supplies List

To construct Mousey, you'll need these tools:

- Rotary (Dremel) tool (required for this project)
- Cut-off wheel for Dremel tool (a piercing or jeweler's saw works well too)
- Needlenose pliers
- Hobby knife
- Wire cutters
- Screwdriver set
- Soldering iron and related soldering tools and supplies
- Breadboard and hook-up wire
- Superglue

- Poster putty
- Recycled component leads
- Ruler, electrical tape, white glue, cellophane tape

FIGURE 8.2

All the parts needed for Mousey the Junkbot.

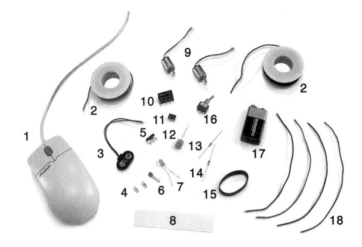

1. Sacrificial computer mouse
2. 22-guage stranded hook-up wire
3. 9V battery snap
4. IR light sensors
5. SP/ST touch switch
6. 2N3904 (or PN2222) transistor
7. LED
8. Plastic piece
9. 2 DC motors
10. 5V DT/DT relay
11. LM386 audio op amp
12. 10μF to 10μF electrolytic capacitor
13. 1kΩ to 20kΩ resistor
14. 1kΩ resistor
15. Rubber band
16. SP/ST toggle switch
17. 9V battery
18. 22-guage solid hook-up wire

Building the Body

Unlike our walker project, we need to build the body of our mousebot first. We're going to freeform its circuitry, as in the last project, but we're going to do this inside of the mouse case. So, we'll prep the case, install the motors, breadboard the circuit, and then install and solder up all of the internal components. Sound like a plan?

tip

If you're unsure about any of these tools or what they do, refer back to Chapter 5, "The Right Tools for the Job."

Alien Mouse Autopsy

The first thing you'll need to do is find a computer mouse to use. It's very likely that if you've been in the digital game for awhile, you have an old mouse (or two) gathering dust someplace. If you're still using a mechanical mouse on your computer (as opposed to an optical one), you need to join the twenty-first century anyway, so order yourself up an optical mouse, yank out that creaky old mechanical model, and there's your robot-in-waiting. If you don't have access to an old mouse (don't forget to ask friends and col-

caution

If you're a minor and are about to undertake this project, make sure you have your parents' permission and guidance. C'mon, you know Mom's just dyin' to build some bots!

leagues), you can buy a brand-new, super-cheap model for under $12. For our mousebot, we used an old (still working) Kensington Mouse-in-a-Box Scroll model. It's a nice choice because it's big enough to fit all of our robot components inside of it. Making sure there's enough room in the case is something you're going to have to be certain of before you unholster your Dremel tool and start eviscerating your mouse. Unscrew the mouse case and eyeball the placement of all of the parts. The main thing you need to be certain of is that the 9V battery and the two DC motors fit inside the case (and that you can still close it), as shown in Figure 8.3.

FIGURE 8.3

Rough layout of what parts need to fit where inside of your mouse case.

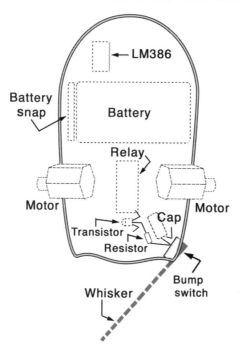

EVENTUAL PARTS PLACEMENT

After you've found a suitable candidate, it's time to remove unwanted innards from the mouse case. There are actually parts we want to keep for use in our robot, so carefully remove all of the mechanics and electronics inside. You should be able to unhook the mouse cable (from its plug-type connector), pop out the scroll wheel (if it has one), and then pry out the PCB. Set all of these parts aside while you work on the mouse case itself.

With all of the above parts removed, what you should have left is the plastic mounts for the two encoder wheels (which are used to translate movement of the mouse ball to cursor movement on your screen), mounts for the scroll wheel (if applicable), and the screw post(s) (which attaches the case top to the bottom). Using your Dremel tool (and a cut-off wheel), carefully remove everything but the screw post, as shown in Figure 8.4.

> **tip**
>
> If you're having trouble finding the screw(s) that hold your mouse together, look under the label, the little nylon feet, or nylon tape strips on the bottom of the mouse. They're likely hidden there. Save the nylon bits so that you can put them back on the mouse after you're finished with the build. They help move the mouse smoothly across the floor.

FIGURE 8.4

Our mouse with all of its innards removed.

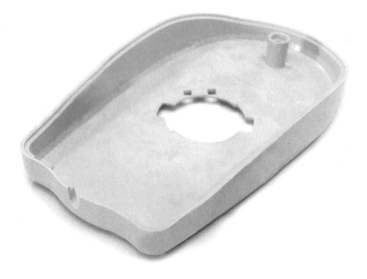

After you have the bottom part of the mouse body cleared out, flip over the top of the mouse case and have a look. It too is likely to have a lot of plastic structure you don't need. Zip all of it off with the Dremel (except the top screw post). You might want to leave on all of the material that holds the two mouse buttons in place. Also, if there are little plastic tabs along the side that help keep the top attached to the bottom, leave those as well.

caution

Cutting away all of this plastic is a bit messy. Work on newspaper or something else disposable. Also, you must wear safety goggles. It's also not a bad idea to wear a paper dust mask too. Plastic dust particles are nasty.

Motor and Switch Placement

Now we're ready to figure out where we want the motors to go and we'll need to cut openings for them in the sides. This project introduces a little hardware-hacking improvisation. Nearly every model of mouse is slightly different in shape, size, and with slightly different internal components. So, your mouse might be different from ours. You're going to have to think through your build and try to work around specific challenges that your mouse design presents. For instance, if you're using a Microsoft IntelliMouse, it has a body that curves to one side. You'll have to make sure that you install your motors perpendicular to the centerline of the body so that the bot can move in a straight line. For the IntelliMouse, you'll also likely have to remove the screw shaft in the back so that you can mount the battery behind the motors.

After you've figured out where the battery will sit, and where the motors should be installed, you're ready to cut the openings for the motors. We *highly* recommend the Solarbotics RM1 motors. They are extremely small, lively, and dirt cheap. These motors are actually the Mabuchi FF-030-PN, so if you can find a pair of these in toys you already have—great. If not, look for them at your local hobby/electronics shop. You obviously can use other DC motors, but they have to be small enough to fit inside the mouse case without horrendous amounts of cutting, and they need to be torquey enough to move your mouse around after all of its innards are assembled.

tip

You can pretty much turn any mouse into our Mousey the Junkbot, but the bigger and more symmetrical the mouse case is, the easier the build will be. Appropriately enough, the common analog Microsoft mouse (with its curved body) is the most difficult to work with.

Keep cutting and test fitting your motors until they can rest comfortably in the case with the lid closed. Taking a cue from BEAM robotics, we're going to use the drive shafts/gears of the motor themselves as our wheels. To do this, we need to angle the motors coming out of the mouse body so that they're at about a 60-degree angle. Make sure the top still fits completely when the motors are in place (see Figure 8.5).

When you're confident you have the right motor placement, you can use superglue (or epoxy, if you prefer) to glue your motors in place. If you'd like to test them out first, you can use poster putty

tip

If you'd like to learn more about DC motors, and even see a dissected version of the motor we're using in this project, check out www.solarbotics.net/ starting/200111_dcmotor/ 200111_dcmotor2.html.

to hold them in place. With the motors temporarily attached, place the mouse on the tabletop, get down at eye level with the table, and make sure that the gears on the motor (our "wheels") are making good, level contact with the table. With this done, you can go ahead and glue them in place.

FIGURE 8.5

Motor placement (and angle) and switch placement are very important for making Mousey work properly.

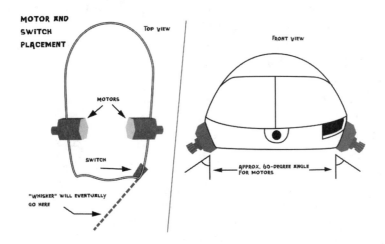

The next thing we need to do is to make an opening in the case bottom for our bump switch. Our mousebot is going to have one giant "whisker" across its front that, when bumped, triggers its mousey, scuttle-away behavior. We'll actually use one of the tiny switches found in our computer mouse for this bump switch. All you need to do is find one of the button switches on the mouse's PCB and desolder it (jump ahead to Figure 8.9). It'll be on the board, under where the buttons were on the top of the mouse, and will look like a teeny plastic box with a (usually) white or gray cylinder sticking out of it (the switch itself).

When you have this removed, test fit it to the left side of your mouse's front end. You want to place it so that, when you attach a long, rectangular piece of plastic across it, it will cover the front end of the bot (refer back to Figure 8.5). Again, placement might have to be different on your robot, depending on the design of your mouse case. The idea is for the whisker to cover as much "real estate" in front of the mouse as possible, so a bump anywhere along the length of the whisker will trigger the switch. When you're confident about the placement, you can cut the opening in the mouse case bottom where the switch will go.

> ## caution
>
> Make sure your motors are totally level and in equal contact with a flat surface when they're finally glued. If one motor makes more contact than the other, your mouse will spin around in circles and you'll have to gouge out the motors and re-glue them (which isn't easy after all of the electronics are installed).

You also can go ahead and cut a 1/4-inch × 2 1/2-inch piece of plastic that will form the bumper. As with the walker project, you can use scrap plastic, but this time, you'll have to be sure that it's very springy plastic. We got great results from a strip of .03 Plasticard stock, which we happened to have on hand. The plastic used in credit cards also works. So the next time a server brings you your maxed-out credit card shredded on a silver platter, save those plastic scraps!

The last mechanical item we need to attend to in the bottom half of the body is putting tires on the motors. This simply involves getting a rubber band the same width as the sprockets on the drive shafts, measuring out the necessary lengths (by wrapping a piece of the band around the sprocket, marking it, and cutting it), and then gluing the tires in place. We recommend using white glue for this bond. Make sure the rubber is glued firmly to the sprocket. The width of the rubber band should be no more than the sprocket, so if it's wider, you'll need to trim it with your hobby knife. For our Mousey tires, we used two pieces of the purple flexible tubing found in the LEGO MINDSTORMS kit (see Chapter 6, "Acquiring Mad Robot Skills"). It fit perfectly, and was just too tempting not to use. If you have the LEGO kit, try some of this tubing; if not, the rubber bands work fine (see Figure 8.6).

> ## tip
>
> We recommend not actually gluing the bump whisker in place until you've soldered on its control wires and components (during the circuit freeforming), as this will make soldering and installing it easier.

FIGURE 8.6

Finished motor installation, switch notch, and LEGO tubing tires shown with battery test-fit.

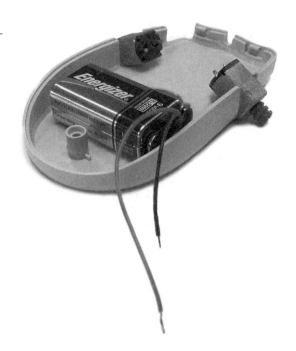

Installing the Control Switch

We have one last piece of major bodywork, and that's making a hole in the top of the mouse case for the power switch. We thought it might be cute (everyone say it together now: "Awwww") if we used a rather large toggle switch and placed it as far back on the mouse case top as possible to make it serve as a tail. The plastic post that accepts the screw from the bottom of the case is right where we *wanted* to put our tail, so we installed the switch in front of this post. If you'd rather the tail be placed in a more tail-appropriate spot, you can remove the screw post on your top and simply tape (or lightly glue) the top and bottom halves together when your robot is finished. To install the switch, all you need to do is drill a hole in the top big enough for the threaded bushing on the switch to fit through. Most switches come with two nuts on the bushing. You simply take one nut off, stick the bushing through the hole you've made in the top, and then tighten down the second nut on the outside of the case to attach your switch (see Figure 8.7). You're done!

caution

Don't put your toggle switch *too* far back on Mousey's behind or your robot might switch itself off when it's in full retreat from an obstacle and it backs into something.

Understanding Mousey's Brain

Where our coat hanger walker made use of the versatile and minimalist Bicore circuit (see Chapter 7, "Project 1: Coat Hanger Walker"), this robot uses a different, but similarly inventive, chip hack. The main component of this bot's control circuit is the LM386 audio operational amplifier. As mentioned previously, the 386 was designed to boost the signal in certain consumer electronic devices that use small speakers. It "listens" to two input signals and compares them. If one signal is lower than the other, the chip is designed to boost that signal to equalize the chip's output. Because these inputs don't have to be sound signals, the op amp can be used to compare other inputs (in our case, light values). If we hook up the output to two DC motors, we suddenly have a little brain that can read the input from two light sensors, compare them, and equalize the output, and use that signal to drive two motors. This gives us a robot that will follow a light source, all the while adjusting power to the two motors to keep itself heading toward that light source.

We then can add further lifelike behavior to Mousey with the addition of electronics that make it become pseudo-photophobic (in other words, it moves away from its light-seeking direction). Our bump whisker will trigger a relay switch that reverses the motors for a few seconds, making Mousey suddenly scuttle away from its previous course.

Figure 8.8 shows the circuit diagram for Mousey's brain. Study this diagram. It might look intimidating at first, but after you've spent some time with it, breadboarded the circuit, and then freeformed it inside the mouse, everything will start to make perfect sense.

A Brief Word About Mousey's Senses

We'll start building Mousey's control circuit in a second—first on the breadboard, and then inside of the mouse case. But before we do, we have to sort out what light-sensing technology we're going to use.

The cool thing about our LM386-based control circuit is that it's sensitive enough to read a wide range of light values. What this means is that we can use nearly any type of light-sensing component. If you have two photodiodes, photoresistors, phototransistors, or even ubiquitous light-emitting diodes (LEDs)—you likely have plenty of these in your techno-junk pile—you can use them as Mousey eyes. Cadmium-sulfide (CdS) sensors (as discussed in the "You Light Up My Cadmium-Sulfide Sensors" section of Chapter 5) also can be used. They're readily available at your local electronics shop.

But we don't have to wade through the aisles of cell phones, boom boxes, and R/C monster trucks at our local Radio Shack to buy the light sensors. What we need is right there on the PCB we pried from our mouse case: two infrared emitters.

note

Relays are nifty, ubiquitous electronic components. If you crack open many of the appliances in your home, you'll find relays lurking within them. The idea behind a relay is simple. A coil inside generates a magnetic field. This field is then used to attract the contacts of a switch (or several switches). So, it's sort of like, instead of you reaching into the circuit and throwing a mechanical switch, the circuit has a clever way of throwing switches by itself. I don't know about you, but when I think of relays and robots, I always think of Robbie the Robot from *Forbidden Planet*. He had about six giant, clacking logic relays visible in his clear-domed head, and somehow, these crude on/off switches (and however many more where stuffed into his body), were enough to give him high-order intelligence and Asimovian laws of behavior. Ah…if only it was that simple.

FIGURE 8.8

Circuit diagram
for Mousey's
brain. A larger
version of this
image is avail-
able at www.
streettech.com/
robotbook.

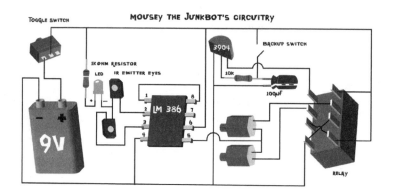

A mechanical computer mouse has an ingenious way
of translating the movement of a mouse's rubber ball
into X- and Y-coordinate movement on your screen.
Inside the mouse are two encoder wheels (basically
the same type of encoders we discussed in the
"Sensors" section of Chapter 4). As you move the
mouse around on the desk, it rotates these wheels
inside (one wheel being perpendicular to the other,
with their drive shafts in contact with the ball). The
encoder wheels have slits in them, and on either
side of these slits are infrared emitters and infrared
detectors. The detectors count the number of light
pulses reaching them (through the moving slits in
the encoder wheels) from the emitters, and then the
on-board chip translates these pulses into X/Y coor-
dinates that it sends to your computer screen (and
that's where your cursor can be found). You would

note

You'll have to use an analog
mouse for this project. The newer
optical mouse only has one red
LED. We need two light sensors.
Because the optical mouse is still
fairly new, you're also much more
likely to have a dead (or neglected)
analog mouse lying around.

think we'd want to use the detector, but it's too weak to be of use in our robot (we
don't want it to be *too* mousey!). We want to find and remove the IR emitters (also
known as IR phototransistors). On most mice, the emitters are clear plastic with a little dome protruding from them while the detectors are solid black. Find the clear
emitters and desolder them from the PCB. Congrats—you're now the proud possessor
of a pair of robot eyeballs (see Figure 8.9).

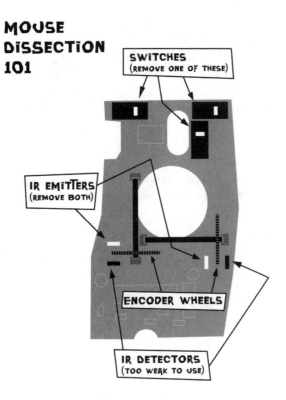

MOUSE DISSECTION 101

SWITCHES
(REMOVE ONE OF THESE)

IR EMITTERS
(REMOVE BOTH)

ENCODER WHEELS

IR DETECTORS
(TOO WEAK TO USE)

Breadboarding the Circuit

With the body of Mousey all prepped, it's time to build the control circuit. We'll breadboard it first, making sure it has the sensitivity and behaviors we're after, and then we'll freeform it right inside of our robot body.

Creating Eyestalks

Before we start hooking up the wiring on our breadboard, we need to give our Mousey eyes some optic nerves. Our computer mouse IR emitters only have two stubby little pins coming out of their packages. Our robot is going to have *eyestalks* that jut from the front of its body. These will not only look cool, but we can use them to adjust Mousey's sensitivity to light by moving the stalks backward and forward and from side to side.

Before we can hook the wires to the eyes, we need to find out which pin on each emitter is positive and which is negative. Time to crank on the ol' digital multimeter (DMM). Set it to Diode Check mode (consult your meter's manual if you're unsure how to do this). Hook one probe to one of the emitter pins and the other probe to the second pin. If the read-out is OL (which indicates no connection), reverse the probes. When connected correctly, you should get a reading of about 1V. Take note of which probe is on which pin. The red probe indicates the anode (or positive) pin and the black probe indicates the cathode (negative) pin. Also, most emitter packages have a 45-degree cut (or chamfer) on one of the top edges. This marks the cathode side. You should still check it with a DMM though because super-cheap parts are sometimes packaged incorrectly.

To create the stalks, cut four 6 1/2-inch pieces of 22-guage solid core hook-up wire. If you have red and black wire, cut two of each color. Solid core wire is better here than stranded wire because it'll give you much stiffer stalks that will hold the shapes you mold them into as you adjust light sensitivity.

Now we're going to do something strange. We're going to reverse our red and black wires. We're going to solder the red wire to the cathode (-) pins on the emitters and the black wires to the anode (+) pins. This is called *reversed biasing* in electronics. It's a technique that will help improve our light sensitivity even more.

When the wires are soldered in place, twist them together and strip some of the jacket off of the other ends (to fit into the tie points of your breadboard). The stalks are so tall because they're going to stick out of the mouse about 4 inches when installed in our actual robot, and the negative wires will need to reach our control chip in the bottom of the mouse (see Figure 8.10).

tip

If you don't have a Diode Check feature on your DMM, set the meter to Volts and attach the two needle probes. If the read-out shows a positive voltage of about 0.6V, the red probe is connected to the positive (anode) emitter pin. If it reads *negative* 0.6V (or thereabouts), the red probe is connected to the negative (cathode) pin.

tip

As in Chapter 7's project, you'll want to use your third hand tool (or a house member who owes you favors) to hold parts while you hold the iron and the solder.

FIGURE 8.10
Our finished
eyestalks, ready
to shed some
light on our
control circuit.

Hooking Up the Op Amp

With all of your electronic components in hand,
you're ready to breadboard the circuit. Here are the
steps to installing the op amp, our main control circuit (each step is illustrated in Figure 8.11):

1. The first thing you'll need to do is install the
   LM386 across the trench on your board. As
   always with ICs, the pin to the immediate
   left of the little dimple (as you hold the chip
   up so that the dimple is on "top" of the
   chip) is Pin 1. Count in a "U" shape down
   the left side (in this case, Pins 1–4), across
   the IC, and then up (Pins 5–8).

2. Now you'll want to connect Pins 1 and 8
   with a piece of hook-up wire. These two pins
   were designed (with the IC wearing its original op amp hat) as gain controllers. By connecting them together, we're boosting the
   gain (in other words, increasing the sensitiv-
   ity of the signals coming from our computer mouse's IR emitters).

We're using our trusty, horizontally
oriented breadboard (with power
buses on the top and bottom) and
installing our chips so that they
face to the left (in other words,
the dimples that indicate the #1
pins are facing to the left side of
the board). Keep this in mind as
we refer to wire and component
placement. Your breadboard might
be laid out differently.

3. The next components to connect are the eyestalks. Take a black wire from
   one of the stalks and put it in a tie-point group for Pin 2 and the other black
   wire in a tie point for Pin 3. Pins 2 and 3 are the op amp's input pins. Now
   take the two red wires for the stalks and plug both of them into the same tie-
   point group someplace above the 386 chip (about five to six tie-point groups
   above the chip). They're on their way to the upper positive power bus, but
   we're going to add some additional electronics before we get there.

4. We found in our experiments using the IR emitters from mice that they're not as sensitive to light as they could be. In *Junkbots, Bugbots, and Bots on Wheels,* Dave Hrynkiw includes a sensitivity-boosting subcircuit that BEAM uber-hacker Wilf Rigter (`wilf.solarbotics.net`) came up with for Herbie-based bots. We decided to add these components, which are nothing more than an LED and a 1kΩ resistor. Adding this booster on the breadboard simply involves plugging the negative (cathode) lead of an LED into the same 5-point group on the board where you have the two red leads for the eyes plugged in, and then jumping the trench and plugging the positive LED lead into the corresponding 5-point group on the other side of the trench. In that same 5-point group, plug in one lead from your 1kΩ resistor and then plug the remaining lead into the positive tie point of your upper power bus.

5. Now to finish up this part of our circuit, all we have to do is connect the power pin of the LM386 (Pin 6) to our upper positive power bus, and the ground (Pin 4) to the lower negative bus.

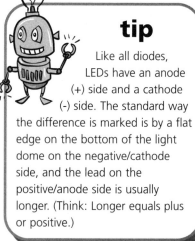

tip

If you're sayin' to yourself right now, "What in blue blazes is *the trench*?" or you otherwise need a refresher course in breadboard fundamentals and terminology, see the "Thumbnail Guide to Breadboards" in Chapter 7.

tip

Like all diodes, LEDs have an anode (+) side and a cathode (-) side. The standard way the difference is marked is by a flat edge on the bottom of the light dome on the negative/cathode side, and the lead on the positive/anode side is usually longer. (Think: Longer equals plus or positive.)

We could just leave Mousey with this much intelligence, hook up the motors (see the following Note), and be done with it. But our robot would be a one-hit wonder. It would only persistently pursue a light source until its battery died (or we toggled off its tail). This would make for a rather boring bot. As we discussed in Part I, "Robot.edu," it's only when a simple robot has a few behaviors (or more) that are triggered by interactions with its environment that we start to see lifelike, even emergent, behaviors. So we're going to give Mousey the Junkbot a little fear reflex.

Creating the Runaway Circuit

To create the circuit for the runaway subroutine, we
basically need one of the switches that we pulled
from the mouse PCB and a 5V DP/DT relay. When
our mouse whisker gets bumped and its switch is
triggered, our runaway circuit switches on the relay
for a set period of time (which we'll discuss in a second), our motors reverse direction, and the robot
backs away from the direction it was heading—in
other words, from the light source it was moving
toward.

Although the relay is the electromagnetic switch
that, when tripped, makes Mousey throw it hard
into reverse, how long the reverse action occurs is
determined by a set of components that together
create an on-board timer. These components consist
of a capacitor, which holds the charge that will
drive the motors in reverse; a resistor, which will
control the rate at which this charge leaves the
capacitor; and a transistor, which will act as a
switch, passing control from the light-following circuitry to the runaway circuitry and back again
(when the capacitor has drained its charge). You
can play around with different cap and resistor values on the breadboard until you find the amount of
reverse motor juice you want. Try capacitors in the 1kΩ to 20kΩ range and resistors
in the 10µF to 100µF range. In both cases, the bigger the value, the longer the discharge time. You want a combination that gives you the backup time you desire. For

note

If you want to create a one-
behavior bot (light following), or
just want to try hooking up the
motors to the LM386 control circuit before we add the circuitry for
the runaway subroutine, here's
how to connect the motors: The
positive wire from one motor
(marked with a silver dimple on
the Solarbotics RM1) connects to
Pin 5 (the output pin) on the chip,
and the negative wire goes to a
negative bus on the board, and on
the other motor, it's reversed. (The
negative wire goes into Pin 5 and
the positive wire goes into a positive bus on the board.)

our Mousey, we used a 10kΩ resistor and a 100µF capacitor. This gave us about 8 seconds of backup time. Because our Solarbotics DC motors are so zippy, Mousey can cover a fair amount of real estate in that time.

Here's the sequence for attaching wires and components to the breadboard to create the runaway circuit (see Figure 8.12 for an illustration of the completed steps):

1. Plug in the relay about six tie-point groups to the right of the LM386, making sure that the dimple on the relay is facing in the same direction as on the 386 (in other words, to the left).

2. Cross a hook-up wire from Pin 8 to Pin 11 and another wire from Pin 6 to Pin 9. It is these two wires, when the relay is engaged, that will reverse the motor connections.

3. Plug in your capacitor so that the positive lead goes into a hole in the tie-point group *above* Pin 1 of the relay and the negative lead goes to the bottom negative power bus. (On electrolytic capacitors, the negative lead is usually marked in some way, often with a stripe and/or a - symbol.) What we're doing here is starting to install our timer subcircuit between the main control circuit and the runaway relay.

note

When referring to the tie-point holes on either side of the relay, we're pretending the relay has 16 pins when it actually only has 8 (4 on each side). But it would be too confusing to say, "tie-point hole 8, counting down from Pin 1, and so forth," so after you plug in the relay, for breadboarding purposes, pretend it has 8 pins on a side corresponding to the 8 tie-point groups on either side.

4. Now connect your resistor in the same tie-point group as the positive lead from the capacitor and jump the trench to the tie-point group directly on the other side of the board, right above the relay.

5. Take your transistor and splay out the three pins on it so that they each will fit in a tie-point hole. Plug it into the breadboard above the relay with the flat side facing toward the trench and position it so that the center pin (the positive pin) is in the same tie-point group as your resistor lead. This puts a negative pin in the tie-point group to the left of this center positive pin and one negative pin to the right of the positive pin. This left-most pin should then be in the tie-point group for Pin 16 of our relay.

6. We now need to hook up the switch itself to our reversing circuit. You don't actually need to attach the touch switch itself. If you're lazy like us (and

confident your cannibalized mouse switch works), all you have to do is plug in a couple of pieces of same-length hook-up wire. The first wire goes into the same bottom tie-point group as the resistor and capacitor. You *must* install this wire in a tie hole *between* the resistor and the cap. The other wire goes to the bottom positive power bus. Bend the tips of the wires so that they can touch, but keep them separated until you want to test the switch. To engage your breadboard switch, all you have to do is touch the two wires together. If you do want to use your actual switch, solder two wires to the poles on the switch and plug the wires into the tie holes indicated previously.

7. The last thing we need to do is connect the power and ground wires on the relay. First connect a wire from Pin 8 on the relay to the bottom positive power bus. Connect another wire from Pin 1 to the bottom positive power bus. Now connect Pin 9 to the top negative bus. And last, but not least, connect the emitter pin on the transistor (the left-most pin with the flat side of the transistor facing the trench) to the top negative power bus.

note

As briefly discussed in Chapter 6, there are two main types of discrete component transistors: PNP (positive/negative/positive) and NPN (negative/positive/negative). The 2N3904/2N2222 transistor used in this project is an NPN type. This refers to the polarity of the three pins that protrude from the transistor package. The three pins are called the *emitter*, the *base*, and the *collector*. If we use the analogy of a switch (because that's how a transistor is frequently used), the collector and emitter are like the poles of the switch and the base is the electronic finger that engages the switch and connects the poles (completing the circuit).

That's it. Look over your cool robot brain! Let's flip the board sideways (so that the left side of the board is now the top). This way, we see a logical layout of our circuitry. At the top, we have the eye sensors and the eye sensitivity booster subcircuit. In the center we have our control chip, and bringing up the rear (appropriately enough), we have the motor-reversing subcircuit with timer components and relay switch.

Connecting the Motors and Power

After you've checked all of the preceding steps to make sure that all components have been properly installed on the board, you're ready to connect the motors and power and see how you did.

FIGURE 8.12

Our breadboard
with control
chip, timer, and
relay circuits
installed. A
larger version
of this image
is available
at www.
streettech.com/
robotbook.

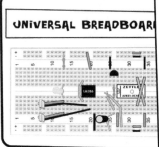

1. Connect the positive wire from one motor to Pin 5 of the LM386 chip. On many
   motors, the positive terminal is marked in some way, often with a dimple or with
   a plus sign (+). Our Solarbotics RM1 motors have silver dimples next to the positive terminals. (Er…and they have a red wire attached to the positive terminal
   and a black wire attached to the negative terminal. That's always a solid clue.)
   Connect the negative motor wire from this first motor to Pin 4 on the relay.

2. Connect the negative wire from the other motor to Pin 5 of the LM386 chip.
   Then connect the positive wire to Pin 13 on the relay. (Notice that the polari-
   ties are reversed from step 1.)

3. Everything should be hooked up by now and we should be ready to power
   our circuit. Because the robot itself will use a 9V battery, that's the power you
   should deliver to the board. You can use the 9V battery snap you bought for
   this project, or you might already have created a special snap with banana
   plugs for your breadboard work (you geek!). That's what we did (being *total*
   geeks…and proud of it!).

Your completed breadboard circuit should look similar to Figure 8.13.

FIGURE 8.13

Finished bread-
board circuit
with motors and
power attached.

So, did it work? If so, congratulations! Get yourself a flashlight and start having fun moving the light beam closer and farther away from Mousey's light sensors. The motors should change speed as the light gets closer to one sensor than the other. Then, when you touch the two wires together that form our makeshift switch (or when you engage the actual switch, if you've attached it to the board), you should hear a distinct "click" as the relay engages, and then the motors should reverse direction for a few seconds (the time determined by your cap/resistor combo) before reverting back to forward motion. If all did *not* go well, see the following trouble-shooting section.

Troubleshooting the Breadboard

If you deliver power to the board and nothing happens, you'll have to put on your propeller beanie to do a little troubleshooting:

- As always with breadboarded circuits, you'll want to first check to make sure that everything's where it should be. Especially check to make sure the capacitor, resistors, and transistor are in the correct tie-point holes. Also make sure the positive and negative wires are going to the correct positive and negative holes on the power bus.

- "She needs more power, captain!" If you hear Scottie yelling this from down in engineering, it might be that only one of the motors is spinning, which means not enough dilithium crystals…er…volts are being delivered to the board. Check your battery with your digital multimeter to make sure you're getting enough juice. Start with a fresh 9V battery so that you know you're delivering all the power you can. Check the power bus itself with your DMM to make sure that the connections from the battery to the binding post, and then to the power bus, are all good.

- Our junkbot's computer mouse emitter eyes should be powerful enough to react to light, especially because we added the sensitivity booster circuit. If the robot doesn't whine like crazy when you shine light in its eyes, check to make sure that your solder joins are healthy on the eyestalks, and that they're plugged into the correct holes. If you're still unhappy with the sensor reaction, or getting no reaction at all, you might want to swap out the eyeballs for a new set. Radio Shack (or an officemate's unattended mouse) to the rescue!

Freeforming Mousey's Control Circuit

Now that we have a light-hungry robot brain, we need to install it in our mouse body so that it can motor around to feed (add your own zombie/*Night of the Living Dead* sound effects here). Obviously, all of the hook-ups will be the same as on the breadboard, but here we'll want to switch to a lighter-gauge and/or a stranded wire. The 22-guage solid core wire used in most breadboard jumper kits is too stiff for most of the connections we'll need to make inside of our mouse case. It makes it too hard to close the lid and puts unwanted stress on our solder joins. Either try a stranded 22-guage wire or a 24-guage (stranded or solid) wire. We used a 22-guage stranded wired and were happy with the results.

Before you start soldering the parts together, test fit all of them inside your case (refer back to Figure 8.3 to see where we placed our components). You'll need to improvise depending on your case design, but you need to find room for the bump switch (which you should have a notch cut for already), the LM386, the relay, and the cap/resistor/ transistor combo. And, of course, the biggest component: the battery. This is the item to test fit first, and then all of the other components can be fit around it. We'll also be installing the resistor/LED sensitivity booster circuit, but this will go along the inside of the mouse-case top from the toggle-switch tail to the eyestalks in front.

Installing the Battery and Relay

When you're happy with the basic placement, heat up the ol' *fire stick* (or *soldering iron*) and follow these building steps for the battery and the relay (the two biggest components):

> **tip**
>
> When figuring out where to place the parts, don't forget to put the top on your mouse to make sure the components still fit with the lid on! Also, keep in mind that you'll have solder connections from the pins on the relay and IC, and lots of wire inside of the case. You'll need to consider the "headroom" for all of this stuff as you plan your layout.

1. Using two-way tape, Velcro tape, or poster putty, install the battery where you want it to go. You don't want to permanently glue the battery in place 'cause you'll want to replace it when your Mousey gets that Alkaline-poor, run-down feeling.

2. Before you install the relay, you might as well solder what you can to it while it's still outside of the mouse case. This makes getting to the pins on it a bit easier. Turn it over onto its back with its pins in the air (also known as "dead bug mode"). Cut short lengths of wire (solid core is best for this) that will connect the switch pins on the relay (the four pins close to each other at the bottom). Solder them in an *X* configuration as shown in Figure 8.14.

3. Solder the emitter pin of your NPN transistor (that's the right-most pin looking at the transistor with the flat side facing you) to the top-left coil pin on the relay. We'll be soldering on the resistor for our timer circuit to the base

pin (the center pin) later on. Finally, on the collector pin (the left-most pin), solder on a length of wire (preferably black, as this is a negative connection). The length depends on the placement of your other components, but about 3–4 inches is probably plenty. This wire will eventually connect to the negative pin (Pin 4) on the LM386 IC. You can always clip off any excess (see Figure 8.14).

4. There are a few more relay hook-ups we can do "out of body." First, solder a short positive (red) wire from the top pin on the right side of the relay to the bottom pin on the right side. (We're connecting our two power pins together here.) Next, solder a negative wire, about 2 inches long, onto the bottom-left pin of the relay. And finally, solder a positive wire, about 3 inches long, onto the bottom-most right pin. These wires are eventually headed to our positive and negative pins on the control chip (see Figure 8.14).

tip

Throughout this project, we're going to be recommending red wire for positive connections and black wire for negative. You don't *have* to do this, but it really does help, especially when learning electronics—to see what's what, polarity-wise. It's also very handy when troubleshooting. Also, keeping everything neat and color-coded makes it look better when you're showing your robot off to everybody: "Hey look at these cool mouse guts I stitched together myself!" [Cue maniacal mad scientist laughter.]

FIGURE 8.14
"Out-of-body" connections on our relay.

RELAY SOLDERING

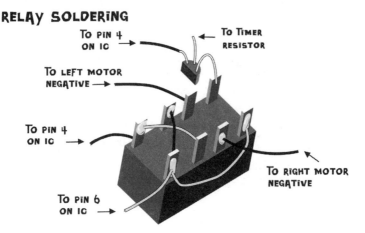

To pin 4 on IC →

← To Timer Resistor

To left motor negative →

To pin 4 on IC →

To Right Motor Negative

To pin 6 on IC →

5. Now you're ready to glue the relay in place and solder its remaining connections. We glued ours right between the two motors (refer to Figure 8.3). Use superglue and glue it in place (in dead bug mode, obviously), and allow it to dry before soldering anything else onto it.

6. Solder the negative wire from the left motor onto the middle pin on the left side. Again, use black wire if you have it. Then solder the negative wire from the right motor onto the right middle pin on the relay (refer to Figure 8.14).

Connecting the Switch Components

One thing that makes placing the relay close to the front (and to the bump switch) a good idea is that we can attach the lead of the timer resistor directly to the capacitor (without need of wires), and then the cap directly to the pole of the bump switch. Again, before we install the switch, we can attach components to it "out of body" for easier soldering (for steps 1–4, see Figure 8.15 for some visual help).

1. Before soldering the timing capacitor to the left-most switch pole, go ahead and solder a 3- to 4-inch negative wire to the negative lead of the capacitor (which should be marked with a stripe or negative symbol).

2. Now solder the positive lead of the cap onto the left-most pole of the bump switch.

3. Solder the timer resistor to the same left bump switch pole as the positive cap lead.

4. Connect a short length of red wire (about 1 inch) to the middle bump switch pin.

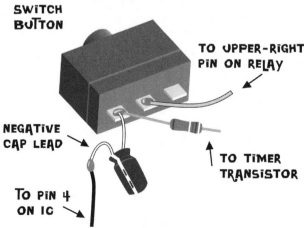

FIGURE 8.15

"Out-of-body" connections on the bump whisker switch.

TOUCH SWITCH CONNECTIONS

SWITCH BUTTON

TO UPPER-RIGHT PIN ON RELAY

NEGATIVE CAP LEAD

TO TIMER TRANSISTOR

TO PIN 4 ON IC

5. Install the switch itself into your mouse body (in the notch you cut earlier). Again, superglue to the rescue!

6. Solder your timer resistor onto the center (base) pin on your transistor.

7. Take the wire you attached to the switch in step 4 and solder it to the right-top pin of the relay (which you've already jumpered with a wire to the bottom-right pin).

Your bump switch/runaway internals are all hooked up and ready to make Mousey afraid—very, very afraid.

Our Motors Are Feeling Powerless

Before we move away from the relay/switch/motor area, we have a few more things we need to do.

1. Solder about 2 inches of positive wire to each of the positive motor terminals (if the motor doesn't have wires already).

2. Solder the stripped ends of these two wires together side-by-side.

3. Finally, solder a third positive wire, about 3 inches long, onto the soldered end of the two motor wires you created in steps 1 and 2 (see Figure 8.16). What we're doing here is making the two positive motor wires into one positive wire that we'll attach to the output pin on our control chip.

note

One of the most frustrating parts of soldering, especially for beginners, is soldering a component or wire onto a pin or lead that already has something soldered to it. Can you guess why? That's right, Billy. Heat melts solder. So you've got to heat the new piece and solder it in place without melting the solder on the existing components. There's no real trick to it—it's just a question of practice and, ironically, having a good, hot iron; a well-tinned tip; and being quick on the draw, or in this case, the withdraw. Practice on other junk components until you work the kinks out of this procedure. It's also always best to make a "mechanical connection" (in other words, twist wires together) if possible, so that they'll stay put when you solder on additional wires.

Installing Our LM386 Control Chip

Find the spot in your mouse case where you decided to install the LM386. You'll want to position it in dead bug mode (with its pins in the air). It doesn't matter how the chip is oriented. Ours has pins 1 and 8 facing towards the robot's rear.

1. Before you glue in your chip, go ahead and bend Pins 1 and 8 toward each other and solder them together.

2. Glue your LM386 IC in place.

FIGURE 8.16

Our positive
motor wire
junction. A
larger version
of this image
is available
at www.
streettech.com/
robotbook.

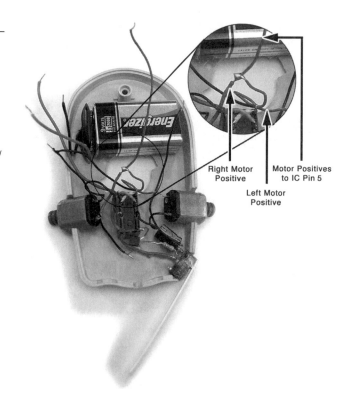

Right Motor
Positive

Motor Positives
to IC Pin 5

Left Motor
Positive

3. Now find the negative wires from the transistor (attached in step 3 of "Installing the Battery and Relay"), the relay (attached in step 4 of "Installing the Battery and Relay"), and the timer cap (attached in step 1 of "Connecting the Switch Components"). Solder all of these negative wires together, side by side. Here we're joining all of the negative wires together on their way to the control circuit and to power.

4. Solder the negative wire from the battery snap onto the 3-wire negative junction you joined previously in step 3.

5. Solder a short piece of negative wire (about 1 inch) to Pin 4 of the 386. Then solder this wire to the uber-negative wire junction created in steps 3 and 4.

6. Solder the positive wire from the relay (attached in step 4 of "Installing the Battery and Relay") to Pin 6 on the chip (our output pin). Solder the motor junction wire to Pin 5 on the IC.

FIGURE 8.17

The LM386 control chip wired and ready for input/output. A larger version of this image is available at www.streettech.com/robotbook.

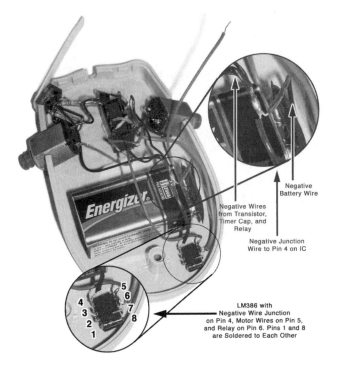

Negative Battery Wire

Negative Wires from Transistor, Timer Cap, and Relay

Negative Junction Wire to Pin 4 on IC

LM386 with Negative Wire Junction on Pin 4, Motor Wires on Pin 5, and Relay on Pin 6. Pins 1 and 8 are Soldered to Each Other

You're now done making all of the connections for the bottom half of our robot. Now we'll make the connections in the top half of the mouse case, install the eyestalk sensor, and then attach the wiring from the top half to the bottom half.

Mousey Gets a Pair of Eyes and a Wag in Its Tail

Set aside that gnarly tangle of wire and components we're calling the bottom half of our bot and let's work on Mousey's lid for a moment.

Installing the Eyestalks

As you install the eyestalks, following the steps listed here, refer to Figure 8.18 for some visual cues.

1. The first thing we want to do is make holes in our mouse top to thread the eyestalks through. The two buttons on most computer mice are separate pieces of plastic that snap onto the mouse's top half and rock a little forward and backward (so that they can engage the actual switches on the PCB underneath). You'll probably want to glue these two pieces onto the rest of the top so that you have a solid upper half. When the glue is dry, mark

where you want to sink your eyestalk holes, and drill out just enough of a hole (using your Dremel tool and a drill bit) to feed your eyestalk wires through. We installed our eyestalks about 1/2 inch from the front edge of the mouse case.

2. Thread the eyestalks through the holes so that about 1 3/4 inches of wire and sensor remains outside of the case. On the inside, clip the two positive wires so that they just reach each other on the underside of the top and overlap each other a bit. Solder them together.

3. Run the negative wires along the top of the case and bend them down wherever your IC is located (ours is in the back of the mouse's bottom half). Don't solder them to the IC just yet. We want to keep the two mouse halves unconnected until the end.

> **tip**
>
> When drilling holes, it's always a good idea to start a pilot hole so that the bit doesn't travel across your material when you go in for the kill…er…drill. We just used the tip of a hobby knife and twisted it in a circle. If you want to, you can even completely "drill" the holes with such a razor knife, but we like the precision of the Dremel.

4. Cut a short piece of red wire (about 1 inch) and strip the ends. On one end, solder on the 1kΩ resistor of our sensitivity booster subcircuit. On the other end of the resistor, solder on the anode side of the LED (with its leads splayed out). We want this wire/resistor/LED combo to fit from the middle pole of our toggle switch to the junction of the two positive eyestalk wires we soldered together previously in step 2.

5. Go ahead and connect the sensitivity booster from the middle switch pole to the positive solder join of the eyestalk wires.

6. Our LED/resistor combo is really only in our circuit to add a *voltage bias*, or in other words, to tweak the sensor output voltage so that it's more in line with what the chip's input is expecting. This will result in the motors turning on and off more gracefully, leading to smoother mouse motion. One cool side benefit to the booster is that it can double as a "power-on" indicator, so we want it showing through the top of the mouse case. Consequently, we're going to have to drill a hole in the top for the dome of the LED to poke through. With the booster circuit in place, you can see where that should be. Gently bend the LED out of the way (so that you don't drill into the top of it!) and sink your hole. Then poke the light up through it. Use a piece of electrical tape to hold the LED in place inside the hole you just cut.

FIGURE 8.18

FIGURE 8.18

Finished inside of mouse top with eyestalk placement, sensitivity booster, and power switch.

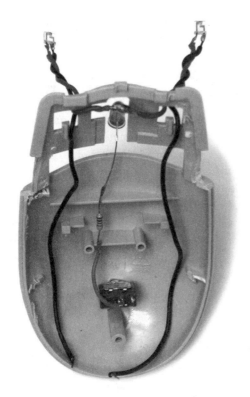

It's All About Connections, Baby

We almost got bot! All we have to do now is make the final connections between power, the switch, and the control chip, and to install our front whisker.

1. Solder the negative eyestalk wires to Pins 2 and 3 on the LM386.

2. Solder the positive (red) wire from the battery snap to the (normally open) outside pole of the toggle switch.

3. Solder a short red wire (about 1 inch) from the center pole of the switch to Pin 6 of the IC.

4. Connect your bumper plastic to the bumper switch. We simply used a couple of layers of cellophane tape. You want to install it so that it's connected to the left side of the switch front, so that it crosses over the face of the switch and across the little cylinder

note

Unfortunately, we can't show you a photo of these final steps because the wires are too short to open up the two halves of the mouse to see the connections. Hopefully, these connections will be clear enough at this point. You're probably such a solder-wieldin', circuit-hackin' fool by this time that you don't even need us anymore. Sniff...sniff.

that's the switch button itself. When you press in the plastic, you should be able to hear an itty-bitty click sound as the switch engages.

It's a Slightly Anxious, Light-Seeking Robot!

That's it! You're done (see Figure 8.19). All you have to do now is connect the battery snap to the 9V battery, screw (or tape) the two mouse halves back together, put Mousey down on the floor, and switch it on. If it takes off like a bat outta hell, congratulations—you've turned your lowly computer mouse into an autonomous, light-seeking robo-critter. Read the following section, "Playing with Mousey," for tips on terrorizing house pets and other cool things you can do with your new robot. If Mousey doesn't move, or travels in circles, or goes forward, but doesn't back up (or other frustrating misbehaviors), see the "Troubleshooting a Wayward Mousey" section later in this chapter.

FIGURE 8.19
Congratulations! It's a baby robot that likes to eat sunshine.

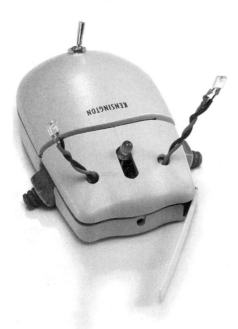

Playing with Mousey

If all went well, how Mousey the Junkbot works should be readily apparent as soon as you flip its tail. The robot should motor off, and eventually, hone in on the brightest area of light in the room. It's best if you make the lighting high contrast.

In other words, it's best if you turn off all lights (close window shades and so forth) and just leave one light on, or leave one sun-soaked window open. Here are some other fun experiments:

- Put Mousey in the hallway and close all doors except one. Make that room as light saturated as possible. Try orienting Mousey in different directions and see if it eventually scuttles into the lit room.

- Use a flashlight as a lure to make Mousey follow you around.

- Use the stiffness of the eyestalks to "tune" the light reception strength. Move the stalks farther apart, closer together, one bent forward, one back, and so on, until you get the "steering" you're looking for.

- Use the flashlight controller and Mousey to drive pets insane! Be careful 'cause agitated pets will attack your little robot and try to rip it to its "discrete components."

Troubleshooting a Wayward Mousey

If you turn on Mousey and nothing happens (cue annoying sound effect, "Wha-wha-WHAAAA"), or if it acts strange (like chasing its own toggle switch), turn it off immediately. Something has gone wrong during the build. Here are a few things to check out:

- The first thing you always want to do (if nothing is happening) is ask yourself the tech support alpha question: "Is it plugged in?" Make sure that the battery is good (new), the battery snap is well seated, and the positive and negative wires from the battery are properly connected.

- Also (if nothing is happening), make sure wires (the bare metal parts or solder joins) are not making contact with other wires or joins they have no business touching (as in positive wires touching negative ones). A clue for such a short circuit is to touch the battery. If it's warm (after you just switched it off), that's a sign of a shorted connection someplace. If you suspect any of this kind of hanky-panky, wrap some electrical tape around the exposed wires.

- The next thing to try to revive a dead Mousey is to go back to the "Freeforming Mousey's Control Circuit" section and carefully double check all of the solder connections. Besides being in the right place, they should all be fat and healthy looking—plump, shiny joins. If any look suspicious, resolder them.

- If you turn on Mousey and it starts spinning around frantically in a circle, it's a sign that you hooked up the motors incorrectly. If Mousey chases its own tail in a clockwise direction, its right motor is running forward and its left motor is running backward. You'll need to reverse the wires on the left motor. If it drives counterclockwise, reverse the right motor connections.

- If you do get a circular motion, before you switch motor wires, make sure that the motors are, in fact, spinning in opposite directions. If they're spinning in the same direction and the mouse is still circling, it might just be that the motors are not level with each other and one is getting more traction than the other. Break the glue bond and reglue the motor so that the same amount of rubber (band) meets the road.

- If you turn on Mousey and it kicks into reverse with a vengeance, you likely have both motors wired incorrectly (dyslexics of the world, unite!). Swap the wiring on both motors.

Further Experiments

There are a number of things you can do to experiment with Mousey. You can obviously try different types of light sensors (photoresistor eyes, such as cadmium-sulfide sensors, and photodiode eyes). You also can connect a trimpot (a variable resistor) to the eyes so that you can better "tune" their sensitivity. You also can add a subcircuit that will enable you to create two proper touch-sensing whiskers on Mousey, instead of the rather un-mouselike front bumper.

These hardware hacks are too much to detail here, but you can find out more about them (and other cool hacks) in Dave Hrynkiw's *Junkbots, Bugbots, and Bots on Wheels* book. Also look for schematics and installation instructions for additional Mousey hacks on our book's Web page (www.streettech.com/robotbook).

tip

If you like the idea of turning computer parts into robots, you don't have to stop at mice. Fran Golden of Goodwin College has plans for building a robot almost entirely from a floppy disk drive (remember those?). Complete plans, pictures, video, and more can be found at ohmslaw.com.leech.dk/ robot.htm.

tip

To find other ideas for hacking your Mousey, and other bots built around the LM386 op amp, do a google.com search on "robot +LM386," "herbie +LM386," and "'Randy Sargent' +robot."

THE ABSOLUTE MINIMUM

With this project, we learned that it really doesn't take much to create an honest-to-goodness robot, with sensors, multiple behaviors, unpredictability, and so forth. Other points worth taking away:

- Techno-junk really is the roboticist's friend. You can transform that drawer full of dead Walkmans, cassette players, modems, and mice into cool robo-critters with just a few extra parts, a soldering iron, and the skills to use them.

- Continuing in the grand cyberpunky tradition of "the street finds its own uses for things," we took a lowly chip originally designed to boost the speaker on your home answering machine and we turned it into a robot brain.

- We also made use of another commonly found electronic component, the 5V relay. Relays are a natty way of creating switches powered by electromagnetic energy. Here we used the relay to make the motors of our robot kick into reverse.

- You didn't learn this in the chapter, but you'll learn it with your completed Mousey: Little robots make great conversation pieces. (Parts to build a robot out of a computer mouse? About $10. Soldering iron and stand? About $15. Watching your co-workers on their hands and knees chasing a tiny robot around the office? Priceless.)

PROJECT 3: BUILDING A DiscROVER

AOL Just Got a Whole Lot Smarter

In this, our final project, we raise the complexity level a little bit, and also increase the versatility of the bot we're building. The other two projects were fairly limited in what we could do to expand upon when we were finished. Because this project is basically a miniature development platform like the ones we discussed in Chapter 4, "Robot Anatomy Class," it offers all sorts of experimental possibilities. This is mainly because we've added a reprogrammable computer component, a *microcontroller unit (MCU)*. Our walker and junkbot were hardware-only critters. To alter them, we had to add, remove, or change hardware components. With the DiscRover, we have an onboard microcontroller that enables us to create programs that control the robot's behavior (see Figure 9.1). All sorts of different sensor systems can then be added and programmed.

The idea for this project came to me during a nap (we old people need naps). I've always wondered what sorts of useful things one could make out of those "Free Trial!" CD-ROMs, sent through the mail by Internet providers on a seemingly daily basis. In bed, it hit me: Turn them into mini robot development platforms! They're perfect. They're plentiful, plastic (easy to drill into, glue on, and so forth), and round—they even have built-in cable pass-thru holes in the center! Using widely available aluminum standoffs, you could stack several layers of discs on top of each other to hold mechanics, power supplies, MCUs, sensors—the works! I was so excited; I bolted (oops...unintended robot-builder pun) out of bed

tip

As with anything built from instructions, it's a good idea to read through this chapter carefully before you begin building. When you're done with the read-through, also make sure to check the book's Web site (www. streettech.com/robotbook) for any instruction clarifications, bug fixes, and so forth. You can also ask questions there if you have them.

to start writing up some specs. About 10 minutes into it, something else hit me. This was *way* too obvious. Somebody must have already thought of this idea (because many a geek ponders the "What to do with all those dang discs?" conundrum). I stopped my sketching and did a Web search. After some creative Googling, I found the robot taxonomy I was looking for: "CDBots." Sure enough, people have used junk CDs to make similar platform robots, they've used them as wheels for bots (just add rubber band tires), and they've even hot-glued dozens of them together, fanned out into large multidisc platform levels. I was slightly crestfallen that this "big idea" wasn't mine, but glad to see that people had finally found something useful to do with the ol' *plastic spam* (as these snail-mail-borne discs are sometimes called).

This project serves as your gateway to future experimentation, and future robots. Like any respectable robot development platform, it provides a basis for new hardware add-ons and new software controls. We'll build the mechanics for the platform, hook up our two controllers (the microcontroller unit and the motor controller), build the first sensor system, and then create our first control program (to run the motors and link motor control to sensor input). From there, we'll point you to excellent resources that show you how to add additional sensors and sensor-control software.

note

We only named our second robot project, Mousey the Junkbot, but think about naming all of your robots. It might sound corny, but it really does add an extra dimension to your "relationship" with your bots. As I was writing this chapter, the Robot Store (www.robotstore.com) sent me its latest offering, the OctoBot Survivor—a self-charging robot—to test out. The bot it sent was named Clarke. My son and I thought it was silly at first, but it actually added to the experience as we updated each other on what sort of trouble Clarke was currently getting himself into: "Clarke's bashing his brains out on the dining room table leg again."

Gathering the Parts

You'll be able to scavenge a few parts for this project (such as the promo CDs that will comprise its structure), but unlike the first two projects, there are a few significant components you'll need to purchase (namely the MCU and the motor controller).

The Parts List

The parts you need to build your DiscRover are listed here and shown in Figure 9.2.

- ■ (2) AOL CD-ROMs (or other junk CDs)
- ■ (2) Hacked servo motors (Solarbotics Part #GM4)

- (1) OOPic-R microcontroller starter kit
- (1) OOPic compiler program (free download, Windows PC required)
- (1) OOPic programming cable (comes with OOPic-R kit)
- (1) Pololu micro dual serial motor controller
- (1) 20-pin IC socket
- (1) 9V snap (comes with OOPic-R kit)
- (1) battery holder for four AAA batteries
- (2) bump switches (taken from computer mouse)
- (2) approximately 1/4-inch × 3 1/8-inch pieces of .03 plastic stock
- (1) toggle power switch
- (1) spool of 22-guage stranded hook-up wire
- (1) set of 65mm plastic wheels (Solarbotics Part #GMPW)
- (4) 1 1/2-inch 4-40 aluminum hex standoffs (male/female type)
- (6) 3/4-inch 4-40 aluminum hex standoffs (male/female type)
- (4) small sheet-metal screws
- A pile of 4-40 screws (preferably Phillips head), washers, and hex nuts (at least 15 of each)
- (6) female crimp pins (designed for 1/4-inch headers)

tip

Solarbotics (www. solarbotics.com) offers convenient parts bundles for the projects in this book. See its ad in the back. Besides these bundles, if you mention, "Gareth says I get a deal!" when you order, you will receive a *one-time* 10% discount off the price of any order worth over $75USD ($100CND). This discount does not apply to existing specials, solar cells, or parts bundles not for this book. For other parts sources, see Chapter 10, "Robot Hardware and Software."

note

We need two servo-type motors hacked (electronics and mechanical stop removed) for continuous rotation. The Solarbotics #GM4 comes already hacked. If you use different servo motors, you'll need to remove the electronics and clip off the stop on the motor's final gear. See the "Hacking the Servo Motor" section of Chapter 7, "Project 1: Coat Hanger Walker," for details.

FIGURE 9.2

All the parts you'll need to build your DiscRover.

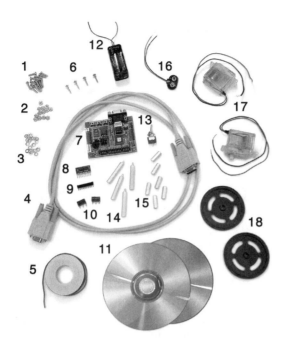

1. 4-40 screws
2. 4-40 hex nuts
3. Washers
4. OOPic programming cable
5. Stranded hook-up wire
6. Servo motor screws
7. OOPic-R microcontroller
8. Pololu motor controller
9. 20-pin IC socket
10. Bump switches
11. Junk CDs
12. AAA battery holder
13. SP/ST power switch
14. 1 1/2-inch hex standoffs
15. 3/4-inch hex standoffs
16. 9V battery snap
17. Motors
18. Wheels

The Tools and Supplies List

Along with the basic parts necessary to build the DiscRover, you'll also need the following tools:

- Needlenose pliers
- Hobby knife
- Wire cutters
- Screwdriver set
- Soldering iron and related soldering tools and supplies
- Breadboard and hook-up wire
- Superglue

tip

Screws are measured by a screw size (commonly #0 through #12) and a threads-per-inch count. So, in our case, a 4-40 screw is a #4-size screw with a 40-thread count. 4-40 screws are a common size for hobbyists and robot builders and can be found at many electronics stores and hobby shops.

- Poster putty
- Rotary (Dremel) tool (optional, but highly recommended)
- Metal ruler
- Electrical tape
- (1) 9V battery
- (4) AAA batteries

tip

If you're unsure about any of these tools or what they do, refer to Chapter 5, "The Right Tools for the Job."

The Brains Behind the Bot

Our previous two robots have been more brawn than brain. Our DiscRover is the exact opposite. It's little more than two motors, two wheels, and two plastic discs. Most of the action in this project is in the microcontroller brain (and the supporting motor controller). To gain a full appreciation for what this robot can do, let's first discuss the controllers we'll be using.

caution

If you're a minor and are about to undertake this project, make sure you have your parents' permission and guidance. C'mon, you know Mom's just dyin' to build some bots!

OOPic: The "Hardware Object"

The microcontroller module we'll use on this project is called the *OOPic* (pronounced *EW-pik*). A *PIC* is a *programmable integrated circuit* or, in other words, an IC that you can download programs into. The *OO* in *OOPic* stands for *object-oriented* (programming).

Object-oriented programming (OOP) is a fascinating and increasingly popular form of programming that treats computer code as discrete objects. Conventional forms of coding have the program functions (operating instructions) in one place and the data someplace else. OOP combines everything an object needs to function (all of its operating instructions and its data) within the object code itself, creating a discrete software entity. This way, it becomes like a physical component, a "part" that

note

The OOPic has a number of different objects already created for talking to popular motor controllers (we'll get into motor controllers in a second). More can be found on the Internet.

you can "plug in" to a program to give the program that added functionality. So, if we want to power a DC motor (or more accurately, a DC motor controller) that's attached to our OOPic controller, all we have to do is find (or create) a motor control object, tell the program to use it (in a statement list at the beginning of the program, called a *Dimension* or *Dim*), define its procedures (what we want our program to do with the object), and we have a motor controller routine.

The OOPic is a unique microcontroller because it is the first to use an object-oriented language to control it. Its developer, Scott Savage, has done a great job of creating a hardware/software package that is relatively easy to use, but has incredible power and versatility. He's even extended the object-oriented paradigm, blurring the lines between soft objects (software code) and hard objects (MCUs and what they're attached to). Whereas OOP tries to get programmers to think in terms of code as something like a real-world object, Scott wants you to also think about electronic components as being virtual entities. The OOPic makes use of *virtual circuits*, a method for linking objects together in a similar way that you might wire together real-world electronic components. By using such virtual circuits, you can have processes running in the background while the main program process is being executed. We won't use or get into virtual circuits in this project, but you'll really want to explore (www.oopic.com/virtcirc.htm) this fascinating aspect of OOPic programming after you've mastered the basics.

The OOPic module we use here is the latest, the OOPic-R starter kit. It sells for $89 and comes with the OOPic module, the 9V battery snap, and the serial programming cable to attach the module to your PC (for transferring programs). The OOPic multi-language compiler program is available as a free download from the OOPic Web site (www.oopic.com). A version of the OOPic-R without the serial cable and battery snap is available for $79. One of the great things about the OOPic-R is that it's a serious little computer with lots of potential. You'll be able to add all sorts of functionality to your DiscRover, and when you're building your next reprogrammable

tip

To get your digits wet with OOPic programming, before you hook your microcontroller to your robot, go to www.oopic.com/hardobj.htm and click on "oDio1 Object" and then "Your First OOPic Application." This is a simple project to build an LED flasher, program it, and download its control software into the OOPic module. You'll need a red LED and a 220Ω resistor for the hardware.

tip

After you've downloaded the OOPic compiler program, launch it, go to the Help menu, and choose Update Help Files. This downloads the most recent manual pages to the Manual directory in your OOPic folder.

bot, you can just unscrew the OOPic-R from this robot and slot it into another one. Have hardware object, will travel!

Figure 9.3 shows many of the features of the OOPic-R. We won't get into too many of them here because they are outside the scope of this beginner's project. This MCU has an excellent manual that downloads with the compiler program and lots more resources can be found on the OOPic Web site. For now, we'll point out only the components of interest to us for the basic operation of the DiscRover.

FIGURE 9.3

The heavily populated OOPic-R module.

OOPIC MICROCONTROLLER

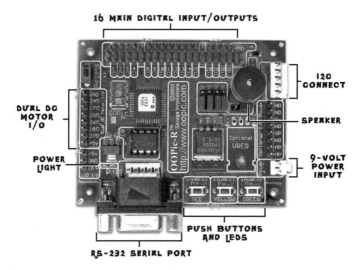

- **Main input/output lines**—The OOPic-R has 16 main bidirectional I/O pins arranged in three groups. The first four pins have analog input capability (see *A/D and D/A converters* in the "Thumbnail Guide to Microcontrollers" in this chapter). The second group of pins (8–15) includes a built-in resistor for connecting hardware directly to the module without the need for adding external resistors. Pins 16–31 have built-in Schmidt Triggers, a type of signal conditioning circuit (for dealing with "noisy" signals), when they're set to input mode.

- **Dual DC motor I/O lines**—This I/O group has two lines (Pins 17 and 18) offering *Pulse-Width Modulation (PWM)* and four other digital motor control lines. This area on the board also offers a 5V power supply and ground. We're going to use one of the main I/O pins for our serial motor controller output, but we will use the 5V power supply and ground pin on this pin group. The OOPic has a number of existing motor control objects. Which one you use depends on what motor controller you go with. Because our controller is serial, we used one of the available serial objects (called oSerialx).

- **EAC (EEPROM Activity Check/Power Good)**—This tiny little LED (marked *EAC* on the module) tells you the status of the EEPROM (see "Thumbnail Guide to Microcontrollers") clock. If this light is on, it means that the microcontroller chip is getting enough power to drive the EEPROM's clock, which means that everything's getting enough juice to run your robot.

- **RS232 port**—The OOPic-R has a serial port so that you can hook it up to the serial port on a PC for downloading programs and controlling the MCU. This port and the built-in Serial Control Protocol enables any device with a serial connection (PC, PocketPC, Palm Pilot) to send programs, debug programs, provide tethered control, and so forth.

- **Push-button switches and LEDs**—We won't use these in our project, but the module includes three programmable buttons and LED lights that can be controlled via the oButton object.

- **Speaker**—Warning tones and other sounds can be programmed into your apps using the oFreq object.

- **I2C network connector**—This 5-pin connector lets you hook the OOPic up to other OOPic modules, computers, or other devices using the I2C networking standard (common in embedded systems). This port also can be used to attach additional devices such as a precision servo motor controller, driver for an LCD screen, voice synthesizer, A/D or D/A converters, and other hardware.

Motorhead

As we discussed in the "Controllers" section of Chapter 4, many robots not only have a central controller (an MCU), but additional controllers for motors and/or power management. Our DiscRover makes use of such a secondary board—in this case, a motor controller. Obviously, the most common motor controller is an on/off switch. But when you want to get more sophisticated than that, such as controlling speed and direction of a motor (make that *two* motors), you need a motor controller. This motor-devoted chip bridges the gap between actuators and the microcontroller unit, devoting itself to juggling motor control signals while the MCU wrestles with other tasks. The most common motor control configuration is called an *H-bridge*. This refers to the circuit layout that usually shows the motor in the center and four switches arranged on either side of it. The speed with which these switches are opened and closed determines how much juice the motors get (which, in turn, determines the motor speed) in a process called *Pulse-Width Modulation*. To change the direction of the motor, the direction of current flow is changed. All of this is the motor controller's job.

There are dozens of commercially available motor controllers out there, but we chose the Pololu micro dual serial motor controller (see Figure 9.4). Some advantages of this unit are that it's extremely small (measuring only .90 inch × .45 inch), affordable ($23), and comes preassembled (many motor controllers are sold as kits). It'll fit nicely on the bottom platform of our discbot.

FIGURE 9.4

The Pololu micro dual serial motor controller.

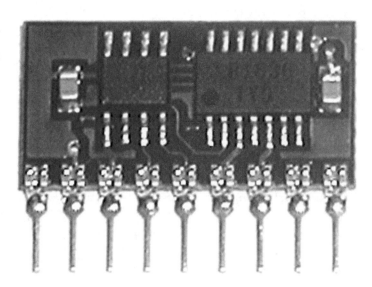

Our hacked servo motors for the DiscRover can't use the same power supply as our OOPic's 9V battery. The motor controller has two power pins, a motor power supply pin, and a logic supply pin. The motor pin can handle up to 9V, but our motors can't. (Actually, they have been pushed to 9V without getting cooked, but they're more comfortable at around 6V.) The logic pin needs between 3 and 5.5V. What we'll do is use a separate 6V supply (four AAA batteries) as our motor power source. To power the logic pin, we'll connect it to the +5V output pin on the OOPic, and then power the OOPic itself via the 9V battery cable that it comes with.

Because we want to be able to fit the motor battery pack between the motors, we'll use a four-battery AAA holder with two batteries on top and two on the bottom (as opposed to a four-battery holder in which all the batteries are side-by-side). The two-over/two-under pack fits nicely between our motors.

note

Astute readers might be wondering why we can't just run our motors off of the +5V output pin on the OOPic. The problem is power spikes. As the motors turn on or motors change directions, you can get power spikes, which cause the controller to reset (interrupting motor operation). It's usually a safer bet to use a dual power supply for motors and microcontrollers.

Thumbnail Guide to Microcontrollers

note

The heart (the soul? the brain?) of our DiscRover is a little computer called a *microcontroller unit (MCU)*. We discussed these miniature wonders a bit in Chapter 4, but because you're going to be spending a lot of time with one in this project, we thought we'd expound a little. We like to expound.

Like a full-blown microcomputer, a microcontroller has a central processing unit (CPU), fixed memory (ROM), Random-Access Memory (RAM), input/output (I/O) pins, an internal clock, and so forth, but unlike a microcomputer, an MCU is designed to fit all of the preceding onto a single chip, or a chip and a very small circuit board (called a *module*). Also unlike a desktop computer, which is designed to receive ongoing input from a person, run multiple programs, and perform varied tasks, an MCU is designed to do one thing over and over again with no, or limited, human input. MCUs are often called *embedded processors* or *embedded systems* and can be found in increasing numbers of consumer devices, from your car to your coffee maker.

Microcontrollers have been a boon to robotics because they enable builders to load sophisticated control programs (created on a desktop computer) into even the tiniest robots. These MCUs can have dozens of sensor inputs and actuator outputs, state timers, analog/digital converters, multiple connection ports, and other goodies all on a module no bigger than a Post-It Note.

People often think that the word *microcontroller* and *microcomputer* are synonymous. They aren't. They're both computers, but the microcontroller is designed for a specific, limited purpose, and with miniaturization in mind. The goal is to "bake" everything, or nearly everything, into the chip, so that it can fit in your phone, microwave oven, or whatever. By contrast, a microcomputer, like the one I'm typing this on, is a bloated slob that, in addition to the CPU, has many microcontrollers on the motherboard to handle various functions. As the old saying goes, things like to expand to fill the available space. This is actually happening with microcontrollers too. To add functionality, MCU manufacturers are starting to build more and more features into the MCU modules that require support controllers and other hardware.

Let's look at a few components usually found on the typical MCU:

■ **Central processing unit (CPU)**—As with any computer, the CPU is the traffic controller for its own internal processes and handling input and output duties. Microcontrollers are usually built with specific functions in mind and will often have tools for these functions hardwired into them. For instance, MCUs designed with motor actuation in mind (as in robot drive trains) often have special motor controller connectors built into them.

■ **Memory**—Most popular MCUs use Flash memory, which enables you to read and write to this memory space over and over again. Like all EEPROM (electrically erasable programmable read-only memory), Flash memory retains the data in it even when the component is powered down. The Flash memory space is where the programs and permanent data are stored. As with your desktop computer, RAM is where all of the temporary data (such as sensor inputs) are stored. The size of the EEPROM or Flash limits the size of the programs you can run on your MCU. The amount of RAM limits the amount of temporary data the controller can handle.

■ **Clocks**—Most MCUs have a clock (also known as a *crystal oscillator*) on the module. Some also have clocks built into the chip. These clocks affect how fast the MCU can execute instructions. Clock speeds are measured, as with microcomputers, in MHz. On MCUs, clock speeds from 4MHz up to 75MHz are common. As with desktop computers, clock speeds can be deceiving. Different MCUs execute instructions in different ways, so a raw comparison of speed is not always the best way to go. Also, a robot you build is not likely to be doing a lot of multiprocess juggling (like a desktop computer does), so your MCU doesn't have to be a speed demon to be effective.

■ **Input**—Physical pins on the MCU module enable you to plug sensors into the controller. In buying an MCU, you always want to make sure you have enough inputs available for the sensors you'll be using (and a few more for future additions). Most MCUs have bidirectional I/O pins that can be programmed for input or output use.

■ **Output**—If there are fixed input and output pins on the MCU (as opposed to bidirectional I/O), there are usually fewer outputs available. Some modules have both regular output pins and pins specifically designed for motor control (with motor control hardware and/or software built in).

■ **A/D and D/A converters**—On an MCU, most I/O is read as digital. This means it

> **tip**
>
> When buying an MCU for a robot project, apply the K.I.S.S. rule (see Chapter 3, "Robot's Rules of Order"). You want to make sure the unit has enough memory, I/O pins, and so forth for your application, but there's no need to go overboard (unless it's a bot with expansion plans in mind). Figure out what your needs are, add a little room for growth, and then shop accordingly. Don't be tempted to get a module that can "do it all," unless you have plans for doing just that. And remember: More functionality = more. Don't spend $100 on an MCU if a $5 chip can do the trick.

will input or output either 0V (in other words, "0," "off," or "low") or 5V (in other words "1," "on," or "high"). If your sensor produces an analog value, it needs to be converted to a digital value. This is done with an A/D converter. Having an A/D converter on-board the module saves you having to add one as an additional piece of hardware. A D/A does the same thing, but in reverse. It takes a digital (on/off) signal and coverts it to an analog output.

- **Ports**—You can't do much with an MCU if you can't talk to it. Microcontrollers use a number of different interfacing methods, from standard parallel and serial cables to more specialized interfaces such as I2C (Inter-Integrated Circuit). I2C is a networking protocol, developed by Philips, which is very common in embedded systems. On the OOPic, and other I2C-equipped MCUs such as the Acroname BrainStem, you can connect multiple MCUs (up to 127 of them), allowing them to talk to each other. I2C is also commonly found in X-10 systems, the network technology commonly found in home automation.

For more information on microcontrollers, check out the Microcontroller FAQ at `www.faqs.org/faqs/microcontroller-faq`.

Trash Recycling Through Bot Building

Now that we know the basic control concepts behind our DiscRover, it's time to put our robot body together, screw down our robot brain, load the control program into it, and start botin' around. First, let's tackle the main structure.

DiscRover's Body

The body of our bot is nothing more than two junk CD-ROMs or CDs you don't want anymore (may we suggest Jewel's last offering?). To make things a lot easier for you, we've created two templates for the CDs you'll need (the bottom and top platforms). You can photocopy these, cut them out, and then use them as a guide to drill the holes as marked. See Figures 9.5 and, later in this chapter, Figure 9.9.

tip

The full-size CD templates for this project are available for download at this book's Web site (www.streettech.com/robotbook).

Building the Bottom Half

To build the first layer of your discbot, attach a printed and cut-out copy of the bottom DiscRover template (see Figure 9.5) to a junk CD. You first need to decide whether you want the printed side of the CD to be upright or not on your finished robot. It adds to the joke of using AOL CDs (or other plastic spam) to build bots if the printed side is showing, but it also looks rather tacky (in our humble opinion). We oriented our discs silver-side up, so the steps that follow will reflect that. When you decide which side should face up, attach the template *to the opposite side*, orient the CD template-side up, and follow these steps:

tip

When it comes to junk CDs/CD-ROMs, older is better. CDs from the '80s and early '90s were made with a much stronger plastic than today's discs, which shatter easily. So dig deep in your closets and attics for those Lionel Ritchie, Cher, and Styx CDs. Come on, you know *Domo Origato, Mr Roboto* is just dying to be made into an actual roboto!

1. Drill the four support holes marked for the 1.50-inch standoffs. These will be used to attach the two CD levels. Use a Dremel tool (or other drill) and a bit big enough to make a hole for a 4-40 screw. Be careful when drilling. CDs are thin, brittle, and can easily split apart. Take your time. You might also want to use the tip of your hobby knife to twist in pilot holes before you drill. This also helps in twisting the paper of the template out of the way so that it doesn't get bound in the drill bit.

FIGURE 9.5

The assembly template for the bottom DiscRover platform. CD-sized templates are available to download from the book's web site.

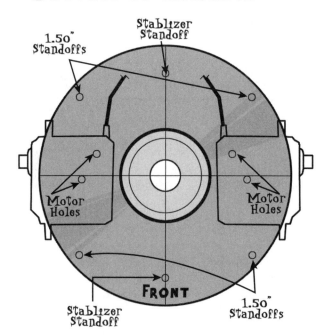

BOTTOM CD TEMPLATE

2. Drill the two holes for the .750-inch under-body stabilizer standoffs. One of these is located in the front of the bot and one in the back to create four points of contact with the floor (the two standoffs and the two drive wheels).

3. Drill the four holes in the CD for anchoring the two hacked servo motors, as indicated on the template. We could have fabricated fancy brackets to hold our motors in place, or superglued them, but we decided—because most of the case on the Solarbotics GM4 motor is empty (and 'cause it's clear, so we can see what's what inside)—we'd just drill right into them to attach to our bottom platform. The trick is in the placement, which we've figured out for you and have indicated on the template.

> ## caution
>
> Obviously, you want to wear goggles or safety glasses while drilling and you want to have something underneath the CD to drill into, such as a piece of wood.

4. After you've finished step 3, place one of the servo motors (as indicated on the marked template front-to-back center line) *on top* of the CD. In other words, the motor is on the top side, the CD is underneath the motor, and the template is underneath the CD. You'll obviously have to eyeball this a bit to see where the motor should be in relationship to the lines indicated on the template underneath. Make sure the motor shafts are lined up with each other (as indicated on the marked template line) and perpendicular to the bot's side-to-side centerline (also marked on the template). When you're confident you have the motor in the right place, stick your hobby knife through one of the motor holes you drilled through the template into the CD. Be sure to hold the motor case firmly against the CD while you twist the knife around to start a pilot hole in the motor case itself.

5. Take the motor away from the CD and carefully drill the two holes you've marked on the servo motor case with the tip of your knife.

6. Do the same procedure as in steps 4 and 5 on the second motor case.

7. Remove the template and screw your two servo motors in place.

8. Place the AAA battery holder in the center of the bottom platform. Affix it to the disc with poster putty. Make sure the power wires for the holder are jutting out of the back.

9. Using Figure 9.6 as a guide, solder the motor wires, power wires, and the two output wires onto the motor controller. Note that the negative battery wire needs to be split so that it eventually goes to both Pin 2 on the motor controller and the ground pin on the OOPic's motor I/O group. Don't bother with

that yet. Just make sure you solder on enough 22-guage stranded wire on the motor controller Pins 2–5 so that they can reach the OOPic when it's mounted on the top disc (about 2 1/2 inches above the bottom platform). We'll be feeding all of the wires from the bottom level through the hole in the center of the top CD, so measure wire accordingly.

10. Take four 1.50-inch standoffs and poke the threaded male ends through the holes you made in the CD. Bolt them on with 4-40 bolts from underneath. These are the standoffs that attach the bottom platform to the top.

11. Tread your two plastic wheels by stretching the rubber-band tires that came with them. Put a 4-40 washer on either side of a wheel, and using the screws that came with your servo horn pack, mount the wheels to the servo motors.

tip

If you don't want to solder directly onto the pins of the motor controller module (we sure didn't), you can use an IC socket. Just take a regular 20-pin socket (like the one we used for the Coat Hanger Walker Bicore) and cut it in half (with a razor knife or wire snips). You end up with two 10-pin sockets. Solder the wires as shown in Figure 9.6, onto 9 adjoining socket pins. If you mess up, no biggie—just use the other 10-pin socket. When done, plug in the motor controller.

FIGURE 9.6

Hook-ups for the motor controller. Consult the Pololu manual for more info.

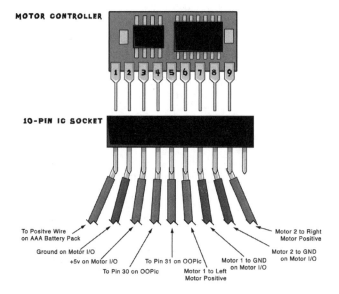

MOTOR CONTROLLER WIRING

MOTOR CONTROLLER

1 2 3 4 5 6 7 8 9

10-PIN IC SOCKET

To Positve Wire on AAA Battery Pack

Ground on Motor I/O

+5v on Motor I/O

To Pin 30 on OOPic

To Pin 31 on OOPic

Motor 1 to Left Motor Positive

Motor 1 to GND on Motor I/O

Motor 2 to GND on Motor I/O

Motor 2 to Right Motor Positive

12. The last thing we need to do on the bottom platform is to attach the two stabilizer standoffs that go underneath. Poke the male ends through the holes in the bottom platform and bolt them in place. The .750-inch standoffs will not reach the floor by themselves. Thread on a washer (or two) and a 4-40 screw and tighten the screw until the stabilizer is just touching the floor (but not raising the wheels off the floor at all). CDs are thin and fairly flexible, so you want to make sure you're not raising the disc up slightly in the center because your stabilizers are too tall. If you're nervous about your stabilizer screw heads scratching the floor, glue on little discs of junk plastic, nylon from computer mouse "feet," or some other smooth substance to the screw heads (see Figure 9.7).

FIGURE 9.7

Our DiscRover thus far, with motors/wheels, motor controller, battery holder, and standoffs. Note use of electrical tape to organize controller wires.

Building the Bump Sensors

There are dozens of different ways you can build bumpers, whiskers, and other switch sensors for the front of your rover. One disadvantage of some schemes using wire whiskers, often fashioned from thick guitar strings (gimme an "E!"), is that you can end up with sensor blind spots. These are places where an object might "get by" a whisker and the bump switch (and the resulting backup routine) will not be engaged. The design we're using is very simple and has only the tinniest of blind spots between the two switches. Here's how we made it:

1. Get two computer mouse touch switches. Most mice have at least three, so you should have two left on your mouse PCB from our last project, "Mousey the Junkbot," in Chapter 8.

2. Cut two rectangles .03 inch (or thereabouts) of plastic stock. Make the height of the strips the exact height of the switch (when it's lying on its side) and make them 3 1/8 inches long.

3. Solder two wires onto the switches (each about 6 inches long). Solder one wire onto the center pin and one onto the appropriate outer pin. On most mouse-device switches, this pin is marked with a C (for *common*).

4. Apply a thin line of superglue to the edge of the switch face on the opposite end of where the switch button itself is located (see Figure 9.8). Using a strip of electrical or other tape, attach the plastic strip so that it crosses from the glued edge across the button on the switch and protrudes from the opposite side. Attached in this fashion, the plastic strips should gently press against the buttons, but not enough to actually depress them unless the strip is moved (but don't test it until the glue is dry!).

note

If you look closely on most switches that have more than two posts on them, they are usually marked with tiny letters. These will likely say C (for *common*), N.O. (for *normally open*), and N.C. for (*normally closed*). Normally open means that power to the switch remains unconnected until it is depressed/engaged. That's what we want for our bump switch, so that's why we want wires on the C and N.O. posts.

FIGURE 9.8

Details of the bump switch assembly.

BUMP SWITCH ASSEMBLY

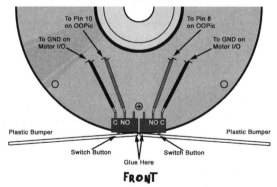

5. Using superglue (or epoxy, if you prefer), glue the switches down onto the places indicated on the bottom CD template (refer to Figure 9.5). They should be side by side, right up against the front edge of the CD, with their two bumpers perpendicular to the bot's wheels (refer to Figure 9.8).

Assembling the Top Half

With the bottom components all in place, you're now ready to build the top platform and attach the four standoffs that will hold the OOPic-R module (see Figure 9.9).

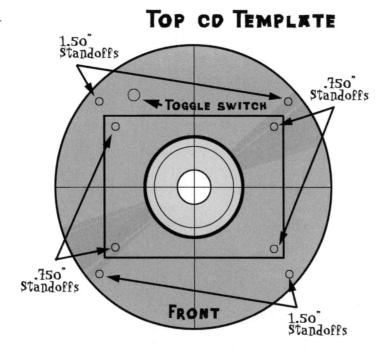

TOP CD TEMPLATE

1.50" Standoffs

.750" Standoffs

TOGGLE SWITCH

.750" Standoffs

1.50" Standoffs

FRONT

1. Using the same technique as with the bottom template, prepare and attach the top CD template to the bottom of the CD you're going to use. Drill the four holes that will attach the main 1.50-inch standoffs from the bottom platform.

2. Drill the four inner holes indicated on the template that will hold the .750-inch standoffs for the OOPic module.

3. Drill the hole where the robot's toggle switch will go (as indicated on the template). When this hole is placed, you can remove the paper template.

4. To install the switch, cut the positive (red) wire on the OOPic's 9V battery snap approximately 3 1/2 inches from the 2-pin connector on the end of the snap. Strip the red wires where you cut them and solder them onto the two posts of the toggle switch (to the C and N.O. posts if there's more than two on the switch). Then install the switch itself (using its two included bolts) in the switch hole you created in step 3. The battery, attached to the battery snap end of the battery wire, will eventually be able to be tucked in, behind the switch, under the OOPic standoffs (which we'll add next). When you finally do attach the battery (don't power anything yet), you can hold it in place with poster putty.

5. Put the male ends of the four .750-inch standoffs into the holes in your top platform and bolt them in place with 4-40 bolts. These will hold the OOPic module.

6. Place the top platform on the main standoffs (with the switch facing toward the back) and screw the top CD into the bottom CD with 4-40 screws. As you're placing this top disc over the standoff holes, thread all of the wires from the motor controller, the bump switches, and the motor battery pack through the center hole.

7. Attach the OOPic-R to its four standoffs on the top platform with 4-40 screws. The serial cable connector on the OOPic should be facing the back of the bot. We're done with the top-half assembly (see Figure 9.10).

FIGURE 9.10

Our DiscRover, built and ready for final wiring.

Making the Final Connections

We're almost ready to download our first control program into the DiscRover. The last mechanical steps we need to make are to connect our motor controller and power wires to the OOPic.

Before we plug our wires into the MCU, we need to attach connectors to the ends of them. So that you don't have a lot of excess wire hanging out all over your robot, it's recommended that you pull each wire to where it needs to go, cut it there (adding a little slack, of course), strip the wire end, and then solder on a connector.

There are all sorts of connector technologies you can use. You can buy connectors made to fit .100-inch *headers* (which is what the pin rows on a PCB like our module are called), but unfortunately, they don't sell them for a single pin (the smallest size is a two-pin connector). The actual metal connector inside of the plastic connector housing is called a *crimp pin*. Using those is probably your best bet. You can buy these (sold as *female crimp pins*) or cannibalize them from connectors in your techno-junk box. If you use these, you'll want to cover the join and the bare metal of the pin with heat-shrink tubing (see Chapter 5 for working with heat shrink). We did something wacky and obsessively techno-junky for our connectors. We took a 10-pin IC socket strip (left over from the 20-pin socket we hacked in half for the motor controller) and cut single-pin connector blocks from this. We soldered the wires onto the socket's pins and then used heat-shrink tubing over the solder connections. The pin sockets are placed so closely on the IC socket package that you have to sacrifice every other one (so you get five—if you're lucky—from the 10-pin strip). We used a second trashed socket (left over from the walker project) to make more connector blocks. You'll end up needing six.

1. Find the positive wire from the left bump switch. This will be the wire connected to the switch's N.O. pin. Test fit its length to Pin 8 on the OOPic's main I/O area. Cut, strip, and solder on a connector, and then, heat shrink over the join. You want to end up with

tip

When test-fitting the wires from the motor controller to the MCU, make sure you leave just enough slack that the controller can sort of "float" comfortably between the bottom and top platforms. We don't want to attach the motor controller because we need to move it out of the way to slide our battery pack in and out of the back end of the bottom platform, when replacing the batteries. Also, make sure there's enough wire on your AAA battery holder to be able to slide it out from the back. You might have to add wire to the wires that came with the battery holder.

note

When using heat-shrink tubing, always remember to slide the piece of tubing onto the wire *before* you solder on the connector! This is disturbingly easy to forget.

a very tight plastic shroud around the connector (if you're using crimp pins) so that the pin (covered in plastic) can fit comfortably in the .100-inch spacing of the header. You don't want any of the crimp pin touching neighboring header pins (but you also don't want the plastic-covered connector so fat that it won't fit). Attach your completed wire to Pin 8.

2. Find the positive wire on the right bump switch. Prepare it exactly as in step 1 and attach it to Pin 10 on the OOPic's main I/O area.

3. Fish out the motor power supply wire that's attached to Pin 1 of the motor controller and attach it to the positive power wire of your motor power supply (the AAA battery pack). Cut this connection so that there's not a lot of excess wire (but keep in mind that you'll need to be able to slide the battery pack out for changing batteries). You can just twist these wires together and slap on some electrical tape.

4. Find the logic power supply from the motor controller (Pin 3). Test fit its length to the +5V power pin on the OOPic's Dual DC Motor I/O group. Prepare the wire with a connector as in step 1. Attach it.

5. Locate the wire attached to the serial input pin (Pin 4) on the motor controller. Prepare the wire with a connector as in step 1. Attach it to Pin 30 on the OOPic's main I/O area. This is our serial output pin where all of the action will happen.

6. Locate the wire attached to Pin 5 of the motor controller. This is the reset pin for the motor controller. By sending a low (0V) signal to it, it will reset the controller. It gets prepped (as explained earlier) and connected to Pin 31.

7. Now we're going to make one big honking ground wire to ground the motor power supply, the motor controller, and our two bump switches to a common ground. Find all of the wires for these. This will be the negative (black) wire on the motor battery pack, the wire attached to Pin 2 on the motor controller, and the two remaining wires from the bump switches. Measure each of these so that it reaches the *GND (ground)* pin on the OOPic's Dual DC Motor I/O group. Cut and strip the ends, but don't attach connectors. Instead, place all four wires side by side and twist them together. Now prepare a short piece of wire (about 1 inch) with a connector on one end and leave stripped, bare wire on the other. Slide a piece of heat-shrink tubing over the four-wire join, solder the short-wire connector to the twisted join, and then slide the tubing over the soldered join and heat shrink the tubing. This uber-ground wire then gets plugged into the GND in the motor I/O group.

That's it! Now all you have to do is plug the 9V battery into its battery snap and load four AAA batteries into the motor power supply. Make sure your robot is switched off before you do this.

Programming Your DiscRover

Hopefully, you took some time while reading the "The Brains Behind the Bot" section earlier in this chapter to explore the OOPic-R and its compiler program. How this MCU works, all of its functions, and its object-oriented programming languages, could comprise a book in itself. When you download the language compiler program, it comes with an excellent manual (found under the Help menu). There are also tons of resources available on the OOPic Web site (www.oopic.com). Before you go any further, spend some time reading at least the first few chapters of the manual and doing the "Your First OOPic Application" project mentioned in the "The Brains Behind the Bot" section. This not only gives you some practical experience in OOPic programming and program loading, but it also lets you know that the serial cable connection between your PC and your OOPic module is working properly.

The OOPic includes a multi-language compiler that can speak three OOP languages: Visual Basic, Java, and C. If you're a beginning OOP programmer, you'll be best served with Microsoft Visual Basic (VB). It's the easiest of the three languages to learn and use (and there are many OOPic programs written in it online). VB is taught in many public schools these days, so through that, or other digital-world travels, you might already know it. The OOPic system is 100% Visual Basic syntax compatible.

Our Motor Controller Code

Now that you've spent some time familiarizing yourself with OOPic programming, let's take a look at our motor controller code and examine some of its key components.

tip

A serial computer cable comes with the OOPic-R starter kit. It not only needs to be plugged into the serial connector on your OOPic module and your PC, but the OOPic software needs to know where to look for it. In the OOPic compiler program, choose Cable Configuration from the Tools menu. With the serial cable connected and the OOPic module powered up, press the Find Serial Cable button. It automatically locates the serial port the cable is connected to and selects the appropriate radio button on the communications port list. Your OOPic-to-PC connection is ready!

tip

You can find tons of resources online about Visual Basic. The first place to start is Microsoft's site (msdn.microsoft.com/vbasic/). Another source is the DevX site (www.devx.com/vb).

When you installed the OOPic program on your PC, it should have loaded a shortcut onto your desktop to the OOPic.exe. Double-click on that icon to launch the program. After you've read through this section first and familiarized yourself with the code we'll use, enter it into the code window on the left side of the program. When you're done, save your code as discRover1.osc.

This isn't the entire control program we will use on our bot; this program only commands the motors to go forward without pause. It does not engage the bump switches. But just to start off more simply, we'll get this program running, and then we'll amend it to add the bump switch control. The entire program is shown in Listing 9.1.

Listing 9.1 This Visual Basic Code Gets Your OOPic Talking to Your Motor Controller, Turning Your Robot's Motors

```
'A motor control program using the Pololu
'micro dual serial motor controller and
'the oSerialx object.

Dim S As New oSerialx
Dim R As New oDio1

Sub Main()
R.IOLine = 31
S.Baud = cv4800
S.IOlineS = 30
S.Operate = cvTrue
R.Direction = cvoutput
 OOPic.Delay = 50

Do
R.Invert 'Invert reset pin

'Motor 1
S.Value = 128 'Start byte
S.Value = 0 'Device type (motor)
S.Value = 0 'Motor and direction
S.Value = 127 'Motor speed (between 0 - 127)
```

```
'Motor 2
S.Value =128 'Start byte
S.Value = 0 'Device type (motor)
S.Value = 3 'Motor and direction
S.Value = 127 'Motor speed (between 0 - 127)
OOPic.Delay = 500
Loop


End Sub
```

Notice that each line in the first paragraph of the listing is preceded by an apostrophe ('). This is called a *comment mark*. The text that follows it is not part of the program code itself, but serves as notes to the programmer/user. Anything following this mark in the remaining line of the code are comments that identify what each line does. Reading these comments left by the coder is a great way of figuring out what everything does and where there might be values you can change (such as motor speed and direction in our preceding code).

```
Dim S As New oSerialx
Dim R As New oDio1
```

This first chunk of actual code is called the Dim statements (*Dim* for *dimension*). These Dim statements name the objects we'll be using and what types of objects they are. In our case, the first object is named s and is identified as a new oSerialx object. Because we want to talk to a serial motor controller, we chose this from the existing library of available serial objects. This one, designated with an x, can use any I/O line we specify in our program. The names in the Dim statement (such as s and R here) are whatever you chose to call them and are used as shorthand later in the code to identify what object you're trying to address.

```
Sub Main()
R.IOLine= 31
S.Baud = cv4800
S.IOlineS = 30
S.Operate = cvTrue
R.Direction=cvoutput
OOPic.Delay =50
```

This section of code is the beginning of our actual program procedure (now that we got notes to mortals out of the way, and everybody introduced themselves and described what they do for a living). Subprocedures in VB always begin with Sub. In this case, the first Sub (our motor control code) is the entire program, so it's followed by Main ( ). The Sub that's the heart of a program is always called Main ( ). The remaining lines of code refer to the beginning Dim statements and describe the program procedures that are related to the objects we named in these statements. So, for instance, the S.Iolines (all references to objects named in the Dim statement are always followed by a period in the program code) is assigning Pin 30 as the serial I/O line to which we'll connect our motor controller serial pin. S.Baud sets the serial baud rate (the speed of data on the serial line), S.Operate is set to cvTrue, which means that on (or true) is the default operation mode of our serial pin (the *cv* stands for *constant value*), and R.Direction tells the bidirectional pin on the OOPic module that it's now an output pin. We've assigned the R dimension to Pin 31 and are using this as the reset pin for our motor controller (refer to Figure 9.6). Finally, the OOPic.delay is an object property that delays execution of the next part of the program for the time indicated (in 1/100ths of a second). Delays like this help a program to catch its breath, so to speak, before moving on to the next part of the code. This helps to reduce errors in code execution.

```
Do
R.Invert 'Invert reset pin

'Motor 1
S.Value = 128 'Start byte
S.Value = 0 'Device type (motor)
S.Value = 0 'Motor and direction
S.Value = 127 'Motor speed (between 0 - 127)

'Motor 2
S.Value = 128 'Start byte
S.Value = 0 'Device type (motor)
S.Value = 3 'Motor and direction
S.Value = 127 'Motor speed (between 0 - 127)
OOPic.Delay = 500
Loop
```

The meat of our motor control program is the Do...Loop (shown in the preceding code). First off, it tells Pin 31 (our reset pin) to reset itself. Then it sends what's called *a four-byte command*. This is represented by the four S.Value statements seen in the preceding code for each motor. The first is the Start byte. This tells our designated motor controller serial line, "Hey, you shiftless I/O pin, wake up and look lively; we've got commands on the way!" The next byte identifies the device type (motor controllers are a 0). This is followed by the motor/direction identifier. This is an important value because this is where you can change directions on your motors if you hook them up and your robot is going in the wrong direction. Values here will likely be 0–3 for addressing two motors. We first used 1 and 2, but that made our bot go backward, so we changed the values to 0 and 3 and everything worked fine. The last of the four-byte command values is the motor speed. It can be any value between 0 and 127. The Loop command at the end of the Do...Loop tells the program to take it from the top and repeat. It will do the ol' shampoo routine (rinse and repeat) until something else in the program barges in to interrupt it. This first Sub procedure (and all Subs) end with the statement (you guessed it), End Sub.

To load this program into your OOPic-R module, you need to do the following:

1. Open the OOPic multi-language compiler program and type in the preceding motor control program. Enter it exactly as shown. You can leave off all of the commented text if you want (everything following an apostrophe), but it's actually a really good idea to keep all of these notes intact (see Figure 9.11).

FIGURE 9.11

The OOPic multi-language compiler program with our motor control code.

2. When you're done, double check to make sure you entered all of the code correctly.

3. Connect your DiscRover to your PC via the OOPic serial programming cable and make sure your OOPic is powered on (by flipping the toggle switch).

4. Press F5 on your PC keyboard. That sends the command to the OOPic program to compile your code and send it to the OOPic module. If it's a new program, it asks you to name it. Let's call it discRover1.osc. osc is the three-letter filename extension used for OOPic programs (it stands for *OOPic Source Code*). The compiler will debug your code and let you know if anything is amiss.

If all goes well, as soon as the motor program hits the module, it will start running and the wheels will turn. That's because we made the motor program our "main" subprocedure. This way, to start our rover going, after its full control program is loaded, all we'll have to do is throw the toggle switch to set it on its way.

Bump Whisker Code

Now that we have our motors powering our rover forward, the next behavior we want to add is obstacle avoidance. With just the creation of two more oDio1 objects and a few additional lines of code, we'll have a DiscRover that can navigate a space. Okay, so crashing into things first and then avoiding them might be a clumsy means of navigation, but we'll start there. After you've got these two routines working successfully, you can add a more refined navigation behavior on top of this contact sensing, such as sonar or infrared distance detection, and the bump whiskers will become a last resort if the distance-sensing system fails.

> **caution**
>
> When powering up a robot, loading programs into it, and so forth, it's always a good idea to have the wheels (or legs) off of the ground in case the robot lurches a bit or drives off the table. While working on the DiscRover, we set it on top of a jumbo coffee mug.

You should have already attached your bump switches to their I/O pins. Check to make sure that they're in Pins 8 (left switch) and 10 (right switch) on the OOPic. For simplicity's sake, we're not going to program separate response behaviors for the two switches. Any trigger of either switch (or both) initiates the obstacle-avoidance routine. This routine reverses the power to our motors for a given time, with one motor turning faster than the other during backup (which changes the forward heading when the main sub is executed again). You can set these values (the reverse motor speeds) to whatever works best for your bot. In Listing 9.2, we have one wheel going twice as fast as the other.

We won't go into the additional bump switch coding line by line. You should be able to figure the code out for yourself, based on our motor control program, the comments in the code, and by using the OOPic manual found under the Help menu in the compiler. There are a few things we should point out. First, check out the code in Listing 9.2.

> **tip**
>
> The OOPic Web site has a good Object Application Note on hooking up switches to an OOPic and use of the oDio1 object, and OOPic's built-in pullup resistor. Check it out here: www.oopic.com/objapp2.htm.

LISTING 9.2   Final Code for Controlling Both the Motors and the
Backup Routine

```
'A simple program to control two DC motors
'(through a 'Pololu Micro Dual Serial Motor Controller) and two
'bump switches.

Dim S As New oSerialx
Dim R As New oDio1
Dim LSwitch As New oDio1
Dim RSwitch As New oDio1

Sub Main()

'Turn internal "pullup resistor" on
OOPic.Pullup = cvTrue

'Assign bump switches to I/O pins 8 and 10
LSwitch.IOLine = 8
LSwitch.Direction = cvInput
RSwitch.IOLine = 10
RSwitch.Direction = cvInput

R.IOLine= 31 'reset pin
S.Baud = cv4800
S.IolineS = 30 'motor controller pin
S.Operate = cvTrue
R.Direction = cvOutput
OOPic.Delay = 50

'Reset the motor controller
R.Invert 'invert reset pin
R.Invert 'invert reset pin

'Run "Do...Loop"
Do
'Check if either switch is depressed
If (LSwitch.Value = 0) or (RSwitch.Value = 0) then

'If depressed, back robot up and turn
```

Listing 9.2   Continued

```
'Motor 1
S.Value = 128 'Start byte
S.Value = 0 'Device type
S.Value = 1 'Motor and direction
S.Value = 120 'Motor speed

'Motor 2
S.Value = 128 'Start byte
S.Value = 0 'Device type
S.Value = 2 'Motor and direction
S.Value = 60 'Motor speed

OOPic.Delay = 200

'Reset the motor controller
R.Invert 'Invert reset pin
R.Invert 'Invert reset pin

End If
'Send robot forward

'Motor 1
S.Value = 128 'Start byte
S.Value = 0 'Device type
S.Value = 0 'Motor and direction
S.Value = 127 'Motor speed

'Motor 2
S.Value = 128 'Start byte
S.Value = 0 'Device type
S.Value = 3 'Motor and direction
S.Value = 127 'Motor speed

OOPic.Delay = 5

Loop

End Sub
```

As you can see from Listing 9.2, we didn't really add that much. We added two new oDio1 objects (our left and right switches) to our Dim statements. We also turned on something called a *pullup resistor*. This is a cool feature of the OOPic. If the module didn't have this feature, we'd have had to add external resistors to the two switch inputs. This resistance is used to help the controller determine what state (1 or 0) the input pin is in. Left idle (with nothing turned on), an input pin can get confused and float in a netherworld of high and low states, even oscillating between the two. The resistor keeps the pin in line. And having this resistance baked in means all we have to do is call it up in software rather than soldering on hardware. Using the pullup resistor means that our input pin stays in a logical high (1) state until the switch is depressed. This is called an *active-low* configuration.

We then assigned our switches to two I/O pins in the 8–15 pin group (which is the I/O group that has the pullup resistor feature available). Then we added an If...Then decision routine to our code. It checks to see if either of the switches is pressed (active-low or 0 mode), and if it is, it goes into a motor reverse routine. If it finds no low signal present on either switch input line, it goes to the motor-forward code we created earlier. The program just keeps looping through this sequence, checking the state of the switches, and making the motors respond accordingly. That's it!

Load 'n Go

You'll want to enter the code outlined in the previous section into a copy of your discRover1.osc program, and give it a different name, such as discRover2.osc. When you're finished entering the new code, double check it, hook up your rover, and compile the program as you did for the motor control program. (When it loads to the rover, it will erase the earlier program. You can only have one program in the MCU at a time.) If you entered the code correctly, you should now have a robot brain that knows how to motor your rover forward, and to back up and turn if it hits an obstacle (see Figure 9.12). If nothing happens, or you get strange behaviors, such as wheels turning in opposite directions, check out the "Troubleshooting the DiscRover" section.

tip

One great way to learn a programming language is to monkey around with it. For our robot programming, create a doodle program (a copy of the preceding code, with a different name) and play around with it. Put your robot "up on blocks" (raised up so the wheels don't touch the ground), and try changing the values in the program and see what happens. Try commenting out lines of code and see which ones "break" the program. Try adding some other objects and functions found in the OOPic manual. Don't be afraid to experiment.

FIGURE 9.12

Our completed DiscRover, ready to spread the word on free trial Internet access!

Troubleshooting the DiscRover

If you've finished your rover, but you're one unhappy bot builder, you might have run into an unfortunate snag along the way. Don't worry—it's likely something simple, such as a bad connection, or you're trying to get your software to talk through the wrong I/O pin, or some similar common mistake. Run through the following troubleshooting list and see if it solves your problem.

The Mechanics

There are few mechanical parts on our rover, but there's still enough complexity that mistakes can happen. Check the following:

- If the bot's getting no power, the first thing you want to do is check the battery supplies. Use your DMM to make sure they're both full/charged. Oh yeah: Did you also switch on the robot? Hey, we *had* to ask! Also, check the tiny LED on the OOPic (marked EAC). If that's lit, it means the processor and Flash memory are operational (which means the OOPic should be doing its thing).

- Also (if there's no power), you want to run through all of your soldered connections to make sure that they're actually connected (and in the right places) and that you have good solder joins. Especially check to see that the motor controller is connected properly. Refer to Figure 9.6 and check to make sure that all of the pins are going where they should.

- If the wheels don't seem to be getting much traction (one or both of them are slipping), it might be that your stabilizer standoffs are too high (raising the wheels off of the ground a tiny bit). Tighten up the screws on the ends of them a few turns (sort of like letting the air out of the tires on a Jeep for over-sand travel).

- If one or both of the bump switches is not working, make sure that they're plugged into the correct I/O pins on the microcontroller (Pins 8 and 10).

If none of these fix your problem, it might be something in your code.

The Code

When you compile your programs in OOPic, the compiler debugs the code before it loads it into the module, so it should point out any syntax errors or other obvious coding problems. So, if it goes through the compile process successfully, you're one step ahead of the game. If it compiles and you still have problems

- Look carefully over the code, especially any values that you've entered. If these are set wrong, it could cause problems.

- If one (or both) of your motors is turning in the wrong direction, you don't need to physically switch the motor wires. This can easily be handled in code. Refer to the "Our Motor Controller Code" section earlier in this chapter. The values to enter will be 0–3. One of those should get your motors going in the right direction.

- If the operations of the motors' forward or reverse sequences don't seem to be smoothly executing, try changing the OOPic.Delay values.

- Try commenting out one or both of the Invert reset pin commands. We found some strange and unpredictable behavior with these resets.

- If all else fails, download the program discRover2.osc from this book's Web site (www.streettech.com/robotbook), open it in the compiler, send it to your rover, and see if that fixes things.

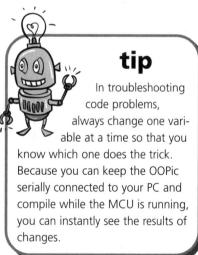

tip

In troubleshooting code problems, always change one variable at a time so that you know which one does the trick. Because you can keep the OOPic serially connected to your PC and compile while the MCU is running, you can instantly see the results of changes.

Further Experimentation

We think the OOPic is a very cutting-edge piece of hardware that has a promising future in the world of small robots and other embedded systems. Learning it, and any of the OOP languages it speaks, should serve you well in your future robot projects. The robot development platform is popular in so many schools and research labs for a reason. The simplicity of the two-wheel differential drive, the ability to quickly add platform layers, the rounded shape (better for bouncing off of obstacles), and so forth all make it a sensible design for a teaching robot. We achieved all of this, in miniature, with our DiscRover. So, what else can you do with it?

tip

Visit OOPic's Web site (www.oopic.com) to see commercial kits and do-it-yourself projects that utilize the OOPic line of controllers.

Add a Prototyping Area

The first thing you're probably going to want to add is a prototyping area. This is a small breadboard that enables you to quickly add and remove sensor components without having to solder or otherwise connect/disconnect anything. Parallax (www.parallax.com) sells a tiny breadboard (1 3/4 inches × 1 1/4 inches) for $4.95 that just fits on the front of your top platform (conveniently located just below the I/O pins of the MCU). It has 34 5-hole tie-point groups, and even has an IC trench! It works exactly the same as a full-size breadboard like the one we detailed in the "Thumbnail Guide to Breadboards" in Chapter 7, "Project 1: Coat Hanger Walker."

Put Another Layer on That Robo-Cake!

If things get too cramped on your two-layer platform, you can easily add a third. Just drill four more holes in the upper platform for another set of 1.50-inch standoffs and four holes in a third CD. This allows you to have a platform devoted to sensors and a prototyping area. If you want to properly get across the whole "Look, I recycled some annoying CD-ROM junk mail!" idea, you can just add a third platform that has the printed side of the CD-ROM showing and no hardware on it. Hilarity will ensue.

Add Higher-Order Sensor Systems

We went with the most basic contact-switch sensing. Now that you've got that working, you can "kick it up a notch" (as Emeril says) with some distance sensors, heat sensors, motion sensors—you name it! The OOPic Web site (www.oopic.com) has project pages that cover the hardware and software required for most of these sensor types.

Make Two DiscRovers Talk to Each Other

CDBot-builder extraordinaire Abe Howell has an excellent how-to article online (www.rentron.com/Robots/CDBot.htm) that shows you how to build a low-cost radio frequency (RF) transmitter/receiver that works with two CDBots and the OOPic. Using this transmitter/receiver system (assuming you've built two DiscRovers), you can get one of them to chase the other around a room.

Build a CD-ROM Boxbot

Now Internet providers are sending you double the amount of bot-building material. It used to be that they were content to just send you the CD-ROM in a paper or cardboard sleeve. But that obviously wasn't getting your attention after several hundred discs, so they started mailing them in nifty metal cases. Drill holes in the four corners of the case top and bottom, connect them with standoffs, and you have a new type of robot body: the CD-ROM boxbot (see Figure 9.13).

FIGURE 9.13

Why let the junk CD-ROMs have all the fun? Make a CD-ROM boxbot!

Take Your OOPic Elsewhere

The great thing about the OOPic is that it's so versatile. It has plenty of I/O pins, several different cable interfacing options, multiple programming languages, different on-board power options, and an active (and growing) user community. After you learn how to use all of its features and the OOPic programming, you can yank the module off of your DiscRover and use it in another bot.

Scott Savage, of Savage Innovations, makers of the OOPic, has a nifty site called simply Robot Projects (www.RobotProjects.com). Here you'll find detailed instructions for using an OOPic controller in a number of different robots. Especially fun is the Dissection section where consumer bots such as the old TOPO, the PooChi robopet, and the Furby get put on the autopsy table. What's not to like? Scott's excellently documented Racing Rover Project is very similar to the Timbot robo-vehicle we covered in Chapter 4.

THE ABSOLUTE MINIMUM

We made a relatively simple robot platform that can now serve as a robot lab for further experimentation. With the building and programming knowledge you've learned in this chapter, and through the online resources pointed to throughout, you should be off and running in the...er...electrifying world of do-it-yourself robotics. If the computer programming in this chapter didn't make you follow Mousey the Junkbot into hiding under the couch, you already know the following:

- It's not plastic junk mail, it's robot building material! We learned how crafty botsmiths are turning promotional CD-ROMs into programmable robots. We did some creative recycling of our own.

- Microcontrollers (MCUs) are tiny computers with growing delusions of grandeur. Their increasing power, versatility, and miniaturization make them the perfect means of robotic control.

- Object-oriented programming is a computer programming approach in which code is constructed into discrete software objects that handle all of the functions and data related to a given computing task. This creates a sort of component-based programming that's very appealing to gearheads like us.

- With a basic robot platform and a powerful MCU, you've got a great base system on which to add all sorts of different sensor systems.

PART III

RESOURCES

Robot Hardware and Software 301

Robot Books, Magazines, and Videos 319

Robots on the Web 333

IN THIS CHAPTER

- Hooking you up with the best places to get robot kits

- Sources for individual components

- Sharing some robot-geek favorite surplus haunts

10

ROBOT HARDWARE AND SOFTWARE

Sourcing the Right Stuff

There are so many choices now in selecting kits, robot parts, microcontrollers, and the software that enables all these things to bring your bots to life. What follows is a list of some of the better retailers, manufacturers, and other sources out there. This is, by no means, an exhaustive list, but this will certainly get you started. To find additional sources, go to some of the "Portals" listed in Chapter 12, "Robots on the Web," and look on their Links, Manufacturers, Retailers, or similar pages.

Robot Kits and Bot-Specific Parts

For the beginner, there's no better way to get your end effectors wet than building a few kits—first, no-soldering-required kits, and then, solder kits and building systems. Here are a few of our "bench-tested" favorites.

Acroname

www.acroname.com

4894 Sterling Drive

Boulder, CO 80301-2350

720-564-0373

Acroname is probably best known for the PPRK (Palm Pilot Robot Kit), which got quite a bit of press a few years ago. Like Evolution Robotics's ER1 (see Chapter 6, "Acquiring Mad Robot Skills"), the PPRK includes a robot body, motors, drive wheels, and everything else except the bot's brains. You provide that by way of a Palm Pilot PDA. The PPRK comes with the robot control software. You build the bot, load the software, plug in your Palm to the top of the robot, and off you go. One of the coolest things about the PPRK is the use of the omni-directional wheels from North American Roller Products (see "Motors and Drive Train" later in this chapter). Acroname also sells parts, other manufacturers' kits, robot books, and more.

Budget Robotics

www.budgetrobotics.com

P.O. Box 5821

Oceanside, CA 92056

760-941-6632

Anyone who's poked his or her nose into the robot hobby knows the name Gordon McComb. Long before Battlebot builders headed to their garages to cannibalize the family mower (in search of 15 minutes of fame in the Battlebox), Gordon McComb was teaching average folks how to build bots. His book, *Robot Builder's Bonanza*, ISBN: 0-0713-6296-7 (see Chapter 11, "Robot Books, Magazines, and Videos"), is legendary. Now Gordon has gone into the robot parts business. Frustrated by the high price of bot kits and parts, which put hobby robotics out of reach for some people, Gordon decided to sell a line of unique and affordable components. He offers several different robot development platforms (complete with wheels and servos), a tank-robot body (with treads and servos), a cheap plastic gripper arm, motors, expanded PVC plastic building material, and other specialty parts. And everything really is "budget" priced. Kudos to Gordon for opening the hobby up to more people.

LEGO MINDSTORMS

www.legomindstorms.com

555 Taylor Road, P.O. Box 1600

Enfield, CT 06083-1600

860-763-3211

You can buy LEGO MINDSTORMS in most toy stores, but you can also order the kits online directly from LEGO. Its Web site is also a portal to a community of LEGO robot builders where you'll find robot plans, RCX control programs, and other useful resources. At its shop.lego.com store, you can order individual components.

Lynxmotion

www.lynxmotion.com

P.O. Box 818

Pekin, IL 61555-0818

866-512-1024

Lynxmotion's unique yellow-colored walking robots have almost become synonymous with hobby robotics. It offers a number of four- and six-legged walkers, along with its Carpet Rovers (two-wheeled development platforms), robotic arms, and a line of Sumo robot chassis (with wheels and motors)—you add the electronics. It also sells robot parts and books.

Mondo-tronics' Robot Store

www.robotstore.com

124 Paul Drive #12

San Rafael, CA 94903

415-491-4600

Roger Gilbertson of the Robot Store has been a tireless champion of hobby robotics for decades. Although other robot retailers have come and gone, the Robot Store has been selling robot kits, building systems, and other robot-related products for 16 years. It sells nearly all of the kits and systems we covered in Chapter 6. It also sells robot-related books, videos, posters, and other fun science and technology-related goodies. It's also the premier consumer source for Shape Muscle Alloy (or Muscle Wire as it is called). It has wire of various diameters, SMA springs, pistons, and more, as well as a great Web site and a free color catalog.

Parallax

www.parallax.com
599 Menlo Drive, Suite 100
Rocklin, CA 95765
888-512-1024

Parallax has probably done more to advance small robot development than anybody. Its BASIC Stamp microcontroller has been the robot brain of record for years now. In addition to the Stamps, it sells full robot kits (such as the BOE-Bot covered in Chapter 6), sensors, robot hardware, and books on Stamp programming.

Solarbotics

www.solarbotics.com
179 Harvest Glen Way N.E.
Calgary, Alberta, Canada T3K 4J4
866-276-2687

As you've probably been able to discern by now, we hold a special place in our high-domed head for BEAM robotics. We like the biological influences, the simplicity and elegance of the designs, and the cleverness of the engineering. Besides Mark Tilden, the creator of the BEAM concept, Dave Hrynkiw has done more to support this path of robot evolution than anyone. His Solarbotics store, both online and via a free print catalog, offers a full complement of kits and parts. The Web site also contains tons of useful information about BEAM and the mechanics and electronics behind it. Dave also sponsors Solarbotics.net, a network of other BEAM builder sites. It's a shame that Mark Tilden doesn't have a bigger Net presence. In light of that, Solarbotics is a godsend to the BEAM enthusiast.

Tamiya

www.tamiyausa.com
2 Orion
Aliso Viejo, CA 92656-4200
800-826-4922

You might know the Japanese company Tamiya from its amazingly detailed military and vehicle scale models, but what you might not know is that it also sells robot parts and kits as well. It has a cool little wall-hugging mouse kit, a solar car, a tracked vehicle chassis, and oodles of useful parts (motors, switches, gear sets, and more).

Building Materials and Supplies

When you get down to designing and building your own bots (or building them from plans), you have no idea what a thrill it is to spec out the parts, hunt them down, order them, and then have the UPS driver deliver them to your front door. Okay, maybe I'm kind of weird that way, but at least it gives *me* a thrill. Here are some of my favorite suppliers.

The Composite Store

www.cstsales.com
16330 Harris Road, Hangar #2
Tehachapi, CA 93561
800-338-1278

If you're interested in composites materials such as carbon composite, carbon/Kevlar hybrid, fiberglass, and similar materials, you'll want to check out The Composite Store. Besides materials, supplies, and tools, it also offers some how-to articles and data on the products it sells. It specializes in sales to hobby and other small-quantity buyers.

Evolution Robotics

www.evolution.com
130 West Union Street
Pasadena, CA 91103
626-535-2814

Evolution Robotics, makers of the ER1 robot kit (see Chapter 5, "The Right Tools for the Job"), also sells the XBeam system, a series of extruded aluminum beams and connectors that can be used to build your own robot frames. The components are not cheap, but they are sturdy and reconfigurable.

Online Metals

www.onlinemetals.com
1138 West Ewing
Seattle, WA 98119
800-704-2157

There's a reason why so many robot links on the Internet point to this site. Online Metals sells a wide range of metals stock and is geared (ah…no pun intended) to the small-quantity market (hey, that's us!). Besides selling aluminum, copper, brass, stainless, and titanium, it also sells plastics and hardware. Its site also includes useful information on the physical characteristics of various metals.

Plastruct

www.plastruct.com
1020 South Wallace Place
City of Industry, CA 91748
800-666-7015

Plastruct is a manufacturer of plastic components for industrial and hobby modeling, from architectural modeling and model making for films to hobby scale modeling. It has all manner of plastic stock and building components, including sheet styrene and tubing, plastic T-beams, I-beams, L-beams—all at various scales. Obviously, given the small size of the components it sells, its use in robotics is mainly limited to miniature and small robots.

Small Parts

www.smallparts.com
13980 N.W. 58th Court, P.O. Box 4650
Miami Lakes, FL 33014-0650
800-220-4242

This amazing retailer could just as well be called Small Orders, 'cause that's its specialty: providing scientists, engineers, researchers, and hobbyists with small quantities of common and hard-to-find industrial building materials. It has everything from shape memory alloy to every type of metal you might need, to dozens of varieties of plastics. Nearly every material type is available in rod, tube, sheet, ball, and other common configurations. It offers a free 300-page catalog and has no minimum order size! If you're anxious to peruse its wares, you can also download its catalog in PDF format.

W.W. Grainger

www.grainger.com
100 Grainger Pkwy.
Lake Forest, IL 60045-5201
888-361-8649

If you have one industrial materials and supply catalog on your desk, it should be the Grainger catalog. This gigantic (free) catalog has just about any material, supply, hardware, and tool you can dream up. You can order online through its print catalog or pick up your booty at nearly 600 retail locations. Be careful going to a Grainger store though—if you're not careful, you'll spend *way* more money than you intended.

Motors and Drive Train

Your robot is nothing but a high-tech paperweight (doorstop?) without motors. We gearheads *love* motors. Here are some of the better-known manufacturers and retailers of high-quality motors and other drive train components.

HiTec RCD

www.hitecrcd.com
12115 Paine Street
Poway CA, 92064
858-748-6948

Probably best known in the robotics community for its line of servo motors, HiTec also deals in everything R/C: radio controllers, motor speed controllers, battery chargers, and more. You can't actually buy products on its site, but it does have manuals and datasheets on its servo motors and other products.

Mabuchi

www.mabuchi-motor.co.jp/english/
430 Matsuhidai
Matsudo-shi, Chiba-ken 270-2280
Japan

If you've...ah...deconstructed a lot of motorized toys in search of useable robot parts, you're no doubt familiar with the name Mabuchi. The DC motors made by this Japanese manufacturer are ubiquitous. You can't buy motors directly from its site (you can get them from mail order sources such as Solarbotics, American Science and Surplus, and elsewhere), but you can access datasheets and other info about its product offerings.

North American Roller Products

www.narp-trapo.com
P.O. Box 2142
Glen Ellyn, IL 60138
630-858-9161

NARP is the manufacturer of the amazing multidirectional rollers (*omni-wheels*) used on Acroname's PPRK (see "Robot Kits and Bot-Specific Parts" earlier in this chapter). These nylon wheels have two rows of barrel rollers all around the hub, which enable the wheels to move forward and back, like a normal wheel, but also sideways and diagonally. The result is 360-degree freedom of movement.

RobotBooks

www.robotbooks.com

RobotBooks is an amazing portal to all things robotic sold online (not just books). The site is run by Carlo Bertocchini (of Battlebots champion Biohazard fame). Among the many things that Carlo sells are the amazing Magmotors. These incredibly "torquey," high-performance, and extremely durable motors are perfect for combat robotics. If you're thinking about building this type of bot, you'll definitely want to check out these motors. Although Carlo doesn't actually sell anything directly (including the Magmotors), his wonderful site is worthy of your support, so buy through him.

Servo City

www.servocity.com

620 Industrial Blvd.

Winfield, KS 67156

877-221-7071

You'd expect a company called Servo City to offer anything and everything related to servo motors, and you'd be right. It carries a complete line of motors from HiTec and Futaba, plus servo connecting horns and arms, gear sets, servo mounts, battery packs, speed controllers, and more. It sells servos in bundles that can save you some bucks if you need more than a couple.

Solarbotics

www.solarbotics.com

179 Harvest Glen Way N.E.

Calgary, Alberta, Canada T3K 4J4

866-276-2687

Dave sells a variety of motors, from motors with gearboxes, to cheap (but surprisingly torquey) small hobby motors from Mabuchi, to pager and vibrating disk motors. He also has servo motors with the control electronics already removed and the motors pre-"hacked" for continuous rotation. And at $12 apiece, they're cheap, too! (See "Robot Kits and Bot-Specific Parts" earlier in this chapter.)

Tamiya

www.tamiyausa.com

2 Orion

Aliso Viejo, CA 92656-4200

800-826-4922

Tamiya offers a number of gearbox motors suitable for robotics. It has a Planetary Gear that enables you to select from eight different gearing ratios (from 4:1 to 400:1). It has a host of other gearbox motors, including a twin-motor gearbox and a six-speed box, and it also sells wheels, gears, pulleys, chains, ball-casters, and other drive train goodies. (See "Robot Kits and Bot-Specific Parts" earlier in this chapter.)

Vantec

www.vantec.com
460 Casa Real Plaza
Nipomo, CA 93444
888-929-5055

If you've watched any episodes of *Battlebots* or *Robot Wars*, or seen a bomb disposal bot doing its jittery job on the evening news, the chances are good that Vantec hardware was lurking in the shadows. It sells heavy-duty servo motors, motor speed controllers, and R/C gear.

Power Systems

Until we have fusion reactors in our back yards and super-efficient fuel cells, we're going to have to rely on the kinds of dry and wet chemical batteries we brought with us from the twentieth century. They're sadly inefficient and short-lived, but there ya go. Maybe with the knowledge you acquire in this book, you can design a nifty self-charging robot that at least is smart enough to feed from its own AC trough.

Battery Mart

www.batterymart.com
1 Battery Drive
Winchester, VA 22601-3673
540-665-0065

This online battery store sells every conceivable type of battery and battery accessory. It also has a FAQ with some useful information on battery power and a battery glossary.

Batteries Plus

www.batteriesplus.com

Batteries Plus is another battery superstore, with thousands of batteries and power-related products available. Unfortunately, it doesn't sell via the Web, but it has over 200 retail outlets nationwide. Check its site for store locations. The site does contain a database of its product offerings.

Robotic Power Solutions

www.battlepack.com

305 9th Street

Carrollton, KY 41008

502-639-0319

These folks sell a line of NiCad and NiMH battery packs optimized for combat robots (but suitable for all robot types). They even sell 24-volt "intercooled" battery packs that have fans in them to keep batteries cool while robot warriors are pushed to the limit. The company also sells chargers and other power accessories. They sell individual cells so that you can build your own battery packs, or they will custom-build packs for you.

Electronics

Take this simple test to see if you're a true electronics geek: Browse through some of the catalogs and Web sites that follow. If your credit card starts to vibrate and mysteriously makes its way from your wallet to your hand, and a little voice inside your head starts chanting, "Buy it, buy it, buy it *all*!" you're a true wire nut (and you're going to have to surrender your card to another responsible adult before it's too late).

Digi-Key

www.digikey.com

701 Brooks Avenue South

Thief River Falls, MN 56701

800-344-4539

If there were one "desert island" catalog for wireheads, it would likely be Digi-Key. It's certainly a mail order catalog that everyone should have on his or her workbench. Besides the thousands of components, supplies, tools, and materials it sells, it also includes lots of technical drawings and information found on manufacturers' datasheets (information critical to understanding what your components are capable of).

Electronix Express

www.elexp.com

365 Blair Road

Avenel, NJ 07001

800-972-2225

The 300-plus-page Electronix Express catalog is crammed full of electronic goodness. It has all the usual stuff, plus a large section of electronic and robot kits.

Future Active

www.future-active.com
41 Main Street
Bolton, MA 01740
800-655-0006

Future Active boasts the largest available-to-sell inventory of electronic components and tools. It has over 150,000 items in stock, as well as a chain of stores. Go to www.activestores.com to see if you have one in your area.

Jameco

www.jameco.com
1355 Shoreway Road
Belmont, CA 94002-4100
800-831-4242

Jameco prides itself on always having everything in stock, having a wide range of offerings, and excellent customer service. Having bought from Jameco for years, I have no arguments with these claims. It has the usual analog and digital electronic components, tools and equipment, but it also has many electronics-related books, cheap electronics kits (great for soldering practice), and even robot kits and parts.

JDR Microdevices

www.jdrmicrodevices.com
1850 South 10th Street
San Jose, CA 95112-4108
800-538-5000

Probably best known as a great source for homebrew computer parts, JDR also sells "discreet" components (capacitors, diodes, LEDs, transistors, and so forth) and lots of both analog and digital integrated circuits.

Mouser Electronics

www.mouser.com
1000 North Main Street
Mansfield, TX 76063-1514
800-346-6873

Mouser has been around since the 1960s, started by Jerry Mouser, a physics teacher who was always trying to hunt down components. Now Mouser does it so you don't have to. Its gigantic (currently 800 pages!) free catalog is a one-stop shop of individual components, test equipment (such as digital multimeters), tools, and more.

Radio Shack

www.radioshack.com
300 West Third Street, Suite 1400
Fort Worth, Texas 76102

Ah...The Shack. Every electronics hobbyist has a love/hate relationship with Radio Shack. It's the place many of us bought our first breadboard and those Forrest Mims books (see Chapter 11), which got us excited about electronics in the first place. Then there was the Radio Shack Color Computer (the "CoCo"), which was this author's first. You never forget your first. But over the years, Radio Shack has gone deeper and deeper into consumer electronics, and in many stores, the components and do-it-yourself areas have gotten smaller and smaller. Still, with a store on nearly every street corner (at least it seems that way), you can't beat the availability. Its Web site used to stink, but now it's pretty good, offering you the ability to order parts online. Also, check the site frequently for its "Virtual Flyers," which list sales in your area.

Remote Control (R/C) Equipment

If you're the brains behind your bot (literally), you're going to need R/C gear. These sources also sell servo motors for use either in radio control or with on-board controllers.

HiTec RCD

www.hitecrcd.com
12115 Paine Street
Poway CA, 92064
858-748-6948

See "Motors and Drive Train" earlier in this chapter.

Tower Hobbies

www.towerhobbies.com
P.O. Box 9078
Champaign, IL 61826-9078
800-637-6050

For years, Tower Hobbies has been a premier source for all types of radio-controlled hobbies (R/C planes, cars, boats, and helicopters), so not surprisingly, it is now frequented by roboticists looking for servo motors, gear sets, and remote control equipment. Its print catalog is $3.

Vantec

www.vantec.com

460 Casa Real Plaza

Nipomo, CA 93444

888-929-5055

See "Motors and Drive Train" earlier in this chapter.

Microcontrollers and Control Software

The following companies sell the most popular microcontrollers and control software used in miniature and small robotics. From them, you should be able to get all of the parts you need to give your robot a brain.

Acroname

www.acroname.com

4894 Sterling Drive

Boulder CO, 80301-2350

720-564-0373

Acroname makes a microntroller module called the BrainStem. This board is built around the PIC 18C252 chip. It has four ports dedicated to servo motor control and a port for controlling the Sharp GP2D02 infrared range finder (a popular navigation sensor). The BrainStem also has five general-purpose digital in/out (I/O) pins and five analog/digital (A/D) inputs. This microcontroller runs a programming language called TEA (Tiny Embedded Application), a subset of the standard ANSI C language.

Basic Micro

www.basicmicro.com

22882 Orchard Lake Road

Farmington Hills, MI 48336

248-427-0040

The BasicATOM microcontroller module, manufactured by Basic Micro, is built on Microchip's popular PIC chips. It sells a number of different modules based around chips with different numbers of I/O pins. All BasicATOM modules have two hardware-based Pulse-Width Modulation (PWM) modules for motor control and four hardware-based analog-to-digital (A/D) converters. They also have two built-in "Capture and Compare" timers, which can count pulses. This capability is great if you're going to be using encoders for wheel rotation sensing (see "Sensors" in Chapter 4, "Robot Anatomy Class"). The BasicATOM uses its own flavor of the BASIC programming language. It is compatible with most of standard BASIC, but not all of it. The ATOMs do come with their own software tools, including a nifty In-Circuit Debugging tool.

Gleason Research

www.gleasonresearch.com

P.O. Box 1494

Concord, MA 01742-1494

800-265–7727

MIT's Handy Board has become a legend in the robotics community. Like the BASIC Stamp, it (and the 68HC11 chip on which it relies) is one of the most ubiquitous microcontroller platforms around. There are instructions available online (see Chapter 12) for putting together your own Handy Board, but you can also buy them, preassembled and tested, from Gleason Research. The Handy Board has motor controllers for up to four motors (each with LED status lights), a built-in Sharp IR decoder (that can receive signals from a standard TV remote), nine digital inputs, and seven analog inputs. The Handy Board uses a language called Interactive C (a subset of the C language). C and its derivatives are probably too "deep geek" for the newbie, but there are a lot of downloadable programs and other software resources available for it. Gleason also sells the Handy Cricket, a newer, smaller sibling to the Handy Board. It can control two motors and two sensors. It has an IR port that enables you to "beam" control programs from your PC (like the LEGO MINDSTORMS kit). The Cricket is programmable in the very easy-to-learn Logo language (popular in grade schools).

Not Quite C

www.baumfamily.org/nqc/

LEGO MINDSTORMS had barely begun moving off of store shelves before hackers began taking apart the RCX microcontroller, figuring out what made it tick, and creating new hardware and software to augment it. One of the first and most enduring user-created tools is David Baum's Not Quite C (NQC) programming language. It's certainly not as easy as the RCX Code that LEGO provides, but it's a lot more powerful. It's also free for download and there are an increasing number of books, Web sites, and other resources devoted to it.

Parallax

www.parallax.com

599 Menlo Drive, Suite 100

Rocklin, CA 95765

888-512-1024

There are numerous events that represent milestones in the small robotics/hobby robotics field (the publishing of Anita Flynn, et al's *Mobile Robots: Inspiration to*

Implementation; the creation of *Robot Wars/Battlebots*; the release of LEGO MINDSTORMS), but few events match the introduction, in 1992, of the BASIC Stamp. This microcontroller suddenly gave the average Joe (and Jane) programmable computer control of electronic circuits. And, it did so using BASIC, one of the most widely known and widely taught programming languages. Not surprisingly, this device became an immediate hit with robot builders. Although newer microcontroller technologies such as the OOPic (see the next section, "Savage Innovations") are raising the bar on what these on-board brains can do, BASIC Stamps still remain a favorite, especially in the educational market. There are hundreds of resources (books, courses, hardware, software, and so forth) available for the Stamp (and its BASIC language).

Savage Innovations

www.oopic.com

Billed as a "hardware object," Savage Innovations' OOPic is an exciting piece of microcontroller technology. The name *OOPic* stands for *Object-Oriented Programmable Integrated Circuit*. It is the first programmable microcontroller to use object-oriented programming. See Chapter 9, "Project 3: Building a DiscRover," for more information on the OOPic and its capabilities. The OOPic Web site has all of the extensive documentation for all of the OOPic objects, the programming manual, software downloads, and of course, the OOPic modules and accessories.

Tools

If you haven't figured it out by now (see Chapter 5, "The Right Tools for the Job"), we love cool tools. Here are a couple of the best sources.

Dremel

www.dremel.com
4915 21st Street
Racine, WI 53406
262-554-1390

The Dremel rotary tool deserves a place in every small robot builder's toolbox. Besides various models of the tool itself, Dremel sells accessories that can turn it into a drill press, a router, a shaper, a grinder, and all sorts of other devices. They don't call it a multitool for nothing.

Micro-Mark

www.micromark.com
340 Snyder Avenue
Berkeley Heights, NJ 07922-1595
800-225-1066

This is, by far, my favorite small tools catalog. It is filled with every imaginable tool for every sort of model making, or any other hobby or activity that involves precision and specialty hand and power tools—like building robots. It has Dremel multi-tools and all the accessories: high-quality pliers, drivers, and metal files; tiny nuts, bolts, and screws; and even building materials such as brass, plastic, and aluminum stock. Its sales and service is always prompt and professional. Check its Web site for special sales and sign up to receive its free print catalog.

Hobby Stores

Your local hobby store is a great source for stock materials (plastic, basswood, and brass), R/C parts, servo motors, small tools, and more. If you don't have a local store, here are a couple of mail order sources.

Hobbytron

www.hobbytron.net
1216 South 1580 West, Suite B
Orem, UT 84058
877-606-8766

This online hobby store has a full complement of hobby products, but its emphasis is on robotics and electronic kits. Check the site frequently for the "Price Mistake of the Day" (the daily sale items).

Tower Hobbies

www.towerhobbies.com
P.O. Box 9078
Champaign, IL 61826-9078
800-637-6050

See "Remote Control (R/C) Equipment" earlier in this chapter.

Surplus

Are you a "surpie" (a surplus fanatic)? If you aren't, you will be after you start getting these catalogs. They're loaded with amazing items you won't find anyplace else, sometimes at unheard of prices. Don't forget to check your Yellow Pages for surplus too. There's nothing like wandering through a surplus yard in person.

All Electronics

www.allelectronics.com
905 S. Vermont Avenue
Los Angeles, CA 90006
818-904-0524

I've been buying surplus from All Electronics for years. Its free, biannual print catalog is a smorgasbord of new and used electronic components and mechanical parts, such as gears, electronics-related hardware, and tools, and then there's the occasional surprise item ("Where's Waldo" mouse pads, anyone?). It's always fun to browse to see what deals you might scare up.

American Science and Surplus

www.sciplus.com
P.O. Box 1030
Skokie, IL 60076
847-647-0011

Hardcore surpies know and love American Science and Surplus. If you don't, you're in for a real treat. Its free catalog, with its smudgy newsprint pages and hand-drawn illustrations, is always a treat to find in your mailbox. Not only is it filled with useful, wonderful, and often bizarre military and industrial surplus and educational toys and kits, but there's a pervasive and wacky sense of fun and good humor that makes reading it a treat. There are plenty of goodies here for the robot enthusiast, from DC motors, gears and wheels, to electronic components and robot kits. And then there's all the other fun stuff. You haven't lived until you see the look on someone's face when you give the gift of a "PooPet" for Christmas (biodegradable yard "art"—of bunnies, ducks, and so forth—made out of pressed animal poo).

Electronic Goldmine

www.goldmine-elec.com
P.O. Box 5408
Scottsdale, AZ 85261
800-445-0697

As its name might suggest, this seller of new and surplus electronic and high-tech goodies trades in some seriously great deals on equally great gear. How about a high-quality, torquey 6V–24V Maxon gearhead motor for $19.95, or a 3V Mabuchi pancake motor for $2 (two for $3)? Check out its Web site frequently for new offerings and sales, or to download a PDF version of its catalog.

IN THIS CHAPTER

- Hooking you up with the best robot books
- The dish on the best robot magazines
- Dialing you in to the best robot videos

11

ROBOT BOOKS, MAGAZINES, AND VIDEOS

Attention Bot-Obsessed Bookworms!

Despite repeated pronouncements over the last few decades of a coming all-electronic, paperless world, it still very much runs on tree meat. The digital age and the Internet revolution have done little to diminish our appetite for books, magazines, and newspapers. In fact, as we do get more of our media through electronic sources, these physical "artifacts" become even more charming and precious, not to mention infinitely more utilitarian. The following selection of books, magazines, and a few videos, are some of the best that tackle the theoretical and practical aspects of robot building and entry-level electronics. Many of these titles are available through robotbooks.com, as well as other online and offline sources. For used copies, bookfinder.com often has many of these titles in its database.

Books (General, Theoretical)

One thing we tried to do with the book you're reading right now is to give you a theoretical basis for thinking about robot building. This is not only helpful in placing your own building efforts within the larger context of robot development, but hopefully, it will also give you some tools for thought in innovating beyond what's come before. Following are some of our favorite books that provide an excellent overview or delve deeper into robot and AI theory.

Battlebots: The Official Guide

Written by: Mark Clarkson
Published by: Osborne, 2003

Unfortunately, this beautifully produced book introducing Battlebots and its most popular robot competitors came out almost exactly at the same time that Comedy Central announced they were canceling the *Battlebots* TV show. Still, this is a handsome and accessible resource for anyone interested in the sport of robotic combat. There are profiles of the build teams, images and diagrams of the bots, detailed competition "battlestats," and even a small how-to section.

Concise Encyclopedia of Robotics

Written by: Stan Gibilisco
Published by: Tab Robotics/McGraw-Hill, 2003

This book promises a lot ("a complete, self-contained reference," "reader-friendly definitions"), but doesn't really deliver. It's geared more toward industrial robots and leaves out a lot of current, cutting-edge robot research. The definitions are also not what one would call "reader-friendly." That said, it's currently the only dedicated robotics dictionary out there, so it still has its uses.

Flesh and Machines

Written by: Rodney Brooks
Published by: Pantheon Books, 2002

In this very straightforward and candid book, MIT AI Lab and robotics pioneer Rodney Brooks lays out his vision of our roboticized future (while providing a glimpse of his personal and professional past). In Brooks's view, we don't need to worry about robots taking over because he believes we will *become* the robots. He sees a gradual, leaky margin between humans and robots (flesh and machines), which will end up with us not being able to tell where we end and where our machines begin. Most of the book details the projects and concepts he has been involved with via his work at MIT's AI Lab and iRobot Corporation. Some have criticized this book for being too rambling and conversational, but I enjoyed its laid-back style and plainspoken English.

Mind Children: The Future of Robot and Human Intelligence
Written by: Hans Moravec
Published by: Harvard University Press, 1990

This controversial book made quite a splash when it was first released in 1990. In it, well-known robotics researcher Hans Moravec paints a picture of a future in which technology overtakes biology as the vehicle of evolution. It is a stunning, and disturbing, thought—a future in which a post-biological human intelligence carries on inside of robotic super bodies that can continuously be upgraded with new hardware and software. Is this really the face of our future? Who can say, but it is certainly fascinating food for thought. Talk about the concepts of this book with others and you can be guaranteed a heated debate.

Mobile Robots: Inspiration to Implementation
Written by: Joseph L. Jones, Anita M. Flynn, and Bruce A. Seiger
Published by: A K Peters Ltd, 1999

It would be a silly overstatement to say that this book changed people's lives, but it certainly changed a lot of people's thinking about robots, robotic intelligence, and what was possible for the average robot builder to accomplish with readily available technology. *Mobile Robots* could just as easily go in the "(Books) Bot Building" section later in this chapter, as it contains detailed information about building two small, mobile robots: The Rug Warrior and TuteBot. But it's more than just a builder's book—it presents a whole way of thinking about robots using, what at that time was called *subsumption architecture*, and is now more commonly referred to as *behavior-based robotics* or *BBR* (see Chapter 2, "Robot Evolution"). The projects (especially the Rug Warrior) are not for beginners, but they are clearly laid out enough that an intermediate builder should have no problem. The book is showing its age (the Rug Warrior uses the 68HC11 chip, which is losing favor among many builders), but the overall information and the how-to-think dimensions of it are still worthwhile—and you can always use the programming ideas and sensor designs in your own, more up-to-date, robots and controllers.

Robo Sapiens: Evolution of a New Species
Written by: Peter Menzel and Faith D'Alusio
Published by: MIT Press, 2000

Very rarely can a book bestow upon you both deep geek and hipster art cred, but *Robo Sapiens* does just that. This breathtakingly photographed, oversized paperback is disguised as a coffee table book, filled with sumptuous photos of cutting-edge (circa 2000) robots, their builders, and their labs. But besides award-winning photographer Peter Menzel's images, there is a wealth of information on robot concepts, design philosophies, and robot specs. There are also authors' field notes and interviews with robot pioneers. It's not only the most beautiful book ever produced on robots, it's also one of the best and most provocative.

Vehicles: Experiments in Synthetic Psychology

Written by: Valentino Braitenberg
Published by: MIT Press, 1986

Anyone serious about exploring robotics, AI, chaos theory, cybernetics, brain research, and a host of related disciplines, needs to read (and reread) this amazing little book. Written by Valentino Braitenberg, director of the Max Plank Institute of Bio-Cybernetics, *Vehicles* presents a series of thought experiments using electro-mechanical vehicles to illustrate them. The first vehicle has one sensor and one motor. Increase the "quality" to which the sensor is attuned (say, temperature) and the motor will run faster. Simple. Straightforward. The second vehicle has two motors and two sensors. Connect the sensors to the motors on the same side and let the sensor quality be light. The vehicle will turn away from the light source because the sensor/motor closest to the light source will be more excited (thereby turning that wheel more, causing the robot to move away from the light). Cross the sensor wires to opposite motors and you have light-following behavior. From there, Braitenberg's vehicles get increasingly more complex, and as they do, we start to see behaviors emerge that look as though intelligence is behind them. *Vehicles* offers a fascinating look at how complex behaviors might arise from simple sets of sensors/actuators. Not surprisingly, this book influenced legions of robot builders, especially the progeny of W. Gray Walter (the BEAM roboticists, Rodney Brooks, and behavior-based robotics, etc.). One of the most amazing things about this book is that, given the rigor and density of the subject matter, it is written in such a warm and engaging style. The book begins with the sentence, "Let the problem of the mind dissolve in your mind." As Keanu Reeves might say, "Whoa!"

Books (Bot Building)

As great as the preceding books are, they would get dirty and take up precious space on your work bench. The following books belong among the resistors, capacitors, and toxic lead fumes (we kid about the fumes, we kid!).

Building Robot Drive Trains

Written by: Dennis Clark and Michael Owings
Published by: McGraw-Hill, 2003

The idea behind the new *Robot DNA* series, of which this book is an early offering, is to cover a single aspect of robot building in substantial detail (see also *Programming Robot Controllers* later in this chapter). This book covers every aspect of robot locomotion in a surprisingly clear and thoughtful manner. The level of detail might scare off some people, but the whole point of the series is to thoroughly explore the subsystem that's being covered. In this case, there's everything from an introduction to various types of locomotion and a look at all the common motor types to motor control circuits and detailed information on wheel rotation encoders. It'll be exciting to see how far this series goes and how a complete library of such subsystem guides might help the serious bot builder.

Build Your Own Combat Robot

Written by: Pete Miles and Tom Carroll
Published by: Osborne, 2002

If you're interested in participating in combat robotics, or just want to peek under the hood of these monstrous machines, this book is a great guide. It offers a brief history of the sport and some first-person accounts and stories about builders. The main focus of the book is, of course, about building robots with bad attitudes. Each chapter covers a subsystem (locomotion, power, construction techniques, and weapons systems) and discusses the trade-offs of different technologies. There's also a color insert showing photos of robots duking it out in the ring. The two authors have a combined 52 years in robot building, so they really know their stuff.

David Baum's Definitive Guide to LEGO MINDSTORMS

Written by: David Baum
Published by: APress, 2000

After MINDSTORMS came out and made such a big and unexpected splash with serious robot enthusiasts, builders started clamoring for any resources they could find. One of the first resources to address the building system on an adult level was this book. Written by David Baum, creator of the NQC programming language, it deftly guides the reader through 14 LEGO robot projects. The successive projects increase in both mechanical and programming complexity, enabling the builder to gradually learn the engineering behind both. The book also covers both LEGO's RCX Code and Baum's own NQC language. The CD-ROM included contains control programs in both RCX and NQC and several versions of NQC (though you'll want to download a more current version from the Web). See the LEGO MINDSTORM section of Chapter 6, "Acquiring Mad Robot Skills," for a bit more on LEGO programming.

Extreme MINDSTORMS

Written by: David Baum, Michael Gasperi, Ralph Hempel, and Luis Villa
Published by: APress, 2000

A followup to Baum's *Definitive Guide*, this book not only delves more deeply into the use of NQC, but also discusses pbForth and LegoOS, two other LEGO-friendly programming environs. One of the niftiest features in the book is a section on creating your own homebrewed sensors for MINDSTORMS robots.

Junkbots, Bugbots & Bots on Wheels

Written by: Dave Hrynkiw and Mark Tilden
Published by: Osborne, 2003

This book is, hands down, one of the best ever written on the subject of robots. Frankly, it would have intimidated us in writing this book, if it weren't for the fact that *Junkbots* deals with a single subgenre of robots: BEAM. There really are no bigger champions of BEAM technology than Dave and Mark and no better hand-holder into this realm than Dave (Mark only contributes the introduction and a very gnarly final chapter on BEAM motor controllers). Living up to the "A is for aesthetics" part of the BEAM initials, this book is extremely handsome, with crisp photos, clear circuit diagrams, fun cartoony illustrations, and an overall design that makes the book very inviting. Dave's writing is also extremely lucid, entertaining, and down to earth. Besides an introduction to basic electronics, tools, shop safety, and other "robots 101" subjects, the book includes seven how-to projects that increase in level of difficulty, from a little spinning solar top to a fairly sophisticated four-legged walker. One great thing about BEAM, and a focus of this book, is the use of technojunk as builder materials. *Junkbots* can teach just about anyone how to turn that drawer full of old Walkmans, pagers, modems, and other castaways into cool robocritters.

Muscle Wires Project Book

Written by: Roger G. Gilbertson
Published by: Mondo-Tronics, 1994

This really isn't a robot book, but it does include a robot project at the end. It is written by Roger Gilbertson of Mondo-Tronics, one of the main commercial distributors of shape memory alloy (sold by Mondo-Tronics under the name Muscle Wire). The book includes background information and technical data on its Muscle Wire, and details fifteen projects, from simple wire-powered levers and lifters to mechanical butterflies, and finally, Boris, a three-segment, six-legged robo-critter that walks without motors. Muscle Wire is a fascinating material that will hopefully find more widespread application as the price of titanium (one of its component alloys) continues to drop.

Programming and Customizing the BASIC Stamp

Written by: Scott Edwards

Published by: Tab Books, 2001

This is the essential starter guide for anyone programming a robot using the BASIC Stamp (BS). The book covers the BS1, BS2, and BS2-SX. There are 12 projects to build in the book, including a robot and a short-range sonar. A CD-ROM included with the book contains BASIC source code, software tools, and all of Edwards's "Stamp Applications" columns from *Nuts & Volts* (see "Magazines" later in this chapter).

Programming Robot Controllers

Written by: Myke Predko

Published by: McGraw-Hill, 2003

Another in the *Robot DNA Series*, this book explores robot programming, specifically, programming the Microchip PICmicro PIC16F627. The book is extremely thorough, not only in exploring the PICmicro, but also in describing different control architectures and sensor systems. An included CD-ROM offers PICmicro datasheets, programming tools, and source code for sample applications.

Robot Builder's Bonanza

Written by: Gordon McComb

Published by: Tab Robotics/McGraw-Hill, 2000

The book that launched a thousand bots, this classic in hobby robotics has been revised with lots of new material on everything that's happened since *Bonanza* first landed on builders' workbenches (back in 1987!). In addition to the old material that's remained relevant (mainly about construction techniques and robot anatomy), there are chapters on microcontrollers, LEGO MINDSTORMS, and currently available sensors. Gordon is a great how-to writer who really makes you feel confident about your ability to realize these projects for yourself.

Robot Builder's Sourcebook

Written by: Gordon McComb

Published by: Tab Robotics/McGraw-Hill, 2003

If there is only one book on your workbench (ahh...besides the book you're currently reading), it should be the phonebook-sized *Robot Builder's Sourcebook*. It is an absolutely exhaustive survey of robot technology sources and anything even remotely related to robotics. There are thousands of listings of products, manufacturers, retailers, Web sites, organizations, journals, and magazines—the list goes on and on. There are also 250 articles and sidebars that detail various aspects of robotic concepts, robot building, robot pioneers, and more. This is truly a phenomenal resource.

Robot Building for Beginners

Written by: David Cook

Published by: APress, 2002

This 568-page tome is an extremely detailed (sometimes maddeningly so) journey into the world of robot building. Dave Cook does an admirable job of guiding the reader through all aspects of building a line-following robot, from learning the basic electronics and assembling required tools to testing the circuits and building the bot. Along the way, some of the detail seems obsessive (he spends 16 pages talking about 9-volt batteries and shows price comparison charts of components in which the prices vary by only a few cents). Still, if you stay awake for the whole thing, and follow his instructions, you'll end up with a really cute little two-wheeled bot that cleverly uses a Ziploc plastic sandwich container as a body.

Robot Sumo: The Official Guide

Written by: Pete Miles

Published by: Osborne, 2002

Robot sumo is one of the fastest growing forms of robotic competition. If you're interested in building a beefy bot that can dig in and muscle other bots out of a ring, this book will help you train for that big match. Written by Pete Miles, of the well-known Seattle Robotics Society (and coauthor of the *Build Your Own Combat Robot* listed earlier in this chapter), it guides you through every aspect of the design and building process. There are so many different ways of building a sumo competitor, and Pete details all of the design trade-offs for each. The book is filled with clear photos and illustrations, helpful charts and formulae, and interesting sidebars. The book also includes the rules for robot sumo and even instructions for building your own sumo ring.

Books (Electronics)

If you want to get deeper into robots after this book (and you will if we've done our job), you'll also need to learn a lot more about electronics. Read these three books and we guarantee (well, not an *actual* guarantee) you'll be comfortable moving on to the next level of electronics wizardry.

Bebop to the Boolean Boogie

Written by: Clive Maxfield

Published by: HighText Publishing, 1995

If you read *There Are No Electrons* (listed later in this chapter) and enjoyed it, this is the perfect book to graduate to. Infused with a similar cocky and mischievous spirit, Clive Maxfield explores more complex electronic concepts, such as digital versus analog, logic gates, and how electronic components are made. No amount of jokes, lively prose, and thoughtful illustrations (all of which this book has) is going to make the subject matter go down easily (this is heady stuff). But after you've read some beginner electronics guides, and done our *Absolute Beginner's* projects, you might be surprised at how many "ah-ha" moments you'll get while reading *Bebop*.

Getting Started in Electronics

Written by: Forrest M. Mims, III
Published by: Radio Shack, 1983

The first thing that strikes you when reading Mims's classic *Getting Started in Electronics* is the fact that the entire book is hand-drawn and handwritten by the author (on graph paper). Anyone who took a mechanical drawing class in school will have fond (or not so fond) memories of the technical lettering and illustrating taught there. The next thing that will hit you is how gosh-darn clear the book is and what a gift Mims has for making electronics easy to understand. Step-by-step, he *draws* you in (okay, that was bad) to the fascinating world of electricity, from atoms and static electricity to direct and alternating currents to digital electronics and logic gates. There are also projects, how-to information (on such things as soldering and breadboarding), and diagrams of 100 basic circuits you can build. If the world runs on electricity, and it is people who put that electricity to work, more of those people probably got their first glimpse of electricity's magic through this book than any-where else.

There Are No Electrons: Electronics for Earthlings

Written by: Kenn Amdahl
Published by: Clearwater Publishing, 1991

The forward to *There Are No Electrons* claims that this little indie-published book was actually one of the inspirations behind all of those beginner's books with the abusive titles (we won't name names). The claim is not surprising. The wacky humor, friendly tone, and engaging storytelling used to explain something as intimidating as electronics inspires many who read this book. You can't help but read it and think: "Wait, I *totally* understand all this stuff! Why couldn't all those *other* books I've read on the subject break it down for me like this?" The truly amazing thing is how much fun you have while you're learning about electronics—and that you're learning it by way of the silliest, most bizarre stories. Amdahl uses little green men (and women), a duck in a motorboat named Bruce, a time-traveling wizard (and his painfully beautiful assistant Belinda), and other fairy tale–like characters to teach electronics concepts. And then there's his assertion that all of the chief electronics pioneers were secretly Norwegian. Why? I can't remember, but in the reality distortion field of this book, it all seems to make sense.

Magazines

If you want to stay abreast of the latest in electronics, digital technologies, and robotics, you'll want to check out the following magazines.

Circuit Cellar

www.circuitcellar.com
P.O. Box 5650
Hanover, NH 03755-5650
860-875-2199
$21.95/12 issues

If you find yourself *really* getting into robotics, especially programmable robotic control, you'll want to check out this venerable electronics magazine. *Circuit Cellar* focuses on embedded systems, and that means microcontrollers, and *that* means robots. It covers every aspect of embedded systems, from the MCUs (microcontroller units) themselves, to sensor systems, display technologies, wireless networking, and more. It also covers a lot of PC technologies (building your own computer, rolling your own hard drive controllers, and so forth). This monthly is a hardware hacker's delight. Warning: Not for the newbie. This is deep computer and electronic engineering (but mere mortals like us with a taste for gadgets can still learn a lot from perusing).

Elektor Electronics

www.elektor-electronics.co.uk
Gibbs Reed Farm
Pashley Road, Ticehurst, nr. Wadhurst
East Sussex, England TN5 7HE
+44-1580-200657

This magazine from the U.K. is geared toward the general electronics hobbyist. Here you'll find articles covering everything from computer hardware to stereo equipment to home-built electronic test equipment. There are articles on all of these (and other) technologies, but the heart of the magazine is the project section. Every issue has complete plans to a half dozen do-it-yourself projects. While *Elektor* overall might be intimidating to a new convert to electronics, there's usually something for all skill levels. The magazine is very pricey to ship to the U.S. (at $83/year), but you can usually find issues on good newsstands.

MIT Technology Review

www.technologyreview.com
One Main Street, 7th Floor
Cambridge, MA 02142
617-475-8000
$28/10 issues

It's not a robot or an electronics magazine, but *MIT Technology Review* is probably the best way to stay on top of all cutting-edge science and technology research (which includes robots). Each issue starts with dozens of small pieces on promising research developments and then segues into more in-depth articles. The magazine is designed and written in a very straightforward, non-intimidating style that makes it accessible to a layperson (while still being "meaty" enough to satisfy the pros). Nearly every issue covers some new development in robots or such related "cyborg" technologies as artificial eyes and organs, neural implants, and other new technological wonders.

Nuts & Volts

www.nutsvolts.com
430 Princeland Court
Corona, CA 92879
909-371-8497
$24.95/12 issues

If you decide to subscribe to one electronics magazine, you'll want to make your check out to *Nuts & Volts*. This is the electronics do-it-yourselfer's dream magazine, with each issue filled with news of new products and services, discussions of emerging technologies, and nifty electronics projects. Although it's not expressly devoted to robotics, it is widely covered (with several monthly columns devoted to the subject), along with microcontrollers, computer control, wireless networking, sensors, and other subjects related to robotics. Like *Elektor Electronics* (listed earlier in this chapter), the content can be a bit daunting to beginners, but it always has some information that appeals to all skill levels.

Real Robots

www.realrobots.co.uk
Eaglemoss Publications Ltd.
5 Cromwell Road, London SW7 2HR, U.K.
08707-270170

It's a magazine! It's a robot kit! It's two, two, two bits in one! *Real Robots* from the U.K. is a unique concept. Each issue comes with new parts (and step-by-step articles) for a robot (called the Cybot). So, as you read about different robot technologies and experiments, you're building a bot to try some of these ideas out for yourself. The glossy full-color magazine is also filled with general articles about robots and related technologies. There are some drawbacks to this idea. Subscribers have to start at issue 1 (they're up to issue 43), which means that the articles not related to the Cybot are a bit dated (things move fast on the cutting edge of today's digital technologies). Also, enthusiastic subscribers can find it maddening having to wait for each new issue to add parts to their robot. Unfortunately, *Real Robots* is not sold in the U.S., but issues (with or without parts) are almost always available on eBay.

Robot Science & Technology

www.robotmag.com
780 S.W. Barrington Drive
Oak Harbor, Washington 98277

As you've probably noticed by now, there aren't many robot magazines out there, including this one. At least, it's not available at the moment. The publishing of this promising magazine has been hit and miss, as the editor/publisher struggles to keep it going. He waits until he has enough ad revenue, or other funding, to put out each new issue. The long-awaited issue 9 may or may not be out by the time you read this. Check the site for details. They still sell back issues.

Videos

We modern media citizens are a visual breed. We like to see our tech in action. Sadly, there aren't that many good videos about robots. This is basically what's available to date. We want our RobotTV!

Battlebots

VHS
$24 each

If you're interested in robot combat (and didn't get enough of the *Battlebots* TV show while it ran on Comedy Central), you might want to check out some of the videos that are available for sale on the Battlebots Web site (www.battlebots.com). *Battlebots: Beginnings (Long Beach 1999)* covers the infamous first gathering of robot warriors under the Battlebots banner. Convened at the Pyramid at Long Beach State University in August 1999, about 60 builders participated. One of the many firsts at this event was a meeting between the creators of Battlebots and TalentWorks, the TV production company that eventually brought the competition to television. *Las Vegas*

1999 Championship was taped by TalentWorks, and was cablecast as a pay-per-view event. It also marked the first appearance of BattleBox announcer Mark Beiro. Both videos contain the robot competitions themselves as well as "pit" segments with the builders and their robots.

BEAM Video Clips

Various formats

www.solarbotics.com

Free

There aren't any BEAM videos currently available, but several are currently in production. Check Solarbotics periodically for updates. In the meantime, you can look at footage of various BEAMbots in action and watch an interview with BEAM creator Mark Tilden on Solarbotics' site. Go to the front page and click on the Videos link in the Galley section.

Extreme Machines: Incredible Robots

VHS

TLC Video, 1999

$19.95

After you've watched *Robots Rising* (listed later in this chapter), you can pop in this cassette and see where robots were a year later. We see MIT's robot COG being put through his (its?) paces, we hear that wacky Hans Moravec rap about robots replacing us, and we get a glimpse of the Honda P-series robot walking up and down stairs. There's also coverage of military robotics circa 1999.

Fast, Cheap & Out of Control

DVD

Columbia/Tristar, 1997

$24.95

This is not a robot film, and might only appeal to those with a bit of an "arty" bent. Shot by the eccentric documentarian Errol Morris (*Thin Blue Line* and the Apple *Switch* TV ad campaign), this 1997 film follows the lives of four obsessively dedicated people. George Mendonca is a topiary gardener and spends his days trying to shape nature to his own desires. Dave Hoover is a lion tamer trying to control the great cats. Ray Mendez spends his days studying the most mundane details of the life of the blind and hairless mole-rat. And finally, Morris trains his restless eye on Rodney Brooks, the outspoken and controversial director of the AI Lab at MIT. Morris always has a way of disarming his subjects and getting the essence of them out, so you see Brooks in a more unfiltered light than in other documentaries (such as *Robots Rising*, later in this chapter). But this film is not just being recommended here because one quarter of it concerns a robot pioneer. The entire film deals with life—understanding it, engineering it, and trying to control it. Every roboticist eventually finds him- or herself looking to the natural world in an effort to sort of reverse engineer it for use in robots. This film trades in that kind of obsessive exploration of life's boundless wonders. Again, this film will not appeal to everyone. Some will find the jumpy camera work, cut-and-paste editing, strange music, and the overall bizarre tone unsettling. Others, like your rather pretentious author, love this sort of thing.

Robotica

VHS
TLC Video
$19.95 each

Robotica, TLC's answer to *Robot Wars* and *Battlebots*, was an intriguing take on robotic competition. Rather than pitting the robots against each other, the bots were challenged by a series of obstacle courses. The two final winners of these matches went on to a sumo-type match on a raised platform (with fire and metal spikes below). Competitors were allowed to alter their robots between matches to better meet each new challenge. Unfortunately, the show's production style (for instance, seriously annoying hosts), and less bot-kicking than the other shows, didn't appeal to a wide enough audience. The series was soon pulled. The first six (50 minute) episodes are available for $19.95 each. Ouch! The show wasn't *that* good. Get your local video store to buy them.

Robots Rising

VHS
Arcwelder Films, 1998
$19.95

This nearly two-hour made-for-TV documentary details the work of the usual suspects (Brooks, Moravec, and Red Whittaker) and takes us into robot labs around the world (principally the U.S. and Japan). Narrated by Linda Hamilton, star of *The Terminator*, this program offers a handsome thumbnail sketch of where cutting-edge robotics was at the end of the twentieth century.

In This Chapter

- Portals
- Robot-specific sites
- Robot communities
- Robot research centers
- Useful bot-building references

12

Robots On The Web

As we discussed in Chapter 2, "Robot Evolution," there has probably been no greater accelerator of robot evolution than the information distribution, communication, and collaboration tool known as the World Wide Web. There are literally thousands of sites and discussion boards dedicated to every type and every aspect of robotics. In the following sections, you will find the proverbial tip of the iceberg. These are sites that either act as gateways (portals) to a host of other sites, or provide a worthwhile summary of a particular area of robotic investigation. We've also listed a few of the best robot *communities* online, places where you can go to hook up with people who share your particular interests. From the sites listed here, like a good little spider, you can probably click your way through to just about every robot-related thread on the Web. Ain't technology grand?

Portals

Every journey begins with a first step—or at least that's what some corndog philoso-pher once said. In online travels, that usually translates to "portals," mega Web sites that offer their own articles, reviews, glossaries, and so forth, and also link to all of the best online material on a given subject. Here are some of the best portals in cyberspace dedicated to all things robotic.

AndroidWorld

Web Site: www.androidworld.com

This has got to be one of the butt-ugliest sites in cyberspace, but you'll find yourself overlooking that unfortunate distinction when you discover the wealth of informa-tion and linkages stored here. If you're keen on robots of the bipedal, humanoid variety, this site will hook you up with just about every professional and amateur project out there. There is so much material here, and so much linked from here, that you could spend weeks traversing all of the connected content.

Botic

Web Site: www.botic.com

Botic looks like it has what you might expect from a robot portal: robot-related news items, product reviews, topic sections, a directory of sites, and so forth. Unfortunately, there's not a lot of content in each area yet. There are so many simi-lar sites built using Web content management templates that look more impressive than the content they offer. That said, there is material here that makes Botic worth a visit. It also sponsors a small network of builders' sites that are linked from the front page.

GoRobotics Network

Web Site: www.gorobotics.net

A similar site to Botic's, but with a more satisfying amount of content. Some nice articles and reviews and a few how-to pieces. It also offers a decent listing of robot clubs and a robot glossary.

MachineBrain

Web Site: www.machinebrain.com

This site has lots of great links to other sites (in categories such as General Robotics, Fighting Robots, Robot Reference), as well as robot technology news, and even a free classifieds section.

NetSurfer Robotics

Web Site: www.netsurf.com/nsr

NetSurfer has been around since the dawn of the Web, offering a series of great electronic newsletters and Web-based "zines" devoted to various educational and scientific subjects. One of these e-pubs covers robots. It's not really a portal, and at this point, it's very infrequently updated, but it does offer robot-related news, book reviews, and links to books, kits, and other robot-related products available online.

Robot Café

Web Site: www.robotcafe.com

This site is dedicated to all those who "live, eat, breath...robots." Please, people, if you start dining on computer chips and shoving nuts and bolts up your nose, you've got something beyond an exciting hobby interest that you need to have a professional take a look at! The two main features of this site are an excellent directory of online robot resources and a gallery of robots with basic information on each bot and links to the robot builder's Web site (if there is one). Anyone can register on the site and post bots.

Robots.Net

Web Site: www.robots.net

Probably the most popular robot portal, Robots.Net is frequently updated with cutting-edge tech news, product reviews, how-to articles, and directory listings. It also has a large gallery of robot projects. Its "Robot of the Day" column on the front page is a great way to see what the gallery has to offer.

Solarbotics.net

Web Site: www.solarbotics.net

If you're interested in exploring BEAM technology, Solarbotics.net is a must-bookmark. The site is not only home to over 40 BEAM-related Web sites, but it's also an invaluable source of information related to BEAM tech, electronics, and other robot technologies. There are tons of goodies here.

TechGeek

Web Site: www.techgeek.com

This is another robot site worth regular visits. It offers news, reviews, a really nice project section detailing site members' building efforts, and robot club listings. There is also a big list of resources.

Robot-Specific Sites

After you have some idea of what types of robots trigger your actuators, you'll want to explore those types in more depth, and bookmark some of these robot-specific sites.

BEAM Online

Web Site: `www.beam-online.com`

Next to Solarbotics.net, BEAM Online is the site you'll want on your radar if you're interested in keeping up with this fascinating area of robotics. The site offers project tutorials, papers (well, at least one really good one detailing how the common BEAM Bicore circuit works), BEAM schematics, a BEAM FAQ, a robot gallery, and much more. The site also contains a search engine to over 40 BEAM Web sites.

Battlebots

Web Site: `www.battlebots.com`

If you're interested in combat robotics, especially Battlebots, this site offers some great resources. It has the full Battlebots rules, profiles of builders and their bots, information and rankings for all Battlebot competitions (from Long Beach 1999 to season 4 on Comedy Central), video clips, some robot-building tutorials, and a store of Battlebots-related books, toys, and other goodies. The creators of Battlebots have also started Battlebots IQ, a high school curriculum that uses robotic competition to teach kids about engineering, design, math, logic, and problem solving. Details and sample curricula are available through the Battlebots site.

Cybot Builder

Web Site: `www.cybotbuilder.com`

This site started out dedicated to the Cybot, a robot kit that comes in installments with a subscription to the U.K.'s *Real Robots* magazine (see Chapter 11, "Robot Books, Magazines, and Videos"). The site has recently been reborn as RobotBuilder.co.uk. It still has lots of info about the Cybot, but it now covers all aspects of hobby robotics. Its forum section can be especially helpful.

Honda Humanoid Project

Web Site: `world.honda.com/robot/`

This official Honda site covers both the P-series robot project and its younger sibling ASIMO. Here you'll find background on the two bots, specs, and movies. The careful reader will notice that on the spec sheets for ASIMO, it says that the "Control Unit" is a wireless connection and the "Operating Section" is a "workstation and portable controller." In other words, a remote brain and a remote control that few people seem to know about. ASIMO is basically a multi-million dollar, high-tech puppet.

Robot Sumo

Web Site: www.robotsumo.com

David Cook, author of *Robot Building for Beginners* (see Chapter 11), has put together an excellent "Illustrated Guide to American Robot Sumo" on his Web site Robot Room. This one-page guide runs through the basic ideas behind robot sumo, the rules, and rules variations, all illustrated with hand-drawn sketches by Dave. There are also links to all of the known rules and classes of the sport.

Robot Wars

Web Site: www.robotwars.co.uk

Like its U.S. counterpart, Battlebots (see previous), U.K.'s Robot Wars has a fairly decent site detailing the competitors (and "house robots") seen on the popular BBC program. If you have Real Player Gold Pass, you can also watch some of the matches. It also sells Robot Wars videos (in PAL format only), toys, video games, and so forth.

Robots Wanted

Web Site: www.robotswanted.com

If your curiousity towards robotics runs toward the retro, the antique robots of yester-year, you'll want to check out Robots Wanted. It actually links to several different sites dedicated to Androbots, Heathkits, RB5Xs, and other fondly remembered favorites. Be sure to check out the gallery and the mailing lists for the various vintage bots.

MINDSTORMS

Web Site: www.legomindstorms.com

LEGO MINDSTORMS's official Web site is not just a place to buy products. LEGO realized early on that MINDSTORMS was about collaborating, showing off builders' handiwork, sharing ideas, and so forth. The site features tutorials, profiles of especially innovative LEGO robot designs, a discussion forum, and software tools. There are also links to other MINDSTORMS-related sites online.

Robot Communities

One of the most exciting things about robot content online is the cool communities of fellow builders with which you can connect. Here are some of the best water holes for discussing robots.

AIBO-Life

Web Site: www.aibo-life.org

This site is where dedicated AIBO owners go to talk obsessively about how much they love their little robot pets, how smart their little mechanical friends are, and how playing with their artificial companions sure beats interacting with silly ol' humans. (*Just kidding*, AIBO owners!) AIBO-Life has discussion forums, real-time AIBO chat, AIBO Q&A, an AIBO image gallery, and listings of gatherings (called "AIBO Lovefests"). If you're not sick to your AIBO stomach yet, then you're probably ready to get an AIBO-Life.

Lugnet

Web Site: www.lugnet.com

Lugnet (which stands for "LEGO Users Group Network") is the place where the international community of LEGO enthusiasts gathers. The site discusses all aspects of LEGOs, not just the robotic variety, but an awesome robotics discussion forum with thousands of postings on various MINDSTORMS-related topics. Besides the discussion areas, the site also contains (or links to) a database of LEGO parts, scans of LEGO building set manuals, and much more. This site is truly an awesome example of how a group of people can come together over a shared interest and devote thousands of collective hours toward building a free resource center and user community.

Solarbotics.net

Web Site: www.solarbotics.net

Solarbotics offers free Web hosting to anyone doing BEAM-related robot projects. There's also a BEAM discussion area. (Refer to the "Portals" section earlier in this chapter.)

Yahoo! Groups

Web Site: groups.yahoo.com

Yahoo! Groups is the home to dozens of discussion groups about robots. Just type in the word **robot** and then whatever area of robotics you fancy—+BEAM, +Sumo, +beginner, and so forth—and you'll see which groups are available. Some are more active than others. Also search for hobby electronics discussions, especially the group "Electronics_101." And if you don't find a group that's talking about what you want to talk about, you can start your own. It's free and it's extremely easy.

Robot Research Centers

A lot of the cutting-edge action in robots is happening in university, corporate, and defense contractor labs. Some of the sites for these developers are little more than "Kilroy pages" (in other words, all they have to say is "we're here!"). Others have surprisingly useful information and cool media (pictures, videos, in-lab Web cams). Here are some of the better ones.

Biorobotics Lab at Case Western University

Web Site: `biorobots.cwru.edu`

The mission of Case Western University's Biorobotics Lab is to study the mechanics of nature to see what of it can be engineered into robo-critters. It studies cockroaches, ants, crickets, and other creepy crawlers, and then creates robots inspired by its investigations. The site shows the evolution of its cricket and cockroach robots, along with a series of bots using *whegs*, a cross between a wheel and an insect leg.

Carnegie Mellon Robotics Institute

Web Site: `www.ri.cmu.edu`

Carnegie Mellon University has almost become synonymous with robotics research, thanks to the decades of innovation coming from its Robotics Institute. One only has to browse through the extensive content on the site to get a sense of the beehive of activity surrounding the Institute. There are some 35 laboratories working on hundreds of projects. Each project has its own page, with pictures, descriptions, links to related papers, and contact information. It's an incredible resource for anyone looking to see what's up with cutting-edge robotics research.

Humanoid Robotics Group

Web Site: `www.ai.mit.edu/projects/humanoid-robotics-group`

MIT's Rodney Brooks and the AI Lab have made the robots Cog, Kismet, and Genghis household names. Okay, only in *very* geeky households, but still… The Lab's robots always seem to get lots of media attention, and with good reason—they're always doing something outside of the robotics/AI mainstream, thinking outside the bot (if you will). The site offers a good indication of why, with tons of current research information, histories, FAQs, videos, papers, and profiles of lab personnel. While you're there marveling at all of the current research, don't forget to check out the "Retired Robots" section, especially the pages for Genghis and the microbot Ant projects.

JPL Robotics

Web Site: robotics.jpl.nasa.gov

Because it doesn't look like we "meatbots" are going to be traveling to other worlds any time soon, most of our hope in extra-planetary exploration lies with our metallic brethren. At JPL's robotics site, you can keep abreast of developments in its space robotics research. It includes pages on current space-related projects for NASA, its non-NASA projects (such as Urbie, an urban recon robot for the military), and you can browse the archive of its completed projects.

Jouhou System Kougaku Lab

Web Site: www.jsk.t.u-tokyo.ac.jp

As we've already discussed elsewhere, Japan has a love affair with robots of the humanoid kind. This site provides a glimpse at what's going on at the JSK Lab at the University of Tokyo. It has pages detailing its research into bipedal walking systems, systems for real-time translation of sensory perceptions into actuation, movement with artificial tendons, and more. Unfortunately, only some of the pages are in English.

Robotics Research in Japan

Web Site: robotics.aist-nara.ac.jp/jrobres/index-e.html

This page provides an excellent master list of all the universities and research centers in Japan devoted to robotics.

Poly-PEDAL Lab

Web Site: polypedal.berkeley.edu

If you need any convincing about the wisdom of studying nature in search of robotic inspirations, you need look no further than the Poly-PEDAL Lab. Professor Bob Full and his students spend their days studying cockroaches, beetles, ants, and geckos. Their gecko research led to a startling discovery. Geckos can climb walls thanks to tiny hairs on their feet that have nanoscale split ends (200 billionths of a meter!) that actually interact atomically with the surface they're scaling. One can only imagine what the practical applications of a technology based upon this design could be, not only for robots, but in all areas of life (can you say Spiderman?). The lab's Web site has information on its various research projects, images of the critters that inspire, research papers, and even a virtual lab tour. Note: *PEDAL* stands for *Performance, Energetics, and Dynamics of Animal Locomotion*. The Poly- part refers to many-footed animals.

Useful Bot-Building References

The Web offers an embarrassment of riches when it comes to free tutorials and reference materials related to robotics, electronics, mechanical engineering, and other subjects of interest to the robot builder. Here are just a few sample sites we found in our bookmark file.

BasicElectronics.com

Web Site: `webhome.idirect.com/~jadams/electronics/`

This site constitutes a free beginners tutorial on electronics. There's a section on theory, one on identifying electronic components, one with reference information and tools (including a resistor color code calculator and an Ohm's law calculator), and other material of interest to electronic newbies and veterans alike.

Battle Robot Tips

Web Site: `www.robotcombat.com/tips.html`

Jim Smentowski, of Battlebots and Backlash fame, has a nice page of tips on building combat robots and a combat robot FAQ.

Dictionary of Measures, Units, and Conversions

Web Site: `www.ex.ac.uk/cimt/dictunit/dictunit.htm`

This amazing site offers an impressive array of conversion calculators and measurements. Length, area, mass, density, pressure, speed, power—the list goes on and on.

Karl Lunt's Homepage

Web Site: `www.seanet.com/~karllunt`

Karl Lunt is well known in the hobby robotics community. He's written numerous articles for *Nuts & Volts* (see Chapter 11) and has written several books on robots. His home page has a lot of useful information and downloadable software tools for various types of robot microcontrollers.

HBRobotics Club Builders Book

Web Site: `www.wildrice.com/HBRobotics/HBRCBuildersBook.html`

The HomeBrew Robotics Club has posted an entire electronic book detailing all of the subsystems found in bots and covering everything from tabletop robots to robots "too big for the kitchen." There are also a few links to other pages related to various aspects of bot building.

How Stuff Works

Web Site: www.howstuffworks.com

If you're endlessly curious about how the world works, this site can help feed your engineer's soul. It covers all types of physical world mechanics (cars, electronics, science), and has now branched out to the soft sciences, covering subjects such as government, entertainment, and economics. It's a great place to start your investigations into any newfound interest.

MatWeb

Web Site: www.matweb.com

A materials composition database that has data (density, tensile strength, melting point, electrical properties, and so forth) on pretty much every kind of metal, plastic, composite, and ceramic. You can search on material type, the manufacturer, the trade name, and other variables.

NASA Tech Briefs

Web Site: www.nasatech.com

Since the 1970s, NASA has offered a free monthly magazine to qualified subscribers within the business, scientific, and engineering sectors. Now you can also view the magazine's content, and other materials via its Web site. The idea of *Tech Briefs* is to share innovations from NASA with the rest of the engineering world. There's a lot of material on lasers, fiber optics, computer-aided design and manufacturing, new materials sciences, and other cutting-edge technologies—and, of course, robots. You can also subscribe to a series of electronic newsletters covering specific technologies.

Online Conversions

Web Site: www.onlineconversion.com

The ultimate site of conversion calculators for converting weights, measures, frequencies, flow rates, force, density, pressure—you name it. It has over 30,000 conversions available!

The Robot Room

Web Site: www.robotroom.com

This site, run by Dave Cook (author of *Robot Building for Beginners*) is a treasure trove of useful information, inspiring ideas, and robot building tips. Definitely worth a visit.

RepairFAQ

Web Site: www.repairfaq.org

Fancy-schmancy Web sites have come and gone; megacorps have tossed millions of dollars away trying to create "sticky" sites (ones with content so gosh-darn interesting and/or useful, you come back again and again), only to have them fail in the end. And through it all has stood the humble, the plain-looking, the free, and the infinitely awesome Sci.Electronics.Repair FAQ. The site has an overwhelming amount of information on repairing consumer electronics, on electronics in general, and on optics and lasers. And this just covers the top-level table of contents. There's so much here, you could get lost. Luckily, there's a site search feature.

Robotics FAQ

Web Site: www.frc.ri.cmu.edu/robotics-faq

This venerable document used to be the first stop many people would make in their explorations of robots and robot resources online. Unfortunately, it hasn't been updated since 1996. It is still a worthwhile destination, though, as it does an admirable job of explaining some of the basic concepts, technologies, and resources related to robots. If some of the links are dead, try doing a Google.com search to see if what you're looking for has simply moved elsewhere.

W.A.R.S.'s Reference Pages

Web Site: www.winnipegrobotics.com/Pages/Robots/Reference/Fasteners.html

The Winnipeg Area Robotics Society has an excellent Web site. Its reference section is especially useful. It has information on robot motors, screws and other fasteners, a resistor color-code chart, the characteristics of different wire gauges, and other material useful to bot builders.

Glossary

AC (Alternating Current) A type of electrical current that moves in two directions, cycling (alternating) back and forth from a negative polarity to a positive one, and back again. Unlike DC current, AC can be "stepped up" to a much higher voltage for efficient long distance transmission, and then "stepped down" again for safe use in devices. Many devices use DC circuits internally, so the AC current from a common wall outlet needs to be transformed to DC before being delivered to the device's electronics. See *DC (Direct Current)*.

Actuator A mechanism used to translate an energy source (such as electricity, air, or a fluid) into motion for purposes of moving part of a device (such as a mechanical arm, leg, or wheel on a robot). The most common forms of actuators in small robotics are the DC (Direct Current) motor and the servo motor. See *DC (Direct Current)*; *Effector (End Effector)*; *Servo Motor*.

Agent A piece of autonomous (or semi-autonomous) software in a computer program or robot control system. An agent has an agenda, a reason for being, and goes about its programmed business (usually inputting and outputting data) unless the control system or other agents override it. See *Augmented Finite-State Machine (AFSM)*.

Analog In electronics, refers to a continuously variable signal (as opposed to an on/off, digital signal). The easiest way to keep analog and digital straight is to think of an undulating wave (analog) versus a set of stair-steps (digital). When shopping for integrated circuits, analog ICs are often called *linear ICs*. One big problem with analog signals is noise (think of pollution and little critters riding in that wave). A lot of effort is spent to try and keep noise out (through filtering, shielding, and special types of connections). See *Digital*.

Android An artificial being made to closely resemble a human, often with lifelike skin and other features so real that it can be passed off as a biological human. The term originally applied to purely mechanical men (*automata*), and the word *robot* was first applied (in Capek's 1921 play *RUR*) to humanoid creatures created through biotechnology. Strangely, at some point (around the late 1930s), the two words switched meanings. There is still much ambiguity about the term, leading to, among other things, robots such as R2-D2 in *Star Wars* being referred to as *droids* (the abbreviated form of android). See *Cyborg*.

Anode The positive pin of a diode or LED (light-emitting diode). See *Cathode*.

Artificial Intelligence (AI) An interdisciplinary branch of computer science that seeks to imbue machines with human-like intelligence. An artificially intelligent machine would be able to learn, reason, adapt, and improve itself. There are at least three "schools" of thought regarding AI. *Hard AI* proponents believe that it is only a matter of time (and processing power) before machine brains can do everything (and more) that human brains can. Those researchers that are said to embrace *soft AI* are not convinced that the human brain can be fully mimicked by machines, but they do believe powerful and smarter-than-today's "thinking machines" are possible. The third "school" has been dubbed *evolutionary AI* (or *nouvelle AI*). Proponents of this approach believe machines might be more likely to reach the level of intelligence sought by hard AI if we "grow" robots in an evolutionary-like manner. As the name implies, this approach looks to nature and human psychology for clues, and emphasizes bottom-up, layered methods of building "smart" machines. See *Behavior-Based Robotics (BBR)*; *Biomimicry*.

Augmented Finite-State Machine (AFSM) A "virtual machine" component in a control program that has its own agency and can operate by itself or via input/output from other AFSMs. Each AFSM can communicate with the "outside world" through sensors and with other machines in the control program network. "Conflicts" between these software machines are resolved through priorities for how the control software should operate, with some machine functions taking over for others at given time intervals or under certain conditions. Developed by Rodney Brooks and MIT's AI Lab. See *Agent*; *Subsumption Architecture*.

Behavior-Based Robotics (BBR) An approach to robotic control that looks to nature and animal behavior for inspiration. The term is open to interpretation, but usually implies a "bottom-up" design philosophy that eschews the building of complex world models/maps of the environment, preferring to program robots with sense-act reflexes, rather than the sense/plan/act architectures of AI. See *Biomimicry*; *Sense-Act (SA)*; *Subsumption Architecture*.

Bicore A common control circuit used in BEAM robotics and developed by Mark Tilden. An integrated circuit is connected in such a way as to create an oscillator that passes a signal back and forth between two "nodes." This signal can then be used to control motors, especially for walking robots that require a back and forth walking gait. See *Nervous Net*; *Node*.

Biomimicry From the Latin *bios*, meaning life, and *mimesis*, meaning to imitate. An approach used in many forms of human engineering (including robotics) in which systems in nature are studied in an attempt to find better approaches to human-made technologies and systems.

Boil the Ocean Slang for trying to tackle all aspects of an ambitious project at once, rather than getting parts of it done first before adding other parts. In other words, you can boil a pot of water, but you can't boil the whole ocean. More behavior-oriented robot builders often accuse hard AI developers and ambitious humanoid engineers of trying to boil the ocean. See *Artificial Intelligence (AI)*; *Scratch Monkey*.

Boolean Logic   Named for the nineteenth-century mathematician George Boole, a form of algebra in which simple operators—AND, NOT, OR, NOR (not or), NAND (not and), and XOR (exclusively or)—are used in logical expressions to yield simple yes/no, true/false answers. If you've ever used a search engine (that uses Boolean operators) and typed in strings such as robots AND projects NOT "Gareth Branwyn", you were using Boolean logic to find every robot project online that I had nothing to do with. (I take offense.) Nobody really knew what to do with Boolean algebra at first until digital electronics came along, which is built upon yes/no, on/off, 1/0 binary switches.

Cathode   The negative side of a diode or LED. Our tech editor Dave Hrynkiw can always keep anode and cathode straight by remembering his high school physics teacher saying: "I feel negatively towards cats." Works for me. See *Anode*.

Chatterbot   A type of computer program that attempts to simulate a "natural" conversation with a person. Such programs have been around since the program ELIZA was created in 1966. These conversational "robots," now popular on the Internet, have become far more sophisticated than earlier efforts, but they're still far from capable of passing the Turing Test.

Circuit Board   See *Printed Circuit Board (PCB)*.

CMOS (Complimentary Metal-Oxide Semiconductor)   A technology for making semiconducting integrated circuits that require less power than other types of digital ICs. An earlier technology, called TTL (Transistor-Transistor Logic), ruled the digital chip world until CMOS fabrication techniques improved, allowing far more circuits to be crammed onto the chips, while consuming far less power.

Compiler   A computer program that takes programming code and translates it from a "high-level" language, meaning one that we mortals can understand, into the lowest-level (machine) language that digital devices can understand. See *Machine Language*.

Conductor   Some atoms have a hard time keeping themselves together and their electrons tend to wander away to hook up with other atoms. Shameless! Materials in which this occurs are known as conductors. They allow electrons to flow freely through them. Metals make good conductors; water is even better. Wood? Not so good. The opposite of a conductor is an insulator, which prevents the flow of an electron current. See *Semiconductor*.

Controller   In robotics, a controller is any device that orchestrates activity between sub-systems (sensors and actuators, actuators and power supply, and so forth). Most robots have at least one main controller, called a *microcontroller unit (MCU)*, and often, separate controllers for motor and power management. See *MCU*.

Cybernetics   The science of control and feedback in living and mechanical systems. Coined by Norbert Weiner of MIT in the late '40s, cybernetics grew out of work on guided missile systems in WWII and the course-correction feedback that was required to keep them on target. Weiner began to see how many complex mechanical and all living systems are controlled through webs of feedback mechanisms. We don't hear much about cybernetics anymore, but the ideas behind it permeate our digital world. In robots, cybernetics, as in control (via the control systems) and feedback loops (via sensors/controllers/actuators chains), is everything. From the Greek *kybernetis*, meaning *steersman*. See *Controller*; *Sensors*.

Cyborg   Short for *Cybernetic Organism*, this is a term for a creature that is part human, part machine. As we become augmented with pacemakers, artificial limbs, designer pharmaceuticals, wearable Internet nodes, and wheels for legs (witness the Segway Human Transport), we're all slowly becoming cyborgs. As every *Star Trek* geek will tell you, the further abbreviated form is *borg*.

DC (Direct Current)   A continuous flow of electricity that moves through a conductor (such as a wire) in only one direction, from the point of highest electrical potential to the lowest. See *AC (Alternating Current)*; *Conductor*; *Semiconductor*.

Dead Bug Mode   An integrated circuit "on its back" (with its pins sticking up in the air) is said to be in "dead bug mode." If it's sitting upright, with its pins down, like the legs of a table, it is said to be in "live bug mode." Be careful for "dead bugs," they can still "sting" like crazy if you sit on them!

Degrees of Freedom (DOF)   In robotics, degrees of freedom refers to planes of movement, either back and forth (linear) or circular (rotational) movement. For instance: A human arm has seven DOF: three in the shoulder, one at the elbow, one rotational at the wrist, and two to move the wrist up and down and side-to-side.

Differential Drive   A simple locomotion system comprised of two wheels on a common axis with each wheel powered independently. Differential drives are probably the most common drive design in robotics. Also shortened to *diff drive*. See *Skid Steering*.

Digital   A property of discrete, binary values (as in on/off, high/low, 1/0), rather than a continuous spectrum of values, as in analog. One of the many advantages of digital technologies over analog is that noise is greatly reduced (because all that's being communicated is "ons" and "offs"). One disadvantage is the loss of information, as when an analog signal is "squared off" as it is converted to a digital one. How much data is lost depends on the "resolution" of the analog-to-digital (A/D) conversion. See *Analog*; *Boolean Logic*.

DIP (Dual In-line Package)   An integrated circuit design in which there are two rows (dual) of parallel pins (in-line) protruding from the bottom of the chip's container (package). Not to be confused with a SIP (Single In-line Package). See *Dead Bug Mode*; *Package*.

Droid   See *Android*.

Dumb Servo   A servo motor that has had its control electronics and mechanical gear-stop removed and is now nothing more than a DC motor and a gearbox. See *Hacking*; *Servo Motor*.

Dustbunny Cowboys   BEAM robot inventor Mark Tilden's nickname for a swarm of dust-wrangling insectoid robots he envisions cleaning homes of the future. Like dung beetles, the bots would wander a room rustling up balls of dust, and then deposit them in a little hole by the brightest window in the house. This "vacuuming" behavior could be created by a simple photophobic/wander aimlessly/photovoric behavior sequence.

Duty Cycle   In electronics, the duty cycle is the "on" time percentage for a device that operates intermittently. In other words, it is the ratio of the actual operating time to the total elapsed time. See *Pulse-Width Modulation (PWM)*.

Effector (End Effector)   In robotics, an effector is any device that interacts with objects in the real world. An effector can be a mechanical hand, a gripper, a power tool, a spray paint nozzle, and so forth. The term *end effector* is often used, especially when referring to mechanical hands and grippers.

Electromotive Force   Anything that causes the flow of electricity from one point to another. Also the energy stored in a power source (such as a battery) for conversion into a moving electrical current. Measured in volts and symbolized by either a *V* or an *E* (for "electromotive").

Evolutionary AI   See *Artificial Intelligence (AI)*.

Force Feedback   The use of sensors/actuators in a robot effector, a joystick, or other manual control device to gauge how much force to apply in response to pressure, or to provide vibration, resistance, or other force effects (think: robot with a firm, but not bone-crushing, handshake). See *Effector*.

Freeforming (Freeform Circuit)   The creation of an electronic circuit by simply soldering components together, rather than soldering them to the pads of a printed circuit board. This technique is used a lot in BEAM robots, in which the robots are often very small and space is at a premium. Dispensing with the PCB also means that the builder is free to improvise (think freeform jazz) and to use the components themselves to construct the robot's body. See *Printed Circuit Board (PCB)*.

Gumby Legs (or "Gumby Wire")   Nickname in BEAM robotics for 10-guage copper wire (or other easily bendable wire) used for legs on robot walkers. Because it's bendable, it can be adjusted or completely reshaped to get a proper walking gait.

H-Bridge   A common circuit design for motor control that uses four switches that are rapidly opened and closed in such a way as to alter the flow of current to the motor, thereby changing its direction and speed. This technique is called *pulse-width modulation*. This circuit type gets its name from the "H" shape made by its layout in a circuit diagram (with the motor depicted in the middle and the four switches around it). See *Pulse-Width Modulation (PWM)*.

Hacking   Using one's technical expertise to innovate, improvise, and/or cobble together solutions to tech problems. The term "hardware hacking" is often used to refer to altering an existing piece of technology to do something new with it, especially something for which it was never intended (such as "hacking a servo" for continuous rotation). See *Open Source; Servo Motor*.

Hard AI   See *Artificial Intelligence (AI)*.

Induction   When an electron current flows through a conductor, it creates a magnetic field around it. When that field is in flux (changing lines of magnetic force) and comes into contact with another conductor (say an unpowered wire), it generates a current flow in that second conductor, in the opposite direction of the first flow. This phenomenon is known as induction. See *Conductor*.

Linux   A Unix-like computer system built upon GNU, another Unix-like OS. Open source software with many different versions available (many of them free of charge), Linux has taken the computer world by storm. Ironically, even though it's been coded collaboratively, by programmers from around the world, it is reputed to be far more robust, less crash-prone, and more code-efficient than Microsoft Windows. Because of Linux's efficiency, and its cheap (or free) cost, it is becoming extremely popular in embedded systems, including robot controllers. Unix is an OS created by Bell Labs scientists in the early 1970s as "get by" software for handling electronic documentation. They never could have imagined that 30 years later, it (and its offspring) would be one of the chief software architectures holding together a globe-spanning "cyberspace." See *Open Source*.

Live Bug Mode   See *Dead Bug Mode*.

Logic Gate   An electronically controlled switch used to carry out operations in Boolean algebra. Such AND, NOT, OR, NOR (not or), NAND (not and), and XOR (exclusively or) gates are created using various types of switching hardware, from relays to steams of water (fluidics) to fiber optics. The most common type of logic switch, however, is the transistor, from discrete components to ICs with millions of switches baked into them. See *Boolean Logic*; *Semiconductor*.

Machine Language   The language your computer (or microcontroller) *really* speaks—its native tongue. All other control languages and programs piled on top of it must be translated into machine language for execution. Machine language is comprised entirely of the binary digits "0" and "1." See *Compiler*.

MCU (Microcontroller Unit)   If you want to be *really* geeky, use μCU, with *μ* being the symbol for the Greek letter *mu*, which is used to mean *micro* (or *mikros* in Greek). See *Controller*.

Meatbot   Computer hacker slang for a human being (as in *meat robot*). Many in the computer hacker and robot builder communities have a playful love/hate relationship with the "mechanics," the strengths and limitations, of the human body, as evidenced by this somewhat demeaning term. Even more demeaning: *meat bag*. See *PEBCAK*.

Mechmaniacs   Short for *mechanical maniacs* and coined by the robot builders at Survival Research Labs to describe themselves and other "mechanical performance artists" who can't get enough of animating machines, and especially, getting them to destroy each other.

Mobots   Short for *mobile robots*. See *Staybots*.

Nanobot   A robot built on a nano-scale (one-billionth of a meter). Nanotechnology engineers see a future in which microscopic robots, both mechanical and bioengineered, repair our bodies, reassemble base proteins into foods (can you say "*Star Trek* replicator?"), and eat pollution in our atmosphere. See *Nanotechnology*.

Nanotechnology   A recent branch of science (still largely theoretical) that deals with engineering machines on a molecular scale. The term was coined by Eric Drexler in the controversial 1986 book *Engines of Creation*. One of the key concepts of nanotech is the creation of nano-scaled robots (called "assemblers") that could be "programmed" to create other devices on the nano-scale. This idea of self-replicating nanobots scares many people, leading to the concept of "gray goo," runaway mutant nanobots that could destroy us all (think: real-world PacMen). Also sometimes referred to as "molecular engineering" by those who want to get away from the hype and sci-fi expectations inherent in the term *nanotechnology*. See *Nanobot*.

Nervous Net   BEAM-creator Mark Tilden's answer to the neural net. Whereas a neural network occurs inside of an artificial brain (in other words, a computer), a nervous net is distributed throughout the body of a robot, creating a sense-act architecture, instead of the sense-plan-act architecture found in AI models of robotic intelligence. See *Neural Network*; *Sense-Plan-Act (SPA)*.

Neural Network   A type of artificial intelligence architecture in which the physiology of the human brain is used as inspiration. A series of artificial nodes (or "neurons") are connected into networks (and networks of networks). This basic structure is really all such AI networks have in common with organic brains, except, like our brains, neural nets are also massively parallel, with large numbers of simple processes happening at the same time and being combined to form a larger networked intelligence. See *Nervous Net*; *Node*.

Node   In computers (and electronics in general), a node is simply a terminal or other point in a network that can create, receive, and/or send data. See *Logic Gate*; *Neural Network*.

Open Source   A movement in the computer sciences to make computer software source code open to programmers who want to improve upon it and distribute their own versions. In the early days of computing, all code was open to examination, improvement, and tinkering, then most of it became proprietary. Open source proponents want to open it up again. Also called the *free software movement*, but that name implies noncommercial, which *open source* does not (you can still sell it as long as you allow users to muck around with it). The open source movement has inspired openness in many areas of high-tech development, including robotics, in which the spirit of software and hardware hacking, and collaboration, is very much alive. See *Hacking*; *Linux*.

Ounce-Inch   An English measurement system used in motors to rate their torque. One ounce-inch means that the motor is delivering one ounce of tangential force one inch from the center of its drive shaft. See *Torque*.

Package   In electronics, the container of an electronic component. Commonly found in the terms *dual in-line package (DIP)* and *single in-line package (SIP)*, two forms of integrated circuits. See *DIP (Dual In-line Package)*.

PEBCAK   Initials stand for *Problem Exists Between Chair and Keyboard*. Computer hacker and tech support slang which refers to the fact that when there's a technical problem with a computer, more often than not, it's the human in the loop (the thing between the chair and the keyboard) that's at fault. See *Meatbot*; *Scratch Monkey*.

Percussive Maintenance   The fine art of whacking the proverbial dust mites out of an electronic device to try and get it working again. Interestingly, studies have shown that bad/intermittent connections are the most failure-prone element in an electronic system, so "percussing" it can often get things running again, at least for awhile. Not to be confused with the *Whack-It Technique*.

Printed Circuit Board (PCB) A device for connecting electronic components together without the need for discrete wires. PCBs are usually made of fiberglass and plastic and have holes drilled in them to accept the electronic components that comprise the circuit the board is designed to carry. Copper foil pads surround the holes (that accept solder) and copper foil tracks (called "traces") connect the holes (and serve as the conductive material between components).

Pulse-Code Modulation (PCM) A means of data transmission in which a pulse train serves as a code for what to do on the receiving end. In servo motor control, pulses are sent down a control wire to the servo's onboard circuitry. The width of the pulse (its time interval) tells the control circuit where the shaft of the motor should be positioned. For instance, a 1.5 millisecond (ms) pulse often signals the center, or 90-degree position.

Pulse-Width Modulation (PWM) A means of controlling the speed of a motor by opening and closing high-speed switches to generate pulses of electricity (rather than a continuous current). By controlling the "on" time percentage (or *duty cycle*) of the signal, the motor speed can be controlled. The speed of the on-time and off-time is fast enough that the pulses smooth out and are not perceptible to us puny humans. See *DC (Direct Current)*; *Duty Cycle*.

Reverse Engineering The process of taking apart an already existing device to figure out how it works so you can then make your own copy or an improved version. See *Hacking*.

Scratch Monkey From the hacker maxim: "Always mount a scratch monkey," which translates to "always proceed with extreme caution when dealing with irreplaceable hardware or software." The apocryphal story goes that a lab monkey died at the University of Toronto when a computer tech, unaware that the monkey's life support hardware was attached to the computer he was working on, powered it down for repairs. See *PEBCAK*.

Semiconductor A wondrous material, such as silicon, that resists an electrical current until a certain threshold voltage is reached, where it then becomes conductive. This property puts semiconductors somewhere between a conductor and an insulator, hence the prefix *semi*. Adding impurities, a process called "doping," can control the degree of the conductivity of the material. A semiconductor doped to create a surplus of electrons is called an N-type (as in "negative"), and one doped to have a shortage of electrons is called a P-type (think: "positive"). See *CMOS (Complimentary Metal-Oxide Semiconductor); Conductor*.

Sense-Act (SA) A type of robot control architecture in which the robot builds no model of its world and the feedback from sensors are fed directly (or as directly as possible) to its actuators. See *Artificial Intelligence (AI); Behavior-Based Robotics (BBR); Sense-Plan-Act (SPA); Subsumption Architecture*.

Sense-Plan-Act (SPA) The most common approach to AI architecture in which sensors on the robot gather data (sense), send that data to a computer for analysis and decision-making (plan), and then the computer sends output to actuators based on its decision on what to do next (act). See *Artificial Intelligence (AI); Neural Network; Sense-Act (SA)*.

Sensors In robotics, any component that reads values from the outside world and sends those values to a controller as data to be processed in some way (usually as feedback that gets passed as output to control actuators). See *Actuator; Controller.*

Servo Motor   A type of motor first created for model vehicles (planes, helicopters, boats) and now used in a host of other applications, including small robots. Servo motors are distinguished by their three main components: a DC motor, a gearbox, and control circuitry. The control circuitry dictates the position of the motor shaft, allowing precise positioning in a 180-degree arc. Servo motors normally have three wires, the standard positive and negative power wires, and a third control wire. Positioning of the servo shaft is determined by the time of electrical pulses on the control wire. To get continuous (360-degree) rotation from a servo, robot builders remove the control circuitry and the mechanical stop on the final servo gear. Also called a *servomotor, servo mechanism,* or simply *servo.* See also *Hacking; Pulse-Code Modulation (PCM).*

Skid Steering   A type of steering common in robots in which the motors/wheels on one side of the bot are denied power (or braked), while the other side continues forward (making the vehicle turn toward the direction of the unpowered side). The degree of the turn is determined by how long power is denied to the stopped side. Also sometimes referred to as *tank steering.*

Soft AI   See *Artificial Intelligence (AI).*

Staybots   The opposite of a *mobot* (or *mobile robot*). Some robots start out life as staybots, built on stationary platforms, but graduate to legs or wheels if they're *really* good (after their control architectures are complete and robust). MIT's famous KISMET and Cog are staybots,

but maybe one day, they'll be allowed to come out and play.

Stuff   The technical term (believe it or not) for the process of assembling ("stuffing") a PCB and its components. Think Thanksgiving and turkey. See *Printed Circuit Board (PCB).*

Subsumption Architecture   A form of real-time robotic control using sensor-triggered behaviors. A series of discrete software "agents" called *Augmented Finite-State Machines* (AFSMs) are programmed so that lower-level processes (such as "go forward") run until sensors (or timers) dictate that a higher-level process wants to subsume lower ones (for instance, to run an obstacle-avoidance routine). Subsumption is a modular way of building robot intelligence because new hardware (sensors and actuators) and software (new AFSMs) can be built upon the already successful hardware/software. It's ground-up intelligence. See *Artificial Intelligence (AI); Behavior-Based Robotics (BBR); Sense-Act (SA); Sense-Plan-Act (SPA).*

Torque   In motors, the rotary force produced on a motor's output shaft. See *Ounce-Inch.*

TTL (Transistor-Transistor Logic)   See *CMOS (Complimentary Metal-Oxide Semiconductor).*

Whack-It Technique   A way of getting stubborn solder or a component off of a desoldered connection. One heats the join and then whacks the PCB against something to dislodge the offending component. Whack with caution. See *Percussive Maintenance; Printed Circuit Board (PCB).*

Whegs   A type of robot locomotion developed at Case Western University that tries to combine the advantages of wheels and legs.

Wheels are easy to engineer and enable vehicles to move quickly. Legs enable creatures to more easily and efficiently traverse varied terrain and obstacles. What do whegs look like? Picture three bicycle spokes on a hub with the ends of the spokes bent (or covered in little booties) to make feet.

Index

Symbols

74HCT240 chips, 193
130-in-1 Electronic Project Labs, 182

A

AC (Alternating Current), 126-127, 176
Acroname
  kits, 302
  microcontrollers/control software, 313
active-low configuration, 293
ActivMedia Robotics
  AmigoBot ePresence, 169
  Web site, 27
actuators, 42, 88
  gears, 88
  hydraulic cylinders, 88
  motors, 88
  muscle wire, 89
  pneumatic cylinders, 89
adaptive robots, 16
Advanced Design, Inc., 161-162
AFSMs (Augmented Finite-State Machines), 43
Age of Automation, 22
AIBO (Artificially Intelligent Robot), 30, 56, 163-164
AIBO-Life Web site, 338
alkaline batteries, 94
All Electronics, 317
alligator clips, 118
Alternating Current (AC), 126-127, 176
aluminum frames, 85-86
Amdahl, Kenn, 67, 176
Amdahl's First Law, 67
American Science and Surplus, 317
American Wire Gauge (AWG), 123
AmigoBot, 169
amorphic robot works Web site, 53
anatomy, 82
Androbots, 27, 97, 171
Android World, 37, 334
antique robots. *See* collectors

Aristotle, 28
Armatron, 173
Artificially Intelligent Robot (AIBO), 30, 56, 163-164
Asimov, Isaac, 21-22
Asimov's Three Laws of Robotics, 62
augmented finite-state machines (AFSMs), 43
auto-ranging, 127
automatons, 28
avoid obstacles behavior, 44
AWG (American Wire Gauge), 123

B

B.I.O.-Bugs, 165-166
back legs (Coat Hanger Walker), 213, 216
Bailey, Clayton, 56
banana plugs, 201
base (epoxy), 123
Basic Micro, 313
BasicElectronics.com, 341
batteries, 94-95, 180
  alkaline, 94
  battery packs, 95
  Coat Hanger Walker, 219
  lead acid, 94
  Li-Ion, 94
  Mousey the Junkbot, 250
  NiCad, 94
  NiMH, 94
  power, 83
  rechargers, 118
  SLA, 95
  testing, 129
Batteries Plus, 309
Battery Mart, 309
Battle Robot Tips Web site, 341
Battlebots, 30, 51, 330, 336
Battlebots: The Official Guide, 320
BBR (Behavior-Based Robotics), 42-43
  recipe, 45-46
  subsumption architecture, 43-45

BEAM (Biological Electronic Aesthetics Mechanics), 46-48, 103
  online, 336
  robot competitions, 48
BEAM Video Clips, 331
Bebop to the Boolean Boogie, 326
Behavior-Based Robotics. *See* BBR
behaviors
  avoid obstacles, 44
  heat sensing, 45
  light seeking, 44
  photovoric, 99
  wander aimlessly, 44
bench vises, 120
Bertocchini, Carlo, 17
Bicore circuits (Coat Hanger Walker), 186, 189
  breadboarding, 190-191
  servo motor, 191-194
  soldering, 194-199
Big Trak, 174
binding posts (breadboards), 200
Biological Electronic Aesthetics Mechanics. *See* BEAM
biomimicry, 15, 40-41
  BBR, 42-43
    recipe, 45-46
    subsumption architecture, 43-45
  BEAM, 46-48
  mutation, 45
Board of Education (BOE), 154
BOB (Brains on Board), 29, 170
bodies
  actuators, 88
    gears, 88
    hydraulic cylinders, 88
    motors, 88
    muscle wire, 89
    pneumatic cylinders, 89
  anatomy, 82
  Coat Hanger Walker, 203
    drive gear, 211
    gears, 204
    idler gear, 208-210
    idler shaft, 205-207
    leg assemblies, 212-213, 216-218
    testing, 219

controllers, 102
  microcontrollers, 103
  nervous nets, 103
  off-board, 103
  Power Controllers, 104
  Speed Controllers, 104
DiscRover, 275
  bottom half, 275-279
  bump sensors, 279-281
  final connections, 282-284
  top half, 281-282
drive trains, 90-91
end effectors, 97-98
frames, 83
  aluminum, 85-86
  carbon composite, 86
  LEGO bricks, 87
  PCB, 87
  plastic, 86
  steel, 85
  titanium, 86
  wood, 84
locomotion, 91
  legs, 93
  tank tracks, 93
  wheels, 92
manipulators, 97-98
Mousey the Junkbot, 231
  bump switch, 235-236
  control switches, 237
  motors, 234-235
  mouse, 232-234
  tires, 236
outer shells, 105-106
power systems, 93
  batteries, 94-95
  fuel cells, 96
  gastrobots, 96
  pressure, 95
  solar, 95-96
sensors, 98
  encoders, 102
  gas, 102
  GPS receivers, 102
  heat, 101
  light, 98-99
  pressure, 98
  sound, 99-100
  strain gauges, 102
  tilt, 102
  vision systems, 100-101
spaghetti, 106
types, 76
  development platform, 81
  embedded robots, 79, 83
  field robots, 77
  humanoids, 79-81
  industrial manipulators, 76-77
  robo-critters, 78
BOE (Board of Education), 154

BOE-Bot kit, 154-155
bolts, 124
Bookfinder.com, 319
books, 320
  bot-building
    Build Your Own Combat Robot, 323
    Building Robot Drive Trains, 323
    David Baum's Definitive Guide to LEGO MindStorms, 323
    Extreme MindStorms, 324
    Junkbots, Bugbots & Bots on Wheels, 324
    Muscle Wires Project Book, 324
    Programming and Customizing the BASIC Stamp, 325
    Programming Robot Controllers, 325
    Robot Builder's Bonanza, 325
    Robot Builder's Sourcebook, 325
    Robot Building for Beginners, 326
    Robot Sumo: The Official Guide, 326
  electronics, 326
    Bebop to the Boolean Boogie, 326
    Getting Started in Electronics, 327
    There Are No Electrons: Electronics for Earthlings, 327
  theoretical
    Battlebots: The Official Guide, 320
    Concise Encyclopedia of Robotics, 320
    Flesh and Machines, 320
    Mind Children: The Future of Robot and Human Intelligence, 321
    Mobile Robots: Inspiration to Implementation, 321
    Robo Sapiens: Evolution of a New Species, 321
    Vehicles: Experiments in Synthetic Psychology, 322
boots (DMMs), 128
bot-building books
    Build Your Own Combat Robot, 323
    Building Robot Drive Trains, 323
    David Baum's Definitive Guide to LEGO MindStorms, 323
    Extreme MindStorms, 324
    Junkbots, Bugbots & Bots on Wheels, 324
    Muscle Wires Project Book, 324
    Programming and Customizing the BASIC Stamp, 325
    Programming Robot Controllers, 325
    Robot Builder's Bonanza, 325

    Robot Builder's Sourcebook, 325
    Robot Building for Beginners, 326
    Robot Sumo: The Official Guide, 326
Botic, 334
brains
  DiscRover, 268
    motorhead, 271-275
    OOPic, 268-271
  Mousey the Junkbot, 238-239
    breadboard, 241
    connections, 257-258
    control circuit, 249-255
    eyestalks, 241-242, 255-256
    motors, 247
    op amps, 243-244
    power, 248
    runaway circuits, 245-247
    senses, 239-240
    troubleshooting, 249
Brains on Board (BOB), 29, 170
Braitenberg, Valentino, 68
Braitenberg's Maxim, 68
breadboards, 116, 151, 200
  Bicore circuits (Coat Hanger Walker), 190-191
  binding posts, 200
  buses, 201
  hook-ups, 204
  horizontal, 203
  Mousey the Junkbot, 241
    eyestalks, 241-242
    motors, 247
    op amps, 243-244
    power, 248
    runaway circuits, 245-247
    troubleshooting, 249
  name origin, 201
  power switches, 202
  power transformers, 203
  tie points, 200
  trench, 201
  troubleshooting, 249
Brooks, Rodney, 18, 67, 154
Brooks's Research Heuristic, 67
Budget Robotics kits, 302
bugs (computer), 69
***Build Your Own Combat Robot*, 323**
Build Your Own Robot kit, 152-153
building materials, 305
  The Composite Store, 305
  Evolution Robotics, 305
  Online Metals, 305
  Plastruct, 306
  Small Parts, 306
  W.W. Grainger, 306
***Building Robot Drive Trains*, 323**

building sets, 159. *See also* kits
  Fischertechnik, 160-161
  LEGO MINDSTORMS Robotics
    Invention System, 159-160
  Robix Rascal Robot Construction Set,
    161-162
  Robotix, 162
bump sensors, 98, 279-281
bump switches, 235-236
bump whiskers, 290-293
buses (breadboards), 201
Bushnell, Nolan, 29
butane micro-torches, 121

C

**cable/satellite combat robotics shows,
  51**
**cadmium-sulfide photoresistor (CdS),
  130**
cameras (teleoperated), 100
capacitance, 127
capacitors, 178-180
Capek, Karel, 19-21
carbon composite frames, 86
**Carnegie Mellon University Institute,
  339**
Carpet Rover 2 Combo kit, 150-151
**Case Western University Biorobotics
  Lab, 339**
CD-ROM box bots, 297
**CdS (cadmium-sulfide photoresistor),
  130**
**Central Processing Units (CPUs), 103,
  273**
chip pullers, 122, 203
Circuit Cellar, 328
circuits
  Bicore. *See* Coat Hanger Walker,
    Bicore control circuit
  control, 249-255
  electronic components. *See* electronic
    components
  integrated, 179
  runaway, 245-247
  virtual, 269
classification system, 16
clocks, 274
co-bots (cooperative robots), 23
Co-Worker, 39
Coat Hanger Walker, 187
  Bicore control circuit, 189
    breadboarding, 190-191

  servo motor, 191-194
  soldering, 194-199
  body, 203
    drive gear, 211
    gears, 204
    idler gear, 208-210
    idler shaft, 205-207
    leg assemblies, 212-213, 216-218
    testing, 219
  experiments, 225
  final assembly, 222-224
  parts, 187-188
  power, 219-221
  tools, 189
cognition box, 42
collectors, 170
  Androbots, 171
  Armatron/Super Armatron, 173
  Big Trak, 174
  Heathkit Heroes, 171-172
  Maxx Steele, 174
  Omnibot, 174
  RB5X, 172-173
combat robotics, 49-51
comments (code), 287
communication, 104-105
communities, 337-338
components (electronic), 177-181
  batteries, 180
  capacitors, 178-180
  diodes, 178-180
  integrated circuits, 179-181
  light-emitting diodes, 178
  motors, 180
  relays, 179
  resistors, 180
  switches, 179
  transistors, 178-180
  variable resistors, 178
  wires, 179
The Composite Store, 305
computer technician repair kits, 123
Concise Encyclopedia of Robotics, **320**
conductors, 176
connections
  DiscRover, 282-284
  jumper wires, 190
  Mousey the Junkbot, 257-258
construction materials, 125
continuity test, 127
**control circuits (Mousey the Junkbot),
  249-250**
  batteries, 250
  LM386 chip, 253-255
  motor wires, 253
  relays, 250-252
  switches, 252-253

control horns, 211
control software, 313
  Acroname, 313
  Basic Micro, 313
  Gleason Research, 314
  Not Quite C, 314
  Parallax, 314
  Savage Innovations, 315
control switches, 237
controllers, 102
  microcontrollers, 103
  motor, 285-290
  nervous nets, 103
  off-board, 103
  Power Controllers, 104
  Speed Controllers, 104
cooperative robots (co-bots), 23
**CPUs (Central Processing Units), 103,
  273**
curing agent, 123
current (I), 65
currents, 176
Cybot Builder Web site, 336
CyBugs kit, 152

D

Dante II, 29
**DARPA (Defense Advanced Research
  Projects Agency), 53-55**
  Distributed Robotics Program, 55
  Grand Challenge, 53
Darwin Awards, 116
data hold, 127
*David Baum's Definitive Guide to
  LEGO MindStorms*, **323**
DC (Direct Current), 126, 176
DC motors, 90, 235
Deep Blue, 30
**Defense Advanced Research Projects
  Agency (DARPA), 53-55**
definitions (robot)
  dictionaries, 14-15
  JIRA, 16
  Robot Institute of America, 15
  roboticists, 17-18
  society, 13
delamination, 86
Demers, Jérôme, 186
desoldering, 137
desoldering pumps, 116
development platform robots, 81
Devol, George, 22

diagonal cutters, 115

Dictionary of Measures, Units, and Conversions Web site, 341

differential steering, 92

Digi-Key, 310

digital multimeters. *See* DMMs

Dim statements, 287

diodes, 180
  check, 127
  functions, 178
  light-emitting, 178

DIP (dual in-line package), 181

Direct Current (DC), 126, 176

DiscRover
  body, 275
    bottom half, 275-279
    bump sensors, 279-281
    final connections, 282-284
    top half, 281-282
  brains, 268
    motorhead, 271-275
    OOPic, 268-271
  experiments, 296
    CD-ROM box bots, 297
    high-order sensors, 296
    layers, adding, 296
    OOPic robots, 298
    prototyping area, 296
    two talking, 297
  loading, 293
  parts, 265-266
  programming, 285
    bump whiskers, 290-293
    motor controller, 285-290
  templates, 275
  tools, 267-268
  troubleshooting, 294-295

disposable gloves, 124

distance detection, 99

DMMs (digital multimeters), 117, 126
  basic features, 126-127
  battery testing, 129
  CdS, 130
  features, 127-128
  functions, 128
  leads, 128
  resistor values, 129-130
  resources, 131

Doerr, Robert, 175

domestic robots, 55
  maids, 57
  mowers, 57
  multitasking, 58
  robo-pets, 55-56

Dremels, 120, 315

drive gears, mounting, 211

drive trains, 90
  DC motors, 90
  servo motors, 90-91
  stepper motors, 91
  suppliers, 307
    HiTec RCD, 307
    Mabuchi, 307
    North American Roller Products, 307
    RobotBooks, 308
    Servo City, 308
    Solarbotics, 308
    Tamiya, 309
    Vantec, 309

dual DC motor I/O lines, 270

dual in-line package (DIP), 181

duty cycles, 90, 127

E

EAC (EEPROM Activity Check), 271

EEPROM (erasable memory space), 103

electricity, 176

Electro/Sparko, 28

electrolytic capacitors, 180

electronic components, 177-181
  batteries, 180
  capacitors, 178-180
  diodes, 178-180
  integrated circuits, 179-181
  light-emitting diodes, 178
  motors, 180
  relays, 179
  resistors, 180
  switches, 179
  transistors, 178-180
  variable resistors, 178
  wires, 179

Electronic Goldmine, 318

electronic screwdrivers, 114

electronics
  books, 326
    Bebop to the Boolean Boogie, 326
    Getting Started in Electronics, 327
    There Are No Electrons: Electronics for Earthlings, 327
  Greenie Theory, 176-181
  learning, 181-182
  overview, 175-176
  suppliers, 310
    Digi-Key, 310
    Electronix Express, 310
    Future Active, 311
    Jameco, 311

    JDR Microdevices, 311
    Mouser Electronics, 311
    Radio Shack, 312

Electronix Express, 310

Elekit, 148

Elektor Electronics, 328

Elmer & Elsie, 23-25

embedded processors (MCUs), 273-275

embedded robots, 79, 83

emergent behavior, 25

encoders, 102

end effectors, 97-98

engines (solarengines), 155

Engleberger, Joseph, 22

epoxy, 208

equipment
  hobby stores, 316
  kits, 302
    130-in-1 Electronic Project Labs, 182
    Acroname, 302
    BOE-Bot, 154-155
    Budget Robotics, 302
    Build Your Own Robot, 152-153
    Carpet Rover 2 Combo, 150-151
    CyBugs, 152
    ER1 Personal Robot System, 157-158
    Hexapod Walkers, 151
    LEGO MINDSTORMS, 303
    Lynxmotion, 303
    Mondo-tronics Robot Store, 303
    Parallax, 304
    Photopopper 4.2b, 156
    Rockit Sound Controlled Robot, 148
    Solarbotics, 304
    SolarSpeeder, 155-156
    Spider 3 Walker, 149
    Sumo-Bot, 153-155
    Tamiya, 304
    WAO-G, 149-150
  microcontrollers/control software, 313
    Acroname, 313
    Basic Micro, 313
    Gleason Research, 314
    Not Quite C, 314
    Parallax, 314
    Savage Innovations, 315
  suppliers, 305
    The Composite Store, 305
    electronics, 310-312
    Evolution Robotics, 305
    motor/drive train, 307-309
    Online Metals, 305
    Plastruct, 306

power systems, 309-310
R/C, 312-313
Small Parts, 306
W.W.Grainger, 306
surplus catalogs, 317
All Electronics, 317
American Science and Surplus,
317
Electronic Goldmine, 318
ER1 Personal Robot System kit,
157-158
erasable memory space (EEPROM),
103
Evolution Robotics, 103, 157-158, 305
Evolution Robots, 85
experiments
Coat Hanger Walker, 225
DiscRover, 296
CD-ROM box bots, 297
high-order sensors, 296
layers, adding, 296
OOPic robots, 298
prototyping area, 296
two talking, 297
Mousey the Junkbot, 260
Extreme Machines: Incredible Robots,
331
Extreme MindStorms, **324**
eyestalks (Mousey the Junkbot),
241-242, 255-256

F

Fast, Cheap & Out of Control, **331**
feedback, 98
feelers, 98
field robotics, 29, 77
final gear (servo motors), 193
fine screwdrivers, 114
Finite-State Machines (FSMs), 44
FIRST (For Inspiration and Recognition
of Science and Technology), 52
Fischer Fischertechnik, 160-161
Flakey, 27
Flesh and Machines, **320**
flux paste, 123
For Inspiration and Recognition of
Science and Technology (FIRST), 52
force feedback, 98
four-byte commands, 288
frames, 83
aluminum, 85-86
carbon composite, 86
LEGO bricks, 87

PCB, 87
plastic, 86
steel, 85
titanium, 86
wood, 84
Frauenfelder, Mark, 18
frequency, 127
Friendly Robotics, 168-169
front legs (Coat Hanger Walker),
217-218
Frye, Jim, 151
FSMs (Finite-State Machines), 44
fuel cells, 96
Furby, 55
Future Active, 311

G

games, 49
combat robotics, 49-51
DARPA Grand Challenge, 53
FIRST, 52
sumo, 51-52
gas sensors, 102
gastrobots, 96
gearboxes, 88
gears, 88
Coat Hanger Walker body, 204
drive, 211
gearboxes, 88
idler, 208-210
General Motors Unimate, 22
General Robotics Web site, 173
Getting Started in Electronics, **181,**
327
Gilbertson, Roger, 17
Gleason Research, 314
Google Web site, 29
GoRobotics Network, 334
GPS (Global Positioning Satellite)
receivers, 102
Greenie Theory (electronics), 176-181
grippers, 97
Guinan, Paul, 28

H

H-bridges, 271
hacksaws, 120
HBRobotics Club Builders Book Web
site, 341
heat guns, 120

heat sensing behavior, 45
heat sensors, 101
heat sinks, 118
heat-shrink tubing, 124
Heathkit Heroes, 171-172
helping hands, 117
hemostats, 121
HERO I, 29, 171
HERO JR., 171-172
Hexapod Walkers kit, 151
historical events, 28-30
HiTec RCD, 307, 312
hobby stores, 316
Hobbytron, 316
holsters (DMMs), 128
Honda
Humanoid Project Web site, 336
P3, 30
hook probes, 128
hook-up wire, 201
hook-ups (breadboards), 204
Hopkins, Johns, 29
horizontal breadboards, 203
horns, 211
hot glue guns, 118
How Stuff Works Web site, 342
Hrynkiw, Dave, 157
human bodies, 75
Humanoid Robotics Group, 339
humanoids, 35-37, 79-81
hydraulic cylinders, 88

I

I (current), 65
I, Robot, **22**
I/O (Input /Output), 103
I2C network connectors, 271
IBM Deep Blue, 30
IC chip extractors, 122
Ideal Maxx Steele, 174
idler gears, 208-210
idler shaft, 205-207
idler wheels, 92
industrial manipulators, 76-77
infrared communication, 105
infrared pyroelectric sensors, 101
Input/Output (I/O), 103
Insectröides Web site, 186

insulators, 176
integrated circuits, 179-181
intelligent mechatronic systems, 16
internal combustion, 93
inverting octal buffers, 189
iRobot Corporation, 154, 167
iRobot-LE, 38
iron stand with sponge, 115

J

Jacquard, Joseph, 28
Jameco, 123, 139, 311
Japanese Industrial Robot Association
(JIRA), 16
JCM Inventures, 152
JDR Microdevices, 311
JIRA (Japanese Industrial Robot
Association), 16
Jouhou System Kougaku Lab, 340
JPL Robotics, 340
jumper kits, 116
jumper wires, 190
Junkbots, Bugbots & Bots on Wheels,
324

K

K.I.S.S. (Keep It Simple, Stupid), 70,
186
Kamen, Dean, 52
Karl Lunt homepage, 341
kits, 148, 302. *See also* building sets
    130-in-1 Electronic Project Labs, 182
    Acroname, 302
    BOE-Bot, 154-155
    Budget Robotics, 302
    Build Your Own Robot, 152-153
    Carpet Rover 2 Combo, 150-151
    CyBugs, 152
    ER1 Personal Robot System, 157-158
    Hexapod Walkers, 151
    LEGO MINDSTORMS, 303
    Lynxmotion, 303
    Mondo-tronics Robot Store, 303
    Parallax, 304
    Photopopper 4.2b, 156
    Rockit Sound Controlled Robot, 148
    Solarbotics, 304
    SolarSpeeder, 155-156
    Spider 3 Walker, 149
    Sumo-Bot, 153-155
    Tamiya, 304
    WAO-G, 149-150

Krogh, August, 68
Krogh Principle, 68

L

lawn mowers, 57
laws
    Amdahl's First Law, 67
    Asimov's Three Laws of Robotics, 62
    Braitenberg's Maxim, 68
    Brooks's Research Heuristic, 67
    Krogh Principle, 68
    Moore's Laws, 63-64
    Moravec timeline, 65
    Ohm's Law, 64
    Tilden's Laws, 62-63
    Turing Test, 66
lead acid batteries, 94
leads (DMMs), 128
Learning Curve (Robotix), 162
learning electronics, 181-182
LEDs (light-emitting diodes), 239, 271
leg assemblies (Coat Hanger Walker
body), 212
    back legs, 213, 216
    front legs, 217-218
LEGO MINDSTORMS Robotics
Invention System, 159-160, 303
    bricks frames, 87
    *MindStorms: Children, Computers,
    and Powerful Ideas*, 160
    Web site, 337
LEGO Users Group Network (Lugnet),
338
Lexan, 86
Li-Ion batteries, 94
light seeking behavior, 44
light sensors, 98-99
light-emitting diodes (LEDs), 239, 271
listings
    bump whisker (DiscRover), 291
    motor controller, 286
Lithium ion (Li-Ion) batteries, 94
LM386 chips, 253-255
loading DiscRover, 293
locomotion, 91
    legs, 93
    tank tracks, 93
    wheels, 92
Loebner, Hugh, 67
Lugnet (LEGO Users Group Network),
338
Lynxmotion, 150-151, 303

M

Mabuchi, 307
MachineBrain, 334
MacMurtie, Chico, 53
Mad Professor, 18
magazines, 328
    Circuit Cellar, 328
    Elektor Electronics, 328
    MIT Technology Review, 329
    Nuts & Volts, 329
    Real Robots, 330
    Robot Science & Technology, 330
magnifying glass, 119
maids, 57
make-life-easier tools, 119
    bench vise, 120
    hacksaw, 120
    heat guns, 120
    hemostats, 121
    IC chip extractor, 122
    magnifying glass, 119
    miter box, 120
    parts cabinets, 121
    parts picker-upper, 119
    rotary tools, 120
    wire-crimping tool, 121
manipulators, 97-98
manually operated manipulators, 16
Marrs, Texe, 171
Mars Pathfinder, 30
Martin, Fred G., 17
MatWeb, 342
Maxx Steele, 174
McCarthy, John, 28
McComb, Gordon, 302
MCU (microcontroller unit), 263
    clocks, 274
    CPUs, 273
    memory, 274
    ports, 275
mechatronics, 16
memory (MCUs), 274
metal file sets, 118
Micro-Mark, 316
micro-rotches, 120
microcomputers, 273
Microcontroller FAQ Web site, 275
microcontroller unit (MCU), 263
microcontrollers, 103, 313
    Acroname, 313
    Basic Micro, 313
    Gleason Research, 314
    microcomputers, compared, 273

Not Quite C, 314
OOPic, 151
Parallax, 314
Savage Innovations, 315
Millions of Instructions Per Second (MIPS), 65
Milton Bradley Big Trak, 174
Mims, III, Forrest M., 181
Mind Children: The Future of Robot and Human Intelligence, 321
MINDSTORMS. *See* LEGO MINDSTORMS Robotics Invention System
MindStorms: Children, Computers, and Powerful Ideas, 160
MIPS (Millions of Instructions Per Second), 65
MIT Technology Review, 329
miter boxes, 120
Mobile Robots: Inspiration to Implementation, 321
modules, 103
Mondo-tronics Robot Store, 303
monolithic capacitors, 180
Moore, Gordon, 63
Moore's Laws, 63-64
Moravec, Hans, 17
    Standford Cart, 29
    timeline, 65
motor controllers, 285-290
motor wires, 253
motorheads, 271-275
motors, 88, 180
    DC, 90, 235
    Mousey the Junkbot, 234-235, 247
    servo, 88-91
        Bicore circuit control, 191-194
        final gear, installing, 193
        mounting brackets, 206
    Solarbotics RM1, 234
    stepper, 91
    suppliers, 307
        HiTec RCD, 307
        Mabuchi, 307
        North American Roller Products, 307
        RobotBooks, 308
        Servo City, 308
        Solarbotics, 308
        Tamiya, 309
        Vantec, 309
mounting drive gears, 211
mounting brackets (servo motors), 206
mounting putty, 124

Mouser Electronics, 311
Mousey the Junkbot
    body, 231
        bump switch, 235-236
        control switches, 237
        motors, 234-235
        mouse, 232-234
        tires, 236
    brain, 238-239
        breadboard. See breadboards, Mousey the Junkbot
        senses, 239-240
    connections, 257-258
    control circuit, 249-250
        batteries, 250
        LM386 chip, 253-255
        motor wires, 253
        relays, 250-252
        switches, 252-253
    experiments, 260
    eyestalks, 255-256
    overview, 227-229
    parts, 229-230
    playing, 258-259
    tools, 230-231
    troubleshooting, 259-260
mowers, 57
multi-language compilers, 285
multitasking robots, 58
muscle wire, 89
Muscle Wires Project Book, 324
musculature, 88
must-have tools, 114
    breadboards, 116
    desoldering pump, 116
    diagonal cutters, 115
    iron stand with sponge, 115
    jumper kit, 116
    needlenose pliers, 115
    safety goggles, 116
    small screwdriver set, 114
    soldering iron, 115
    wire cutters/strippers, 115
mutation, 45

N

N-type transistors, 181
NASA Tech Briefs, 342
needlenose pliers, 115
nervous nets, 103
NetSurfer Robotics, 335
NiCad (nickel cadmium) batteries, 94
NiMH (nickel metal hydride) batteries, 94
nitinol, 89

Nomadic Research Labs, 71
North American Roller Products, 307
numerically controlled robots, 16
nuts, 124
Nuts & Volts, 329

O

object-oriented programming (OOP), 268
off-board control, 103
Ohm, George, 64
Ohm's Law, 64
omni-wheels, 307
Omnibot, 174
Online Conversions Web site, 342
Online Metals, 305
OOP (object-oriented programming), 268
OOPic
    DiscRover, 268-271
    microcontroller, 151
    multi-language compiler, 285
    programming example, 269
    Web site, 269
OOPic-R, 269
    dual DC motor I/O lines, 270
    EAC, 271
    I2C network connectors, 271
    LEDs, 271
    main input/output lines, 270
    push-button switches, 271
    RS232 ports, 271
    speakers, 271
op amps, 243-244
organizing workshop, 133
origins (robots)
    co-bots, 23
    Flakey, 27
    R.U.R., 19-21
    rogue robot plot, 21
    Shakey, 25-27
    short stories, 21-22
    tortoises, 23-25
    Unimation, 22-23
Out of Control: The Rise of Neo-Biological Civilization, 41
out-of-the-box bots, 163
    AmigoBot ePresence, 169
    B.I.O.-Bugs, 165-166
    Robomower, 168-169
    Roomba, 167
    Sony AIBO, 163-164
    Sony SDR-4X, 164-165
outer shells, 105-106

OWI/Movit kits, 148
Rockit Sound Controlled Robot, 148
Spider 3 Walker, 149
WAO-G, 149-150

P

P-type transistors, 180
P3, 30
PackBot, 54
Palm Pilot Robot Kit (PPRK), 302
Papert, Seymour, 160
Parallax
kits, 154-155, 304
microcontrollers/control software, 314
Web site, 154, 296
parts
cabinets, 121
Coat Hanger Walker, 187-188
DiscRover, 265-266
Mousey the Junkbot, 229-230
picker-uppers, 119
Pauline, Mark, 52
PCB (Printed Circuit Board) frames, 87
PEDAL (Performance, Energetics, and Dynamics of Animal Locomotion), 340
performance art, 52-54
The Personal Robot Book, 171
pets (AIBO), 30
Photopopper 4.2b kit, 156
photovoltaic cells, 96
photovoric behaviors, 99
PICs (programmable integrated circuits), 268
plastic frames, 86
plastic spam, 265
Plastruct, 306
playback robots, 16
playing Mousey the Junkbot, 258-259
Plexiglass, 86
plies, 86
pneumatic cylinders, 89
Poly-PEDAL Lab, 340
portals, 334
AndroidWorld, 334
Botic, 334
GoRobotics Network, 334
MachineBrain, 334
NetSufer Robotics, 335
Robot Café, 335
Robots.Net, 335

Solarbotics.net, 335
TechGeek, 335
ports (MCUs), 275
Positronic Robot series, 21
poster putty, 124
pot time, 123
potentiometers, 166
pots, 166
power
batteries, 83
Coat Hanger Walker, 219
battery packs, 219
final assembly, 222-224
power switches, 221
Mousey the Junkbot, 248
power switches
breadboards, 202
Coat Hanger Walker, 221
power systems, 93
batteries, 94-95
fuel cells, 96
gastrobots, 96
pressure, 95
solar, 95-96
suppliers, 309-310
power transformers, 203
PowerControllers, 104
PPRK (Palm Pilot Robot Kit), 302
precision screwdrivers, 114
Predator UAV (Unmanned Arial Vehicle), 54
pressure sensors, 98
pressure systems, 95
Printed Circuit Board (PCB) frames, 87
programmable integrated circuit (PICs), 268
programmable manipulators, 16
programming
comments, 287
DiscRover, 285
bump whiskers, 290-293
motor controller, 285-290
troubleshooting, 295
Programming and Customizing the BASIC Stamp, 325
Programming Robot Controllers, 325
prototyping areas. *See* breadboards
proximity detection, 99
pullup resistors, 293
push-button switches, 271
PWM (Pulse-Width Modulation), 90, 270

R

R (resistance), 65
R/C (Remote Control), 312-313
R.U.R., 19-21
Radio Shack, 312
130-in-1 Electronic Project Labs, 182
Armatron/Super Armatron, 173
RAM (Random-Access Memory), 103
RB Robot Corporation, 172-173
RB5X, 172-173
RCX, 87
Real Robots, 330
relays, 179, 239, 250-252
remote control (R/C), 312-313
RepairFAQ, 343
research centers, 339
Carnegie Mellon University Institute, 339
Case Western University Biorobotics Lab, 339
Humanoid Robotics Group, 339
Jouhou System Kougaku Lab, 340
JPL Robotics, 340
Poly-PEDAL Lab, 340
Robotics Research in Japan, 340
resistance (R), 65, 127
resistors, 176, 180
pullup, 293
values, 129-130
variable, 178
resources
BasicElectronics.com, 341
Battle Robot Tips, 341
Dictionary of Measures, Units, and Conversions, 341
DMMs, 131
HBRobotics Club Builders Book, 341
How Stuff Works, 342
Karl Lunt homepage, 341
MatWeb, 342
NASA Tech Briefs, 342
Online Conversions, 342
RepairFAQ, 343
Robot Room, 342
Robotics FAQ, 343
W.A.R.S. Reference Pages, 343
retro-robotics, 170
Androbots, 171
Armatron/Super Armatron, 173
Big Trak, 174
Heathkit Heroes, 171-172
Maxx Steele, 174
Omnibot, 174
RB5X, 172-173

reusable putty, 124

reversed biasing, 242

RF transmitter/receivers Web site, 297

rheostats, 176

Ristow, Christian, 53

Robert L. Doerr Robots Wanted Web site, 175

Robix Rascal Robot Construction Set, 161-162

Robix Web site, 161

Robo Sapiens: Evolution of a New Species, 321

robo-critters, 78

robo-pets, 55-56

robo-vac, 58

RoboForce, 174

Robomower, 58, 168-169

Robot Builder's Bonanza, 325

Robot Builder's Sourcebook, 325

Robot Building for Beginners, 326

Robot Café, 335

Robot Institute of America, 15

Robot Projects Web site, 298

Robot Room, 342

Robot Science & Technology, 330

The Robot Store, 89, 148

Robot Sumo: The Official Guide, 326

Robot Sumo Web site, 337

robot warriors, 54-55

Robot Wars, 50, 337

Robot Wars Extreme, 51

robot-specific Web sites, 336
    Battlebots, 336
    BEAM Online, 336
    Cybot Builder, 336
    Honda Humanoid Project, 336
    MINDSTORMS, 337
    Robot Sumo, 337
    Robot Wars, 337
    Robots Wanted, 337

RobotBooks, 308

robotic arms, 29

robotic performance art, 52-54

Robotic Power Solutions, 310

robotic presence, 33-34, 37-39

Robotica, 332

roboticists rules, 69, 72
    artist/scientist, 70
    breaks, 72
    build early, build often, 72
    deconstructionist, 70
    generalist, 69

K.I.S.S., 70
    master of many trades, 71
    methodical/patient, 71
    neatness, 71
    thinking outside the bot, 70
    working knowledge, 72

Robotics FAQ Web site, 343

Robotics Research in Japan, 340

Robotix building set, 162

robots
    definitions
        dictionaries, 14-15
        JIRA, 16
        Robot Institute of America, 15
        roboticists, 17-18
        society, 13
    humanoid, 35-37
    origins
        co-bots, 23
        Flakey, 27
        R.U.R., 19-21
        rogue robot plot, 21
        Shakey, 25-27
        short stories, 21-22
        tortoises, 23-25
        Unimation, 22-23

Robots Rising, 332

Robots Wanted Web site, 175, 337

Robots.Net, 335

Rockit Sound Controlled Robot Kit, 148

rogue robot plot, 21

Roomba, 57, 167

rotary tools, 120

rotors, 91

RS232 ports, 271

rules (roboticists), 69-72
    artist/scientist, 70
    breaks, 72
    build early, build often, 72
    deconstructionist, 70
    generalist, 69
    K.I.S.S., 70, 186
    master of many trades, 71
    methodical/patient, 71
    neatness, 71
    thinking outside the bot, 70
    working knowledge, 72

runaway circuits, 245-247

S

safety, 144-145

safety goggles, 116

satellite/cable combat robotics shows, 51

Savage, Scott, 298

Savage Innovations, 315

screws, 124

scrubby pads, 124

SDR (Sony Dream Robot), 165

SDR-4X, 164-165

sealed lead acid (SLA) batteries, 95

sensate robots, 16

sense-act control system, 43

sense-plan-act control systems, 42

sensors, 98
    bump, 279-281
    encoders, 102
    gas, 102
    GPS receivers, 102
    heat, 101
    light, 98-99
    pressure, 98
    sound, 99-100
    strain gauges, 102
    tilt, 102
    vision systems, 100-101

sequential manipulators, 16

Servo City, 308

servo horns, 211

servo motors, 88-91
    Bicore circuit control, 191-194
    mounting brackets, 206

Shakey, 25-27

Shape Memory Alloy (SMA), 89

shells, 105-106

short stories, 21-22

should-have tools, 117
    battery rechargers, 118
    digital multimeter. *See* DMMs
    heat sink, 118
    hot glue guns, 118
    metal file set, 118
    third hand, 117

Sintra, 86

skid steering, 92

SLA batteries, 95

SMA (Shape Memory Alloy), 89

Small Parts, 306

small screwdriver sets, 114

smart robots, 16

software, 43, 106

Sojourner Rover, 30

solar power systems, 95-96

Solarbotics
    kits, 155-156, 304
    LMPs, 214
    motors, 308

RM1 motors, 234
Web site, 155, 335, 338
solarengines, 96, 155
SolarSpeeder kit, 155-156
soldering, 122, 132
Bicore circuit control, 194-199
exercise, 137-139
existing components, 253
instructions, 134-136
tools, 133
ventilation, 133
soldering irons, 115
sonar sensors, 99
Sony AIBO, 30, 163-164
Sony Dream Robot (SDR), 165
Sony SDR-4X, 164-165
Sony Web site, 164
souls, 106
sound sensors, 99-100
space
Mars Pathfinder, 30
robotic arms, 29
spaghetti, 106
spam, 265
speakers, 271
speech, 104-105
Speed Controllers, 104
Spelletich, Kal, 53
Spider 3 Walker kit, 149
sports, 49
combat robotics, 49-51
DARPA Grand Challenge, 53
FIRST, 52
sumo, 51-52
SRI International, 25
stand (DMMs), 128
Stanford Cart, 29
**Stanford Research Institute Problem
Solver (STRIPS), 26**
**Stanford Research Institute (SRI)
International, 25**
statements (Dim), 287
steel frames, 85
stepper motors, 91
strain gauges, 102
**STRIPS (Stanford Research Institute
Problem Solver), 26**
subsumption architecture, 43
avoid obstacles behavior, 44
heat sensing behavior, 45
light seeking behavior, 44
wander aimlessly behavior, 44

Sugarman, Peter, 68
Sugarman Caution, 68
sumo, 51-52
Sumo-Bot kit, 153-155
Super Armatron, 173
Super Science Stories, **21**
superglue, 123
suppliers, 305
The Composite Store, 305
electronics, 310
Digi-Key, 310
Electronix Express, 310
Future Active, 311
Jameco, 311
JDR Microdevices, 311
Mouser Electronics, 311
Radio Shack, 312
Evolution Robotics, 305
kits. *See* kits
microcontrollers/control software, 313
Acroname, 313
Basic Micro, 313
Gleason Research, 314
Not Quite C, 314
Parallax, 314
Savage Innovations, 315
motors/drive trains, 307
HiTec RCD, 307
Mabuchi, 307
*North American Roller Products,
307*
RobotBooks, 308
Servo City, 308
Solarbotics, 308
Tamiya, 309
Vantec, 309
Online Metals, 305
Plastruct, 306
power systems, 309-310
R/C, 312-313
Small Parts, 306
tools, 315-316
W.W. Grainger, 306
supplies, 122
Coat Hanger Walker, 189
construction materials, 125
DiscRover, 267-268
flux paste, 123
heat-shrink tubing, 124
Mousey the Junkbot, 230-231
poster putty, 124
screws/nuts/bolts/washers, 124
scrubby pads, 124
solder, 122
superglue, 123
techno-junk box, 125
two-part epoxy, 123
wire, 123

surplus catalog companies, 317
All Electronics, 317
American Science and Surplus, 317
Electronic Goldmine, 318
Survival Research Labs, 52
switches, 179, 252-253

T

Tab Robotics kits, 152
Build Your Own, 152-153
Sumo-Bot, 153-155
Tamiya
kits, 304
motors, 309
tank tracks, 93
tantalum capacitors, 180
TechGeek, 335
techno-junk box, 125
teleoperated cameras, 100
teleautomaton, 28
Tesla, Nicola, 28
testing
batteries, 129
Coat Hanger Walker, 219
Turing, 66
theoretical books
Battlebots: The Official Guide, 320
Concise Encyclopedia of Robotics,
320
Flesh and Machines, 320
*Mind Children: The Future of Robot
and Human Intelligence*, 321
*Mobile Robots: Inspiration to
Implementation*, 321
*Robo Sapiens: Evolution of a New
Species*, 321
*There Are No Electrons: Electronics
for Earthlings*, 176, 327
*Vehicles: Experiments in Synthetic
Psychology*, 322
third hands, 117
Thorpe, Marc, 17
Three Laws of Robotics, 62
three-claw parts holders, 119
throwbots, 54
tie pins (breadboards), 200
Tilden, Mark, 62
Tilden's Laws, 62-63
tilt sensors, 102
Timbot, 108-109
timeline, 28-30, 65
tires (Mousey the Junkbot), 236
titanium frames, 86

Tomy Company Omnibot, 174
tools, 114
  chip pullers, 203
  Coat Hanger Walker, 189
  computer technician repair kits, 123
  DiscRover, 267-268
  make-life-easier, 119
    bench vise, 120
    hacksaw, 120
    heat guns, 120
    hemostats, 121
    IC chip extractor, 122
    magnifying glass, 119
    miter box, 120
    parts cabinet, 121
    parts picker-upper, 119
    rotary tools, 120
    wire-crimping tools, 121
  Mousey the Junkbot, 230-231
  must-have, 114
    breadboards, 116
    desoldering pump, 116
    diagonal cutters, 115
    iron stand with sponge, 115
    jumper kit, 116
    needlenose pliers, 115
    safety goggles, 116
    small screwdriver set, 114
    soldering iron, 115
    wire cutters/strippers, 115
  should-have, 117
    battery rechargers, 118
    digital multimeter. See DMMs
    heat sink, 118
    hot glue guns, 118
    metal file set, 118
    third hand, 117
  soldering, 133
  suppliers, 315-316
  supplies, 122
    construction materials, 125
    flux paste, 123
    heat-shrink tubing, 124
    poster putty, 124
    screws/nuts/bolts/washers, 124
    scrubby pads, 124
    solder, 122
    superglue, 123
    techno-junk box, 125
    two-part epoxy, 123
    wire, 123
Topo I, 29, 170
Topo II, 170
tortoises, 23-25
Tower Hobbies, 312, 316
transistors, 178-181
trenches (breadboards), 201
troubleshooting

  breadboards, 249
  DiscRover, 294-295
  Mousey the Junkbot, 249, 259-260
Turing, Alan, 66
Turing Test, 66
two-part epoxy, 123
two-way wireless communication, 104

U - V

ultrasonic range sensors, 100
Unimation (Universal Automation), 22-23
V (voltage), 65
values (resistors), 129-130
Vantec, 309, 313
variable resistors, 178
VB (Visual Basic), 285
***Vehicles: Experiments in Synthetic Psychology*, 322**
ventilation system
  building, 139-143
  soldering, 133
Victorian Era robots Web site, 28
videos, 330
  Battlebots, 330
  BEAM Video Clips, 331
  Extreme Machines: Incredible Robots, 331
  Fast, Cheap & Out of Control, 331
  Robotica, 332
  Robots Rising, 332
virtual circuits, 269
vision systems, 100-101
Visual Basic (VB), 285
voice synthesis, 105
voice-dependent speech recognition, 104
voice-independent speech recognition, 105
voltages, 65, 176, 256

W

W.A.R.S. (Winnipeg Area Robotics Society), 343
W.W. Grainger, 306
WABOT-1, 29
Walter, W. Grey, 23-25
wander aimlessly behavior, 44
WAO-G kit (Wise Argent Orb), 149-150

warriors, 54-55
Waseda University, 29
washers, 124
Web sites
  Acroname, 302, 313
  ActivMedia, 27
  All Electronics, 317
  American Science and Surplus, 317
  AmigoBot, 169
  amorphic robot works, 53
  Android World, 37, 334
  Basic Micro, 313
  BasicElectronics.com, 341
  Batteries Plus, 309
  Battery Mart, 309
  Battle Robot Tips, 341
  Battlebots, 336
  BEAM robot competitions, 48
  BEAM Video Clips, 331
  Bookfinder.com, 319
  Botic, 334
  Budget Robotics, 302
  Carnegie Mellon University Institute, 339
  Case Western University Biorobotics Lab, 339
  Christian Ristow, 53
  Circuit Cellar, 328
  Clayton Bailey, 56
  communities, 337-338
  The Composite Store, 305
  DARPA Distributed Robotics Program, 55
  Darwin Awards, 116
  DC motors, 235
  Dictionary Measures, Units, and Conversions, 341
  Digi-Key, 310
  DiscRover templates, 275
  Dremel, 315
  Electronic Goldmine, 318
  Electronix Express, 310
  Elektor Electronics, 328
  Evolution Robotics, 103, 157, 305
  Evolution Robots, 85
  FIRST, 52
  Fischertechnik, 160
  Friendly Robotics, 168
  fuel cells, 96
  Future Active, 311
  gastrobots, 96
  General Robotics, 173
  Gleason Research, 314
  Google, 29
  GoRobotics Network, 334
  HBRobotics Club Builders Book, 341
  Heathkit Heroes, 172
  HiTec RCD, 307, 312
  Hobbytron, 316

How Stuff Works, 342
Hugh Loebner, 67
Humanoid Robotics Group, 339
Insectröides, 186
Jameco, 123, 139, 311
JCM Inventures, 152
JDR Microdevices, 311
Jouhou System Kougaku Lab, 340
JPL Robotics, 340
Kal Spelltich, 53
Karl Lunt homepage, 341
LEGO MINDSTORMS, 159, 303
Lynxmotion, 150, 303
Mabuchi, 307
MachineBrain, 334
MatWeb, 342
Micro-Mark, 316
Microcontroller FAQ, 275
MIT Technology Review, 329
Mondo-tronics Robot Store, 303
Mouser Electronics, 311
NASA Tech Briefs, 342
NetSurfer Robotics, 335
Nomadic Research Labs, 71
North American Roller Products, 307
Not Quite C, 314
Nuts & Volts, 329
Online Conversions, 342
Online Metals, 305
OOPic, 269
Parallax, 154, 296, 304, 314
Plastruct, 306
Poly-PEDAL Lab, 340
Radio Shack, 312
Real Robots, 329
RepairFAQ, 343
RF transmitter/receivers, 297
Robix, 161
Robot Café, 335
Robot Projects, 298
Robot Room, 342
Robot Science & Technology, 330
The Robot Store, 89, 148
robot-specific, 336-337
RobotBooks, 308
Robotic Power Solutions, 310
Robotics FAQ, 343
Robotics Research in Japan, 340
Robots Wanted, 175
Robots.Net, 335
Roomba, 167
Savage Innovations, 315
Servo City, 308
Small Parts, 306
Solarbotics, 155, 304, 308, 335
Sony, 164
Sony AIBO, 163
Survival Research Labs, 52
Tab Robotics, 152
Tamiya, 304, 308

TechGeek, 335
Timbot, 108
Tower Hobbies, 312, 316
Vantec, 309, 313
VB, 285
Victorian Era robots, 28
W.A.R.S. Reference pages, 343
W.W. Grainger, 306
Wiha, 114
Wowwee Toys, 165
Xcelite, 114
XYTronic, 115

Westinghouse Electro/Sparko, 28

wheels, 92, 307

whiskers, 98

Whittaker, Red, 29

WiFi (Wireless Fidelity), 104

Wiha, 114

Winnipeg Area Robotics Society (W.A.R.S.), 343

Wireless Fidelity (WiFi), 104

wires, 123, 179
    crimping tools, 121
    cutters/strippers, 115
    hook-up, 201
    jumper, 190
    motor, 253
    receiving sockets, 200

Wirz, Ben, 154

Wise Argent Orb (WAO-G), 149-150

wood frames, 84

workshop, 131-132
    organizing, 133
    ventilation system, 139-143

Wowwee Toys, 165-166

X- Y- Z

Xcelite, 114
XYTronic, 115

Yahoo! Groups, 338

Zeroth Law, 62

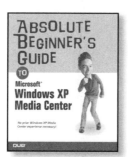